T0296442

CAMBRIDGE STUDIES IN
ADVANCED MATHEMATICS 30

REPRESENTATIONS AND COHOMOLOGY I

Already published

1 W.M.L. Holcombe *Algebraic automata theory*
2 K. Petersen *Ergodic theory*
3 P.T. Johnstone *Stone spaces*
4 W.H. Schikhof *Ultrametric calculus*
5 J.-P. Kahane *Some random series of functions, 2nd edition*
6 H. Cohn *Introduction to the construction of class fields*
7 J. Lambek & P.J. Scott *Introduction to higher-order categorical logic*
8 H. Matsumura *Commutative ring theory*
9 C.B. Thomas *Characteristic classes and the cohomology of finite groups*
10 M. Aschbacher *Finite group theory*
11 J.L. Alperin *Local representation theory*
12 P. Koosis *The logarithmic integral I*
13 A. Pietsch *Eigenvalues and s-numbers*
14 S.J. Patterson *An introduction to the theory of the Riemann zeta-function*
15 H.J. Baues *Algebraic homotopy*
16 V.S. Varadarajan *Introduction to harmonic analysis on semisimple Lie groups*
17 W. Dicks & M. Dunwoody *Groups acting on graphs*
18 L.J. Corwin & F.P. Greenleaf *Representations of nilpotent Lie groups and their applications*
19 R. Fritsch & R. Piccinini *Cellular structures in topology*
20 H Klingen *Introductory lectures on Siegel modular forms*
22 M.J. Collins *Representations and characters of finite groups*
24 H. Kunita *Stochastic flows and stochastic differential equations*
25 P. Wojtaszczyk *Banach spaces for analysts*
26 J.E. Gilbert & M.A.M. Murray *Clifford algebras and Dirac operators in harmonic analysis*
27 A. Frohlich & M.J. Taylor *Algebraic number theory*
28 K. Goebel & W.A. Kirk *Topics in metric fixed point theory*
29 J.F. Humphreys *Reflection groups and Coxeter groups*
30 D.J. Benson *Representations and cohomology I*
31 D.J. Benson *Representations and cohomology II*
32 C. Allday & V. Puppe *Cohomological methods in transformation groups*
33 C. Soulé et al *Lectures on Arakelov geometry*
34 A. Ambrosetti & G. Prodi *A primer of nonlinear analysis*
35 J. Palis & F. Takens *Hyperbolicity, stability and chaos at homoclinic bifurcations*
37 Y. Meyer *Wavelets and operators I*
38 C. Weibel *An introduction to homological algebra*
39 W. Bruns & J. Herzog *Cohen-Macaulay rings*
40 V. Snaith *Explicit Brauer induction*
41 G. Laumon *Cohomology of Drinfield modular varieties I*
42 E.B. Davies *Spectral theory and differential operators*
43 J. Diestel, H. Jarchow & A. Tonge *Absolutely summing operators*
44 P. Mattila *Geometry of sets and measures in Euclidean spaces*
45 R. Pinsky *Positive harmonic functions and diffusion*
46 G. Tenenbaum *Introduction to analytic and probabilistic number theory*
47 C. Peskine *An algebraic introduction to complex projective geometry I*
48 Y. Meyer & R. Coifman *Wavelets and operators II*
49 R. Stanley *Enumerative combinatories*
50 I. Porteous *Clifford algebras and the classical groups*
51 M. Audin *Spinning tops*
52 V. Jurdjevic *Geometric control theory*
53 H. Voelklein *Groups as Galois groups*
54 J. Le Potier *Lectures on vector bundles*
55 D. Bump *Automorphic forms*
56 G. Laumon *Cohomology of Drinfeld modular varieties II*
60 M. Brodmann & R. Sharp *Local cohomology*

Representations and Cohomology

I. Basic representation theory of finite groups and associative algebras

D. J. Benson

University of Georgia

PUBLISHED BY THE PRESS SYNDICATE OF THE UNIVERSITY OF CAMBRIDGE
The Pitt Building, Trumpington Street, Cambridge CB2 1RP, United Kingdom

CAMBRIDGE UNIVERSITY PRESS
The Edinburgh Building, Cambridge CB2 2RU, United Kingdom
40 West 20th Street, New York, NY 10011-4211, USA
10 Stamford Road, Oakleigh, Melbourne 3166, Australia

First published 1995
First paperback edition 1998

A catalogue record for this book is available from the British Library

ISBN 0 521 36134 6 hardback
ISBN 0 521 63653 1 paperback

Transferred to digital printing 2004

To Christine Natasha

Contents

Introduction

These two volumes have grown out of about seven years of graduate courses on various aspects of representation theory and cohomology of groups, given at Yale, Northwestern and Oxford. The pace is brisk, and beginning graduate students would certainly be advised to have at hand a standard algebra text, such as for example Jacobson [**128**].

The chapters are not organised for sequential reading. Chapters 1, 2, 3 of Volume I and Chapter 1 of Volume II should be treated as background reference material, to be read sectionwise (if there is such a word). Each remaining chapter forms an exposition of a topic, and should be read chapterwise (or not at all).

The centrepiece of the first volume is Chapter 4, which gives a not entirely painless introduction to Auslander–Reiten type representation theory. This has recently played an important rôle in representation theory of finite groups, especially because of the pioneering work of K. Erdmann [**101**]–[**105**] and P. Webb [**204**]. Our exposition of blocks with cyclic defect group in Chapter 6 of Volume I is based on the discussion of almost split sequences in Chapter 4, and gives a good illustration of how modern representation theory can be used to clean up the proofs of older theorems.

While the first volume concentrates on representation theory with a cohomological flavour, the second concentrates on cohomology of groups, while never straying very far from the pleasant shores of representation theory. In Chapter 2 of Volume II, we give an overview of the algebraic topology and K-theory associated with cohomology of groups, and especially the extraordinary work of Quillen which has led to his definition of the higher algebraic K-groups of a ring.

The algebraic side of the cohomology of groups mirrors the topology, and we have always tried to give algebraic proofs of algebraic theorems. For example, in Chapter 3 of Volume II you will find B. Venkov's topological proof of the finite generation of the cohomology ring of a finite group, while in Chapter 4 you will find L. Evens' algebraic proof. Also in Chapter 4 of Volume II, we give a detailed account of the construction of Steenrod operations in group cohomology using the Evens norm map, a topic usually treated from a topological viewpoint.

One of the most exciting developments in recent years in group cohomology is the theory of varieties for modules, expounded in Chapter 5 of Volume II. In a sense, this is the central chapter of the entire two volumes, since it shows how inextricably intertwined representation theory and cohomology really are.

I would like to record my thanks to the people, too numerous to mention individually, whose insights I have borrowed in order to write these volumes; who have pointed out infelicities and mistakes in the exposition; who have supplied me with quantities of coffee that would kill an average horse; and who have helped me in various other ways. I would especially like to thank

Ken Brown for allowing me to explain his approach to induction theorems in I, Chapter 5; Jon Carlson for collaborating with me over a number of years, and without whom these volumes would never have been written; Ralph Cohen for helping me understand the free loop space and its rôle in cyclic homology (Chapter 2 of Volume II); Peter Webb for supplying me with an early copy of the notes for his talk at the 1986 Arcata conference on Representation Theory of Finite Groups, on which Chapter 6 of Volume II is based; David Tranah of Cambridge University Press for sending me a free copy of Tom Körner's wonderful book on Fourier analysis, and being generally helpful in various ways you have no interest in hearing about unless you happen to be David Tranah.

There is a certain amount of overlap between this volume and my Springer lecture notes volume [**17**]. Wherever I felt it appropriate, I have not hesitated to borrow from the presentation of material there. This applies particularly to parts of Chapters 1, 4 and 5 of Volume I and Chapter 5 of Volume II.

THE SECOND EDITION. In preparing the paperback edition, I have taken the liberty of completely retypesetting the book using the enhanced features of LaTeX 2$_\varepsilon$, AMS-LaTeX 1.2 and XY-pic 3.5. Apart from this, I have corrected those errors of which I am aware. I would like to thank the many people who have sent me lists of errors, particularly Bill Crawley–Boevey, Steve Donkin, Jeremy Rickard and Steve Siegel.

The most extensively changed sections are Section 2.2 and 3.1 of Volume I and Section 5.8 of Volume II, which contained major flaws in the original edition. In addition, in Section 3.1 of Volume I, I have changed to the more usual definition of Hopf algebra in which an antipode is part of the definition, reserving the term bialgebra for the version without an antipode. I have made every effort to preserve the numbering of the sections, theorems, references, and so on from the first edition, in order to.avoid reference problems. The only exception is that in Volume I, Definition 3.1.5 has disappeared and there is now a Proposition 3.1.5. I have also updated the bibliography and improved the index. If you find further errors in this edition, please email me at djb@byrd.math.uga.edu.

Dave Benson, Athens, September 1997

CONVENTIONS AND NOTATIONS.

- All groups in Volume I are finite, unless the contrary is explicitly mentioned.
- Maps will usually be written on the left. In particular, we use the left notation for conjugation and commutation: ${}^gh = ghg^{-1}$, $[g,h] = ghg^{-1}h^{-1}$, and ${}^gH = gHg^{-1}$.
- We write G/H to denote the action of G as a transitive permutation group on the left cosets of H.
- We write $H \leq_G K$ to denote that "H is G-conjugate to a subgroup of K". Similarly $h \in_G K$ means "h is G-conjugate to an element of K". Thus we write for example $\bigoplus\limits_{g \in_G G}$ to denote a direct sum over conjugacy classes of elements of G.
- The symbol $\square$ denotes the end of a proof.
- We shall use the usual notations $O_p(G)$ for the largest normal p-subgroup of G, $O^p(G)$ for the smallest normal subgroup of G for which the quotient is a p-group, $G^{(\infty)}$ for the smallest normal subgroup of G for which the quotient is soluble, $\Phi(P)$ for the *Frattini subgroup* of a p-group P, i.e., the smallest normal subgroup for which the quotient is elementary abelian, $Z(G)$ for the centre of G, $\Omega_1(G)$ for the subgroup of an abelian p-group G generated by the elements of order p, and so on. The *p-rank* $r_p(G)$ is defined to be the maximal rank of an elementary abelian p-subgroup of G.
- If H and K are subgroups of a group G, then $\sum_{HgK}$ will denote a sum over a set of double coset representatives g of H and K in G.
- We shall write ${}_\Lambda M_\Gamma$ to denote that M is a Λ-Γ-bimodule, i.e., a left Λ-module which is simultaneously a right Γ-module in such a way that $(\lambda m)\gamma = \lambda(m\gamma)$ for all $\lambda \in \Lambda$, $m \in M$ and $\gamma \in \Gamma$.
- If G is a group of permutations on the set $\{1, \dots, n\}$ and H is another group, we write $G \wr H$ for the *wreath product*; namely the semidirect product of G with a direct product of n copies of H. Thus elements of $G \wr H$ are of the form $(\pi; h_1, \dots, h_n)$ with $\pi \in G$, $h_1, \dots, h_n \in H$ and multiplication given by

$$(\pi'; h'_1, \dots, h'_n)(\pi; h_1, \dots, h_n) = (\pi'\pi; h'_{\pi(1)}h_1, \dots, h'_{\pi(n)}h_n).$$

- If X is a set with a right G-action and Y is a set with a left G-action, then we write $X \times_G Y$ for the quotient of $X \times Y$ by the equivalence relation $(xg, y) \sim (x, gy)$ for all $x \in X$, $g \in G$, $y \in Y$.

CHAPTER 1

Background material from rings and modules

Group representations are often studied as modules over the group algebra (see Chapter 3), which is a finite dimensional algebra in case the coefficients are taken from a field, and a Noetherian ring if integer or p-adic coefficients are used. Thus we begin with a rather condensed summary of some general material on rings and modules. Sources for further material related to Chapters 1 and 3 are Alperin [**3**], Curtis and Reiner [**64, 65, 66**], Feit [**107**] and Landrock [**148**]. We return for a deeper study of modules over a finite dimensional algebra in Chapter 4.

Throughout this chapter, Λ denotes an arbitrary ring with unit, and M is a (left) Λ-module.

1.1. The Jordan–Hölder theorem

DEFINITION 1.1.1. A **composition series** for a Λ-module M is a series of submodules

$$0 = M_0 < M_1 < \cdots < M_n = M$$

with M_i/M_{i-1} irreducible.

A module M is said to satisfy the descending chain condition (D.C.C.) on submodules if every descending chain of submodules eventually stops, and the ascending chain condition (A.C.C.)if every ascending chain of submodules eventually stops. A module satisfying A.C.C. on submodules is said to be **Noetherian**.

LEMMA 1.1.2 (Modular law). *If A and $B \supseteq C$ are submodules of M then*

$$(C + A) \cap B = C + (A \cap B).$$

PROOF. Clearly $C + (A \cap B) \subseteq (C + A) \cap B$. Conversely if $b = c + a \in (C + A) \cap B$ then $a = b - c \in A \cap B$ and so $b \in C + (A \cap B)$. □

THEOREM 1.1.3 (Zassenhaus isomorphism theorem). *If $U \supseteq V$ and $U' \supseteq V'$ are submodules of M then*

$$\frac{(U + V') \cap U'}{(V + V') \cap U'} \cong \frac{U \cap U'}{(U' \cap V) + (U \cap V')} \cong \frac{(U' + V) \cap U}{(V' + V) \cap U}.$$

PROOF. It suffices to prove the first isomorphism. The kernel of the composite map

$$U \cap U' \hookrightarrow ((U + V') \cap U') \twoheadrightarrow ((U + V') \cap U')/((V + V') \cap U')$$

is

$$(U \cap U') \cap ((V + V') \cap U') = U' \cap (V + V') \cap U = (U' \cap V) + (U \cap V')$$

by two applications of the modular law. □

THEOREM 1.1.4 (Jordan–Hölder). *Given any two series of submodules*

$$0 = M_0 \leq \cdots \leq M_r = M$$
$$0 = M'_0 \leq \cdots \leq M'_s = M$$

of a Λ-module M, we may refine them (i.e., stick in extra terms) to series of equal length

$$0 = L_0 \leq \cdots \leq L_n = M$$
$$0 = L'_0 \leq \cdots \leq L'_n = M$$

so that the factors L_i/L_{i-1} are a permutation of the factors L'_j/L'_{j-1} (up to isomorphism). Thus the following conditions on M are equivalent.

(i) *M has a composition series.*

(ii) *Every series of submodules of M can be refined to a composition series.*

(iii) *M satisfies A.C.C. and D.C.C. on submodules.*

PROOF. Between M_i and M_{i+1} we insert the terms $(M'_j + M_i) \cap M_{i+1}$, and between M'_j and M'_{j+1} we insert the terms $(M_i + M'_j) \cap M'_{j+1}$. Now use the Zassenhaus isomorphism theorem. □

REMARK. It follows that the length of a composition series, if one exists, is an invariant of the module. It is called the **composition length** of the module.

A module M is said to be **uniserial** if it has a unique composition series. This is the same as saying that M has a unique minimal submodule M_0, M/M_0 has a unique minimal submodule M_1/M_0, and so on.

EXERCISE. Suppose that M is a module of finite composition length. Show that the submodules of M satisfy the distributive laws

$$(A + B) \cap C = (A \cap C) + (B \cap C)$$
$$(A \cap B) + C = (A + C) \cap (B + C)$$

if and only if M has no subquotient isomorphic to a direct sum of two isomorphic simple modules.

Read about Birkhoff's theorem for distributive lattices of finite length in Aigner [**1**]. In effect, this says that for a module with the above property, one can draw a diagram, whose vertices represent the composition factors, and whose edges describe how they are "glued together." In fact, there is a generalisation of Birkhoff's diagrams to modular lattices of finite length (Benson and Conway [**22**]), but this method quickly becomes cumbersome, as there are usually many more vertices than composition factors. For further discussions of diagrams for modules, see Alperin [**2**] and Benson and Carlson [**21**].

1.2. The Jacobson radical

DEFINITION 1.2.1. The **socle** of a Λ-module M is the sum of all the irreducible submodules of M, and is written $\mathrm{Soc}(M)$. The **socle layers** of M are defined inductively by $\mathrm{Soc}^0(M) = 0$, $\mathrm{Soc}^n(M)/\mathrm{Soc}^{n-1}(M) = \mathrm{Soc}(M/\mathrm{Soc}^{n-1}(M))$.

The **radical** of M is the intersection of all the maximal submodules of M, and is written $\mathrm{Rad}(M)$. The **radical series** or **Loewy series** of M is defined inductively by $\mathrm{Rad}^0(M) = M$, $\mathrm{Rad}^n(M) = \mathrm{Rad}(\mathrm{Rad}^{n-1}(M))$. The nth **radical layer** or **Loewy layer** is $\mathrm{Rad}^{n-1}(M)/\mathrm{Rad}^n(M)$.

The module M is said to be **completely reducible** or **semisimple** if $M = \mathrm{Soc}(M)$. This is equivalent to the condition that every submodule has a complement, by Zorn's lemma. If M satisfies D.C.C. then M is completely reducible if and only if $\mathrm{Rad}(M) = 0$. In this case, M is a finite direct sum of irreducible modules.

The **head** or **top** of M is $\mathrm{Head}(M) = M/\mathrm{Rad}(M)$.

If M has **socle length** n (i.e., $\mathrm{Soc}^n(M) = M$ but $\mathrm{Soc}^{n-1}(M) \neq M$) then M also has **radical length** n (i.e., $\mathrm{Rad}^n(M) = 0$ but $\mathrm{Rad}^{n-1}(M) \neq 0$) and $\mathrm{Soc}^j(M) \supseteq \mathrm{Rad}^{n-j}(M)$ for all $0 \le j \le n$.

The **annihilator** of an element $m \in M$ is the set of all elements $\lambda \in \Lambda$ with $\lambda m = 0$. It is a left ideal, which is maximal if and only if the submodule generated by m is irreducible. The annihilator of M is defined to be the intersection of the annihilators of the elements of M. It is a two sided ideal I which is **primitive**, meaning that Λ/I has a faithful irreducible module. We define $J(\Lambda)$, the **Jacobson radical** of Λ to be the intersection of the maximal left ideals, or equivalently the intersection of the primitive two sided ideals.

We claim that $J(\Lambda)$ consists of those elements $x \in \Lambda$ such that $1 - axb$ has a two sided inverse for all a, $b \in \Lambda$, so that it does not matter whether we use left or right ideals to define $J(\Lambda)$. If $x \in J(\Lambda)$ then $1-x$ cannot be in any maximal left ideal (since otherwise 1 would be!) so it has a left inverse, say $t(1-x) = 1$. Then $1-t = -tx \in J(\Lambda)$ so t has a left inverse, and is hence a two sided inverse for $1-x$. Applying this with axb in place of x shows that $1-axb$ has a two sided inverse. Conversely, if $1-axb$ has a two sided inverse for all a, $b \in \Lambda$, and I is a maximal left ideal not containing x, then we can write 1 as ax plus an element of I, contradicting the invertibility of $1-ax$.

If we let Λ act on itself as a left module, we call this the **regular representation** ${}_\Lambda\Lambda$. Since submodules are the same as left ideals, we have $J(\Lambda) = \mathrm{Rad}({}_\Lambda\Lambda)$. We say that Λ is **semisimple** if $J(\Lambda) = 0$. Note that $\Lambda/J(\Lambda)$ is always semisimple.

LEMMA 1.2.2. *If $a \in J(\Lambda)$ then $1 - a$ has a left inverse in Λ.*

PROOF. Since $1 = a + (1-a)$ we have $\Lambda = J(\Lambda) + \Lambda(1-a)$. If $\Lambda(1-a) \neq \Lambda$, then by Zorn's lemma there is a maximal left ideal I with $\Lambda(1-a) \subseteq I$. By definition of $J(\Lambda)$ we also have $J(\Lambda) \subseteq I$, and so $\Lambda \subseteq I$. This contradiction proves the lemma. □

LEMMA 1.2.3 (Nakayama). *If M is a finitely generated Λ-module and $J(\Lambda)M = M$ then $M = 0$.*

PROOF. Suppose that $M \neq 0$. Choose $m_1, \ldots, m_n$ generating M with n minimal. Since $J(\Lambda)M = M$, we can write $m_n = \sum_{i=1}^{n} a_i m_i$ with $a_i \in J(\Lambda)$. By Lemma 1.2.2, $1 - a_n$ has a left inverse b in Λ. Then $(1 - a_n)m_n = \sum_{i=1}^{n-1} a_i m_i$, and so $m_n = b(\sum_{i=1}^{n-1} a_i m_i)$, contradicting the minimality of n. □

LEMMA 1.2.4. *If Λ is semisimple and satisfies D.C.C. on left ideals, then every Λ-module is completely reducible. Conversely if Λ satisfies D.C.C. on left ideals and ${}_\Lambda\Lambda$ is completely reducible then Λ is semisimple.*

PROOF. If Λ is semisimple then $\mathrm{Rad}({}_\Lambda\Lambda) = J(\Lambda) = 0$ and so ${}_\Lambda\Lambda$ is completely reducible. Choosing a set of generators for a module displays it as a quotient of a direct sum of copies of ${}_\Lambda\Lambda$, and hence every module is completely reducible. Conversely if ${}_\Lambda\Lambda$ is completely reducible then ${}_\Lambda\Lambda \cong \Lambda/J(\Lambda) \oplus J(\Lambda)$ and so as Λ-modules,

$$J(\Lambda) = J(\Lambda).(\Lambda/J(\Lambda) \oplus J(\Lambda)) = J(\Lambda).J(\Lambda)$$

so that by Nakayama's lemma $J(\Lambda) = 0$. □

PROPOSITION 1.2.5. *If M is a finitely generated Λ-module then $J(\Lambda)M = \mathrm{Rad}(M)$.*

PROOF. If M' is a maximal submodule of M then by Nakayama's lemma we have $J(\Lambda)(M/M') = 0$ so that $J(\Lambda)M \subseteq M'$, and hence $J(\Lambda)M \subseteq \mathrm{Rad}(M)$. Conversely $M/J(\Lambda)M$ is completely reducible by Lemma 1.2.4 and so $\mathrm{Rad}(M/J(\Lambda)M) = 0$, which implies that $\mathrm{Rad}(M) \subseteq J(\Lambda)M$. □

DEFINITION 1.2.6. A ring Λ is said to be **Noetherian** if it satisfies A.C.C. on left ideals, and **Artinian** if it satisfies D.C.C. on left ideals. A Λ-module is Noetherian/Artinian if it satisfies A.C.C./D.C.C. on submodules.

THEOREM 1.2.7. *If Λ is Artinian then*

(i) *$J(\Lambda)$ is nilpotent.*

(ii) *If M is a finitely generated Λ-module then M is both Noetherian and Artinian.*

(iii) *Λ is Noetherian.*

PROOF. (i) Since Λ satisfies D.C.C. on left ideals, for some n we have $J(\Lambda)^n = J(\Lambda)^{2n}$. If $J(\Lambda)^n \neq 0$, then again using D.C.C. we see that there is a minimal left ideal I with $J(\Lambda)^n I \neq 0$. Choose $x \in I$ with $J(\Lambda)^n.x \neq 0$, and in particular $x \neq 0$. Then $I = J(\Lambda)^n.x$ by minimality of I, and so for some $a \in J(\Lambda)^n$, we have $x = ax$. But then $(1-a)x = 0$, and so $x = 0$ by Lemma 1.2.2.

(ii) Let $M_i = J(\Lambda)^i M$. Then M_i/M_{i+1} is annihilated by $J(\Lambda)$, and is hence completely reducible by Lemma 1.2.4. Since M is a finitely generated module over an Artinian ring, it satisfies D.C.C., and hence so does M_i/M_{i+1}. Thus M_i/M_{i+1} is a finite direct sum of irreducible modules, and so it satisfies A.C.C. It follows that M also satisfies A.C.C.

(iii) This follows by applying (ii) to the module ${}_\Lambda\Lambda$. □

The following proposition shows that whether a ring homomorphism to an Artinian ring is surjective can be detected modulo the square of the radical.

PROPOSITION 1.2.8. *Suppose that Λ is an Artinian ring and Λ' is a subring of Λ such that $\Lambda' + J^2(\Lambda) = \Lambda$. Then $\Lambda' = \Lambda$.*

PROOF. We show that $\Lambda' + J^n(\Lambda) = \Lambda' + J^{n+1}(\Lambda)$ for $n \geq 2$, so that by induction and part (i) of the previous theorem we deduce that $\Lambda' = \Lambda$. If $x \in J^{n-1}(\Lambda)$ and $y \in J(\Lambda)$, choose $x' \in J^{n-1}(\Lambda) \cap \Lambda'$ such that $x - x' \in J^n(\Lambda)$ and $y' \in J(\Lambda) \cap \Lambda'$ such that $y - y' \in J^2(\Lambda)$. Then

$$\begin{aligned} xy &= x(y-y') + (x-x')y' + x'y' \\ &\in J^{n-1}(\Lambda)J^2(\Lambda) + J^n(\Lambda)(J(\Lambda)\cap\Lambda') + (J^{n-1}(\Lambda)\cap\Lambda')(J(\Lambda)\cap\Lambda') \\ &\subseteq J^{n+1}(\Lambda) + \Lambda'. \end{aligned}$$

□

EXERCISES. 1. If Λ is a Noetherian ring, show that a Λ-module is finitely generated if and only if it is Noetherian. Deduce that every submodule of a finitely generated Λ-module is finitely generated.

2. Give an example of a simple ring (i.e., one with no non-trivial two sided ideals) which is not Noetherian.

1.3. The Wedderburn structure theorem

LEMMA 1.3.1 (Schur). *If M_1 and M_2 are irreducible Λ-modules, then for $M_1 \not\cong M_2$, $\mathrm{Hom}_\Lambda(M_1, M_2) = 0$, while $\mathrm{Hom}_\Lambda(M_1, M_1) = \mathrm{End}_\Lambda(M_1)$ is a division ring.*

PROOF. Clear. □

DEFINITION 1.3.2. An **idempotent** in Λ is a non-zero element e with $e^2 = e$.

Note that if $e \neq 1$ is an idempotent then so is $1 - e$, and we have ${}_\Lambda\Lambda = \Lambda e \oplus \Lambda(1-e)$.

LEMMA 1.3.3. (i) *If M is a Λ-module and e is an idempotent in Λ then*

$$eM \cong \mathrm{Hom}_\Lambda(\Lambda e, M).$$

(ii) *We have an isomorphism of rings $e\Lambda e \cong \mathrm{End}_\Lambda(\Lambda e)^{\mathrm{op}}$ (Λ^{op} denotes the opposite ring to Λ, where the order of multiplication has been reversed).*

PROOF. (i) Define maps $f_1 : eM \to \mathrm{Hom}_\Lambda(\Lambda e, M)$ by $f_1(em) : ae \mapsto aem$ and $f_2 : \mathrm{Hom}_\Lambda(\Lambda e, M) \to eM$ by $f_2 : \alpha \mapsto \alpha(e)$. It is easy to check that f_1 and f_2 are mutually inverse.

(ii) This follows by applying (i) with $M = \Lambda e$. It is easy to check that f_1 and f_2 reverse the order of multiplication. □

THEOREM 1.3.4. *Let M be a finite direct sum of irreducible Λ-modules, say $M = M_1 \oplus \cdots \oplus M_r$, with each M_i a direct sum of n_i modules $M_{i,1} \oplus \cdots \oplus M_{i,n_i}$ isomorphic to a simple module S_i, and $S_i \not\cong S_j$ if $i \neq j$. Let $\Delta_i = \mathrm{End}_\Lambda(S_i)$. Then Δ_i is a division ring, $\mathrm{End}_\Lambda(M_i) \cong \mathrm{Mat}_{n_i}(\Delta_i)$, and $\mathrm{End}_\Lambda(M) = \bigoplus_i \mathrm{End}_\Lambda(M_i)$ is semisimple.*

PROOF. By Lemma 1.3.1, Δ_i is a division ring. Choose once and for all isomorphisms $\theta_{ij} : M_{ij} \to S_i$. Now given $\lambda \in \mathrm{End}_\Lambda(M_i)$, we define $\lambda_{jk} \in \Delta_i$ as the composite map

$$S_i \overset{\theta_{ik}^{-1}}{\cong} M_{ik} \hookrightarrow M_i \xrightarrow{\lambda} M_i \twoheadrightarrow M_{ij} \overset{\theta_{ij}}{\cong} S_i.$$

The map $\lambda \mapsto (\lambda_{jk})$ is then an injective homomorphism $\mathrm{End}_\Lambda(M_i) \to \mathrm{Mat}_{n_i}(\Delta_i)$. Conversely, given (λ_{jk}), we can construct λ as the sum of the composite endomorphisms

$$M_i \twoheadrightarrow M_{ik} \overset{\theta_{ik}}{\cong} S_i \xrightarrow{\lambda_{jk}} S_i \overset{\theta_{ij}^{-1}}{\cong} M_{ij} \hookrightarrow M_i.$$

Finally, $\mathrm{End}_\Lambda(M) = \bigoplus_i \mathrm{End}_\Lambda(M_i)$ since if $i \neq j$, Lemma 1.3.1 implies that $\mathrm{Hom}_\Lambda(S_i, S_j) = 0$. □

THEOREM 1.3.5 (Wedderburn–Artin). *Let Λ be a semisimple Artinian ring. Then $\Lambda = \bigoplus_{i=1}^r \Lambda_i$, $\Lambda_i \cong \mathrm{Mat}_{n_i}(\Delta_i)$, Δ_i is a division ring, and the Λ_i are uniquely determined. The ring Λ has exactly r isomorphism classes of irreducible modules M_i, $i = 1, \ldots, r$, $\mathrm{End}_\Lambda(M_i) \cong \Delta_i^{\mathrm{op}}$, and $\dim_{\Delta_i^{\mathrm{op}}}(M_i) = n_i$. If Λ is simple then $\Lambda \cong \mathrm{Mat}_n(\Delta)$.*

PROOF. By Lemma 1.2.4, ${}_\Lambda\Lambda$ is completely reducible. By Lemma 1.3.3 with $e = 1$, $\Lambda \cong \mathrm{End}_\Lambda({}_\Lambda\Lambda)^{\mathrm{op}}$. The result now follows by applying Theorem 1.3.4 to ${}_\Lambda\Lambda$. Note that the opposite ring of a complete matrix ring is again a complete matrix ring, over the opposite division ring. □

REMARKS. (i) Wedderburn has shown that every division ring with a finite number of elements is a field.

(ii) If Λ is a finite dimensional algebra over a field k, then each Δ_i for $\Lambda/J(\Lambda)$ in the above theorem has k in its centre. If for each i we have $\Delta_i = k$, then k is called a **splitting field** for Λ. This is true, for example, if k is algebraically closed, since in this case there are no finite dimensional division rings over k (apart from k itself).

Finally, the following special case of the Skolem–Noether theorem is often useful.

PROPOSITION 1.3.6. *Suppose that* V *is a vector space over a field* k. *Then every* k*-linear automorphism of* $\mathrm{End}_k(V)$ *is inner (i.e., effected by a conjugation in* $\mathrm{End}_k(V)$*).*

PROOF. Since the regular representation of $\mathrm{End}_k(V)$ is a direct sum of copies of V, it follows that $\mathrm{End}_k(V)$ has only one isomorphism class of simple modules. Thus if $f : \mathrm{End}_k(V) \to \mathrm{End}_k(V)$ is an automorphism, then f defines a new representation of $\mathrm{End}_k(V)$ on V, which is therefore conjugate to the old one. Thus f is conjugate to the identity map. □

EXERCISE. If Λ is a finite dimensional algebra over k, show that some finite extension k' of k is a splitting field for Λ (i.e., for $k' \otimes_k \Lambda$). If k is algebraically closed, then k is a splitting field for Λ.

1.4. The Krull–Schmidt theorem

DEFINITION 1.4.1. A (not necessarily commutative) ring E is said to be a **local ring** if it has a unique maximal left ideal, or equivalently a unique maximal right ideal. This maximal ideal is automatically two-sided (see the remarks in Section 1.2) and consists of the non-invertible elements of E. The quotient by the unique maximal ideal is a division ring.

It is easy to see that E is local if and only if the non-invertible elements form a left ideal.

DEFINITION 1.4.2. A Λ-module M has the **unique decomposition property** if

(i) M is a finite direct sum of indecomposable modules, and

(ii) Whenever $M = \bigoplus_{i=1}^m M_i = \bigoplus_{i=1}^n M_i'$ with each M_i and each M_i' non-zero indecomposable, then $m = n$, and after reordering if necessary, $M_i \cong N_i$.

A ring Λ is said to have the unique decomposition property if every finitely generated Λ-module does.

THEOREM 1.4.3. *Suppose that* M *is a finite sum of indecomposable* Λ*-modules* M_i *with the property that the endomorphism ring of each* M_i *is a local ring. Then* M *has the unique decomposition property.*

PROOF. Let $M = \bigoplus_{i=1}^m M_i = \bigoplus_{i=1}^n M_i'$ and work by induction on m. Assume $m > 1$. Let α_i and β_i be the composites

$$\alpha_i : M_i' \hookrightarrow M \twoheadrightarrow M_1$$

and

$$\beta_i : M_1 \hookrightarrow M \twoheadrightarrow M_i'.$$

Then $\mathrm{id}_{M_1} = \sum \alpha_i \circ \beta_i : M_1 \to M_1$. Since $\mathrm{End}_\Lambda(M_1)$ is a local ring, some $\alpha_i \circ \beta_i$ must be a unit. Renumber so that $\alpha_1 \circ \beta_1$ is a unit. Then $M_1 \cong M_1'$.

Consider the map $\mu = 1 - \theta$, where θ is the composite

$$\theta : M \twoheadrightarrow M_1 \xrightarrow{\alpha_1^{-1}} M_1' \hookrightarrow M \twoheadrightarrow \bigoplus_{i=2}^m M_i \hookrightarrow M.$$

Then $\mu M_1' = M_1$, and $\mu(\bigoplus_{i=2}^m M_i) = \bigoplus_{i=2}^m M_i$, so μ is onto. If $\mu(w) = 0$, then $w = \theta(w)$ and so $w \in \bigoplus_{i=2}^m M_i$. But then $\theta(w) = 0$.

Thus μ is an automorphism of M with $\mu M_1' = M_1$, and so

$$\bigoplus_{i=2}^{n} M_i' = M/M_1' \cong M/M_1 = \bigoplus_{i=2}^{m} M_i. \qquad \square$$

LEMMA 1.4.4 (Fitting). *Suppose that M has a composition series (i.e., satisfies A.C.C. and D.C.C. on submodules, see Theorem 1.1.4) and $f \in \mathrm{End}_\Lambda(M)$. Then for large enough n, $M = \mathrm{Im}(f^n) \oplus \mathrm{Ker}(f^n)$.*

PROOF. By A.C.C. and D.C.C. on submodules of M, there is a positive integer n such that for all $k \geq 0$ we have $\mathrm{Ker}(f^n) = \mathrm{Ker}(f^{n+k})$ and $\mathrm{Im}(f^n) = \mathrm{Im}(f^{n+k})$. If $x \in M$, write $f^n(x) = f^{2n}(y)$. Then $x = f^n(y) + (x - f^n(y)) \in \mathrm{Im}(f^n) + \mathrm{Ker}(f^n)$. If $f^n(x) \in \mathrm{Im}(f^n) \cap \mathrm{Ker}(f^n)$ then $f^{2n}(x) = 0$, and so $f^n(x) = 0$. $\square$

LEMMA 1.4.5. *Suppose that M is an indecomposable module with a composition series. Then $\mathrm{End}_\Lambda(M)$ is a local ring.*

PROOF. Let $E = \mathrm{End}_\Lambda(M)$, and choose I a maximal left ideal of E. Suppose that $a \notin I$. Then $E = Ea + I$. Write $1 = \lambda a + \mu$ with $\lambda \in E$, and $\mu \in I$. Since μ is not an isomorphism, Lemma 1.4.4 implies that $\mu^n = 0$ for some n. Thus $(1 + \mu + \cdots + \mu^{n-1})\lambda a = (1 + \cdots + \mu^{n-1})(1 - \mu) = 1$, and so a is invertible. $\square$

THEOREM 1.4.6 (Krull–Schmidt). *Suppose that Λ is Artinian. Then Λ has the unique decomposition property.*

PROOF. Suppose that M is a finitely generated indecomposable Λ-module. Then by Theorem 1.2.7 M has a composition series, and so by Lemma 1.4.5, $\mathrm{End}_\Lambda(M)$ is a local ring. The result now follows from Theorem 1.4.3. $\square$

EXERCISE. Suppose that $\mathcal{O}$ is the ring of integers in an algebraic number field. Show that the Krull–Schmidt theorem holds for finitely generated $\mathcal{O}$-modules if and only if $\mathcal{O}$ has class number one.

1.5. Projective and injective modules

DEFINITION 1.5.1. A module P is said to be **projective** if given modules M and M', a map $\lambda : P \to M$ and an epimorphism $\mu : M' \to M$ there exists a map $\nu : P \to M'$ such that the following diagram commutes.

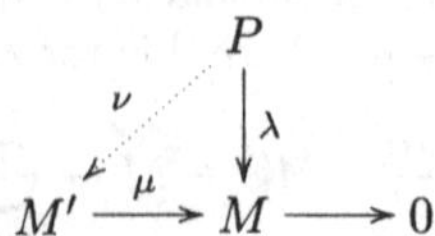

A module I is said to be **injective** if given two modules M and M', a map $\lambda : M \to I$ and a monomorphism $M \to M'$, there is a map $\nu : M' \to I$ such that the following diagram commutes.

$$\begin{array}{ccccc} 0 & \longrightarrow & M & \xrightarrow{\mu} & M' \\ & & \downarrow{\scriptstyle\lambda} & \swarrow{\scriptstyle\nu} & \\ & & I & & \end{array}$$

LEMMA 1.5.2. *The following are equivalent.*
(i) *P is projective.*
(ii) *Every epimorphism $\lambda : M \to P$ splits.*
(iii) *P is a direct summand of a free module.*

PROOF. The proof of this lemma is left as an easy exercise for the reader. □

Note that if P is a projective left Λ-module and

$$\cdots \to M_n \to M_{n-1} \to M_{n-2} \to \cdots$$

is a long exact sequence of right Λ-modules then the sequence

$$\cdots \to M_n \otimes_\Lambda P \to M_{n-1} \otimes_\Lambda P \to M_{n-2} \otimes_\Lambda P \to \cdots$$

is also exact. A left module with this property is called **flat**. Similarly a right Λ-module with the above property with respect to long exact sequences of left Λ-modules is called flat.

Since every module M is a quotient of a free module, it is certainly a quotient of a projective module. If Λ is Artinian, and P_1 and P_2 are minimal projective modules (with respect to direct sum decomposition) mapping onto a finitely generated module M, then we have a diagram

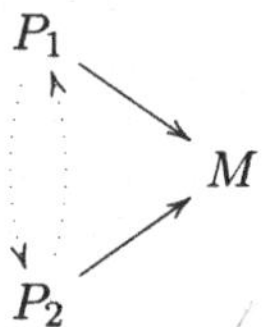

If the composite map $P_1 \to P_2 \to P_1$ is not an isomorphism then by Fitting's lemma P_1 has a summand mapping to zero in M and so P_1 is not minimal. Applying this argument both ways round, we see that $P_1 \cong P_2$. This module is called the **projective cover** P_M of M. We write $\Omega(M)$ for the kernel, so that we have a short exact sequence

$$0 \to \Omega(M) \to P_M \to M \to 0.$$

Even when Λ is not Artinian, we have the following.

LEMMA 1.5.3 (Schanuel). *Suppose that $0 \to M_1 \to P_1 \to M \to 0$ and $0 \to M_2 \to P_2 \to M \to 0$ are short exact sequences of modules with P_1 and P_2 projective. Then $M_1 \oplus P_2 \cong P_1 \oplus M_2$.*

PROOF. Let X be the submodule of $P_1 \oplus P_2$ consisting of those elements (x, y) where x and y have the same image in M (the **pullback** of $P_1 \to M$ and $P_2 \to M$). Then we have a commutative diagram with exact rows and columns

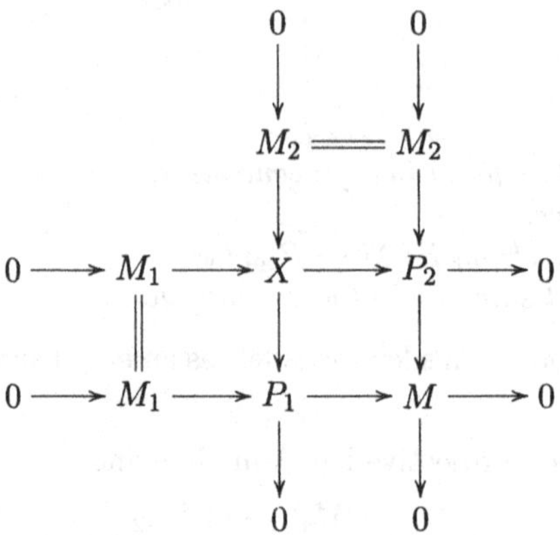

The two sequences with X in the middle must split since they end with a projective module, and so we have $M_1 \oplus P_2 \cong X \cong M_2 \oplus P_1$. □

Thus if we define $\tilde{\Omega}(M)$ to be the kernel of *some* epimorphism $P \to M$ with P projective, Schanuel's lemma shows that $\tilde{\Omega}(M)$ is well defined up to adding and removing projective summands.

If $\alpha : M_1 \to M_2$ is a module homomorphism then we may lift as in the following diagram

$$\begin{array}{ccccccccc}
0 & \longrightarrow & \tilde{\Omega}(M_1) & \longrightarrow & P_1 & \longrightarrow & M_1 & \longrightarrow & 0 \\
 & & \downarrow & & \downarrow & & \downarrow & & \\
0 & \longrightarrow & \tilde{\Omega}(M_2) & \longrightarrow & P_2 & \longrightarrow & M_2 & \longrightarrow & 0
\end{array}$$

and obtain a map $\tilde{\Omega}(\alpha) : \tilde{\Omega}(M_1) \to \tilde{\Omega}(M_2)$ which is unique up to the addition of maps factoring through a projective module. For a discussion of the right functorial setting for $\tilde{\Omega}$, see Section 2.1.

The discussion of injective modules is achieved by means of a dualising operation as follows.

LEMMA 1.5.4. *Every Λ-module may be embedded in an injective module.*

PROOF. If M is a left Λ-module, the dual abelian group

$$M^o = \mathrm{Hom}_{\mathbb{Z}}(M, \mathbb{Q}/\mathbb{Z})$$

is a right Λ-module in the obvious way, and vice-versa. There is also an obvious injective map $M \hookrightarrow M^{oo}$. If P is projective, then the dual P^o is injective, as is easy to see by applying duality to the definition. Thus if P is a projective right Λ-module mapping onto M^o then $M \hookrightarrow M^{oo} \hookrightarrow P^o$ is an embedding of M into an injective left Λ-module. $\square$

If $M \to I$ is an embedding of M into an injective module I then we write $\tilde{\Omega}^{-1}(M)$ for the cokernel.

Injective modules are better behaved than projective modules in the sense that for any ring Λ and any module M there is a unique minimal injective module I (with the obvious universal property) into which M embeds. This is called the **injective hull** of M. A proof of this statement, which is the Eckmann–Schöpf theorem, may be found in Curtis and Reiner [**64**], Theorem 57.13. If $M \to I$ is the injective hull of M, we write $\Omega^{-1}(M)$ for the cokernel.

Exercise. (Broué) Suppose that Λ is a k-algebra. Write Ω_Λ for the kernel of the multiplication map $\Lambda \otimes_k \Lambda \to \Lambda$, so that Ω_Λ is a Λ-Λ-bimodule (usually called the degree one differentials). Show that if M is a Λ-module then $\Omega_\Lambda \otimes_\Lambda M$ is a Λ-module of the form $\tilde{\Omega}(M)$.

1.6. Frobenius and symmetric algebras

Suppose that Λ is an algebra over a field k. If M is a left Λ-module, then the vector space dual $M^* = \mathrm{Hom}_k(M, k)$ has a natural structure as a right Λ-module, and vice-versa. If M is finite dimensional as a vector space, which it usually is because we are normally interested in finitely generated modules, then there is a natural isomorphism $(M^*)^* \cong M$. If M is injective then M^* is projective, and vice-versa, since duality reverses all arrows.

In general, projective and injective modules for a ring are very different. However, there is a special situation under which they are the same.

Definition 1.6.1. We say a finite dimensional algebra Λ over a field k is **Frobenius** if there is a linear map $\lambda : \Lambda \to k$ such that

(i) $\mathrm{Ker}(\lambda)$ contains no non-zero left or right ideal.

We say that Λ is **symmetric** if it satisfies (i) together with

(ii) For all $a, b \in \Lambda$, $\lambda(ab) = \lambda(ba)$.

We say that a ring Λ is **self injective** if the regular representation ${}_\Lambda\Lambda$ is an injective Λ-module.

Proposition 1.6.2. (i) *If Λ be a Frobenius algebra over k, then $(\Lambda_\Lambda)^* \cong {}_\Lambda\Lambda$. In particular Λ is self injective.*

(ii) *Suppose that Λ is self injective. Then the following conditions on a finitely generated Λ-module M are equivalent:*

(a) *M is projective* (b) *M is injective*
(c) *M^* is projective* (d) *M^* is injective.*

Proof. (i) We define a linear map $\phi : {}_\Lambda\Lambda \to (\Lambda_\Lambda)^*$ via $\phi(x) : y \to \lambda(yx)$. Then if $\gamma \in \Lambda$,

$$(\gamma(\phi(x)))y = (\phi(x))(y\gamma) = \lambda(y\gamma x) = (\phi(\gamma x))y,$$

so ϕ is a homomorphism. By the defining property of λ, ϕ is injective, and hence surjective by comparing dimensions.

(ii) It follows from self injectivity that M is projective if and only if M^* is projective, so that (a) and (c) are equivalent. We have already remarked that (a) $\Leftrightarrow$ (d) and (b) $\Leftrightarrow$ (c) hold for all finite dimensional algebras. $\square$

It follows from the above proposition that if P is a projective indecomposable module for a Frobenius algebra then not only $P/\mathrm{Rad}(P)$ but also $\mathrm{Soc}(P)$ are simple. In general they are not isomorphic, but in the special case of a symmetric algebra we have the following:

THEOREM 1.6.3. *Suppose that P is a projective indecomposable module for a symmetric algebra Λ. Then* $\mathrm{Soc}(P) \cong P/\mathrm{Rad}(P)$.

PROOF. Let e be a primitive idempotent in Λ with $P \cong \Lambda e$. Let $\lambda : \Lambda \to k$ be a linear map as in Definition 1.6.1. Then $\mathrm{Soc}(P) = \mathrm{Soc}(P).e$ is a left ideal of Λ and so there is an element $x \in \mathrm{Soc}(P)$ with $\lambda(x.e) \neq 0$. By the symmetry, $\lambda(e.x) \neq 0$ and so $e.\mathrm{Soc}(P) \neq 0$. But $e.\mathrm{Soc}(P) \cong \mathrm{Hom}_\Lambda(P, \mathrm{Soc}(P))$ by Lemma 1.3.3 (i), and so there is a non-zero homomorphism from P to $\mathrm{Soc}(P)$, which therefore induces an isomorphism from $P/\mathrm{Rad}(P)$ to $\mathrm{Soc}(P)$. $\square$

REMARK. We shall see in Section 3.1 that the group algebra of a finite group over a field of any characteristic is an example of a symmetric algebra.

EXERCISES. 1. Show that for a module M over a self injective algebra we have

$$M \cong \Omega\Omega^{-1}(M) \oplus (\text{projective}) \cong \Omega^{-1}\Omega(M) \oplus (\text{projective}).$$

In particular, as long as M has no projective summands, M is indecomposable if and only if $\Omega(M)$ is indecomposable.

2. Show that a finite dimensional algebra Λ is self injective if and only if for each simple Λ-module S with projective cover P_S, $\mathrm{Soc}(P_S)$ is simple, and whenever $S \not\cong S'$, $\mathrm{Soc}(P_S) \not\cong \mathrm{Soc}(P_{S'})$.

3. Show that a finite dimensional self injective algebra Λ is Frobenius if and only if for each projective indecomposable Λ-module P, $\dim_k \mathrm{Soc}(P) = \dim_k P/\mathrm{Rad}(P)$.

4. Show that if Λ is a finite dimensional symmetric algebra then so is $\mathrm{Mat}_n(\Lambda)$.

1.7. Idempotents and the Cartan matrix

Recall that an **idempotent** in a ring Λ is a non-zero element e with $e^2 = e$. If $e \neq 1$ is an idempotent then so is $1 - e$.

DEFINITION 1.7.1. Two idempotents e_1 and e_2 are said to be **orthogonal** if $e_1 e_2 = e_2 e_1 = 0$. An idempotent e is said to be **primitive** if we cannot write $e = e_1 + e_2$ with e_1 and e_2 orthogonal idempotents.

There is a one–one correspondence between expressions $1 = e_1 + \cdots + e_n$ with the e_i orthogonal idempotents, and direct sum decompositions ${}_\Lambda\Lambda = \Lambda_1 \oplus \cdots \oplus \Lambda_n$ of the regular representation, given by $\Lambda_i = \Lambda e_i$. Under this correspondence, e_i is primitive if and only if Λ_i is indecomposable.

PROPOSITION 1.7.2. *Two idempotents e and e' are conjugate in Λ if and only if $\Lambda e \cong \Lambda e'$ and $\Lambda(1-e) \cong \Lambda(1-e')$.*

PROOF. If e and e' are conjugate, say $e\mu = \mu e'$ with μ invertible, then $(1-e)\mu = \mu(1-e')$ and so μ induces an isomorphism form Λe to $\Lambda e'$ and from $\Lambda(1-e)$ to $\Lambda(1-e')$. Conversely if $\Lambda e \cong \Lambda e'$ and $\Lambda(1-e) \cong \Lambda(1-e')$, then by Lemma 1.3.3 there are elements $\mu_1 \in e\Lambda e'$, $\mu_2 \in e'\Lambda e$, $\mu_3 \in (1-e)\Lambda(1-e')$ and $\mu_4 \in (1-e')\Lambda(1-e)$ such that

$$\mu_1\mu_2 = e \qquad \mu_2\mu_1 = e'$$

$$\mu_3\mu_4 = 1-e \qquad \mu_4\mu_3 = 1-e'.$$

Letting $\mu = \mu_1 + \mu_3$ and $\mu' = \mu_2 + \mu_4$, we have $\mu\mu' = \mu'\mu = 1$ and $e\mu = \mu_1 = \mu e'$. □

Under the circumstances of the above proposition, we say e and e' are **equivalent**. Note that if the Krull–Schmidt theorem holds for finitely generated Λ-modules, then $\Lambda e \cong \Lambda e'$ implies $\Lambda(1-e) \cong \Lambda(1-e')$ since ${}_\Lambda\Lambda \cong \Lambda e \oplus \Lambda(1-e)$.

THEOREM 1.7.3 (Idempotent Refinement). *Let N be a nilpotent ideal in Λ, and let e be an idempotent in Λ/N. Then there is an idempotent f in Λ with $e = \bar{f}$.*

If e_1 is equivalent to e_2 in Λ/N, $\bar{f}_1 = e_1$ and $\bar{f}_2 = e_2$, then f_1 is equivalent to f_2 in Λ.

PROOF. We define idempotents $e_i \in \Lambda/N^i$ inductively as follows. Let $e_1 = e$. For $i > 1$, let a be any element of Λ/N^i with image e_{i-1} in Λ/N^{i-1}. Then $a^2 - a \in N^{i-1}/N^i$, and so $(a^2-a)^2 = 0$. Let $e_i = 3a^2 - 2a^3$. Then e_i has image e_{i-1} in Λ/N^{i-1}, and

$$e_i^2 - e_i = (3a^2 - 2a^3)(3a^2 - 2a^3 - 1) = -(3-2a)(1+2a)(a^2-a)^2 = 0.$$

If $N^r = 0$, we take $f = e_r$.

Note that in this proof, if Λ happens to be an algebra over a field k of characteristic p, we can instead take $e_i = a^p$ if we wish.

Now suppose that e_1 is conjugate to e_2, say $\bar{\mu}e_1 = e_2\bar{\mu}$ for some $\mu \in \Lambda$. Let $\nu = f_2\mu f_1 + (1-f_2)\mu(1-f_1)$. Then $\nu f_1 = f_2\nu$, and $1 - \nu = f_2\mu + \mu f_1 - 2f_2\mu f_1 = (f_2\mu - \mu f_1)(1-2f_1) \in N$ so that $1 + (1-\nu) + (1-\nu)^2 + \cdots$ is an inverse for ν. □

COROLLARY 1.7.4. *Let N be a nilpotent ideal in Λ. Let $1 = e_1 + \cdots + e_n$ with the e_i primitive orthogonal idempotents in Λ/N. Then we can write $1 = f_1 + \cdots + f_n$ with the f_i primitive orthogonal idempotents in Λ and $\bar{f}_i = e_i$. If e_i is conjugate to e_j then f_i is conjugate to f_j.*

PROOF. Define idempotents f'_i inductively as follows. $f'_1 = 1$, and for $i > 1$, f'_i is any lift of $e_i + e_{i+1} + \cdots + e_n$ to an idempotent in the ring $f'_{i-1}\Lambda f'_{i-1}$. Then $f'_i f'_{i+1} = f'_{i+1} = f'_{i+1} f'_i$. Let $f_i = f'_i - f'_{i+1}$. Clearly $\bar{f}_i = e_i$. If $j > i$, $f_j = f'_{i+1} f_j f'_{i+1}$, and so $f_i f_j = (f'_i - f'_{i+1}) f'_{i+1} f_j f'_{i+1} = 0$. Similarly $f_j f_i = 0$. □

Now for the rest of this section, suppose that Λ satisfies D.C.C. on left ideals. Then by the Wedderburn Structure Theorem 1.3.5, we may write $\Lambda/J(\Lambda) = \bigoplus_{i=1}^r \mathrm{Mat}_{n_i}(\Delta_i)$. Write S_i for the simple Λ-module corresponding to the ith matrix factor. Then the regular representation of $\Lambda/J(\Lambda)$ is isomorphic to $\bigoplus_{i=1}^r n_i S_i$. This decomposition corresponds to an expression $1 = e_1 + e_2 + \cdots$ in $\Lambda/J(\Lambda)$ with the e_i orthogonal idempotents. Lifting to an expression $1 = f_1 + f_2 + \cdots$ in Λ as in the above corollary, we have a direct sum decomposition

$$_\Lambda\Lambda = \bigoplus_{i=1}^r n_i P_i$$

with $P_i/J(\Lambda)P_i \cong S_i$. By the Krull–Schmidt theorem, every projective indecomposable module is isomorphic to one of the P_i.

LEMMA 1.7.5. $\mathrm{Hom}_\Lambda(P_i, S_j) \cong \begin{cases} \Delta_i & \textit{if } i = j \\ 0 & \textit{otherwise.} \end{cases}$

PROOF. P_i has a unique top composition factor, and this is isomorphic to S_i. □

LEMMA 1.7.6. *$\dim_{\Delta_i} \mathrm{Hom}_\Lambda(P_i, M)$ is the multiplicity of S_i as a composition factor of M.*

PROOF. Use the previous lemma and induction on the composition length of M. Since P_i is projective, an exact sequence

$$0 \to M' \to M \to S_j \to 0$$

induces a short exact sequence

$$0 \to \mathrm{Hom}_\Lambda(P_i, M') \to \mathrm{Hom}_\Lambda(P_i, M) \to \mathrm{Hom}_\Lambda(P_i, S_j) \to 0.$$ □

Dually we have:

LEMMA 1.7.7. *Suppose that I_S is the injective hull of a simple Λ-module S, and $\Delta = \mathrm{End}_\Lambda(S)$. Then $\dim_\Delta \mathrm{Hom}_\Lambda(M, I_S)$ is equal to the multiplicity of S as a composition factor of M.* □

Combining these lemmas, we have the following:

THEOREM 1.7.8 (Landrock [**147**]). *Suppose that S and T are simple modules for a finite dimensional algebra Λ over a splitting field k. Then the multiplicity of T as a composition factor in the nth Loewy layer of the projective cover P_S is equal to the multiplicity of the dual S^* (which is a right Λ-module) as a composition factor in the nth Loewy layer of the projective cover P_{T^*}.*

PROOF. Since k is a splitting field, each Δ_i is equal to k. Since

$$\mathrm{Rad}^n\mathrm{Soc}^n I_T = 0 \quad \text{and} \quad \mathrm{Soc}^n(P_S/\mathrm{Rad}^n P_S) = P_S/\mathrm{Rad}^n P_S,$$

we have

$$\begin{aligned}\mathrm{Hom}_\Lambda(P_S/\mathrm{Rad}^n P_S, I_T) &= \mathrm{Hom}_\Lambda(P_S/\mathrm{Rad}^n P_S, \mathrm{Soc}^n I_T)\\ &= \mathrm{Hom}_\Lambda(P_S, \mathrm{Soc}^n I_T).\end{aligned}$$

By Lemma 1.7.7, the dimension of the left hand side is equal to the multiplicity of T as a composition factor in the first n Loewy layers of P_S. By Lemma 1.7.6, the dimension of the right hand side is equal to the multiplicity of S in the first n socle layers of I_T. The dual of I_T is P_{T^*}, so this is equal to the multiplicity of S^* in the first n Loewy layers of P_{T^*}. The theorem follows by subtraction. □

DEFINITION 1.7.9. The **Cartan invariants** of Λ are defined as

$$c_{ij} = \dim_{\Delta_i} \mathrm{Hom}_\Lambda(P_i, P_j),$$

namely the multiplicity of S_i as a composition factor of P_j. The matrix (c_{ij}) is called the **Cartan matrix** of the ring Λ.

In general, the matrix (c_{ij}) may be singular, but we shall see in Corollary 5.3.5 that this never happens for a group algebra of a finite group. In fact, we shall see in Corollary 5.7.2 and Theorem 5.9.3 that the determinant of the Cartan matrix of a group algebra over a field of characteristic $p > 0$ is a power of p.

Finally, the following general fact about idempotents is often useful.

LEMMA 1.7.10 (Rosenberg's lemma). *Suppose that e is an idempotent in a ring Λ, $e\Lambda e$ is a local ring (cf. Lemmas 1.3.3, 1.4.5 and Theorem 1.9.3), and $e \in \sum_\alpha I_\alpha$, where I_α is a family of two-sided ideals in Λ. Then for some α we have $e \in I_\alpha$.*

PROOF. Each $eI_\alpha e$ is an ideal in the local ring $e\Lambda e$, and so for some value of α we have $eI_\alpha e = e\Lambda e$. □

1.8. Blocks and central idempotents

DEFINITION 1.8.1. A **central idempotent** in Λ is an idempotent in the centre of Λ. A **primitive central idempotent** is a central idempotent not expressible as the sum of two orthogonal central idempotents. There is a one–one correspondence between expressions $1 = e_1 + \cdots + e_s$ with e_i orthogonal central idempotents and direct sum decompositions $\Lambda = B_1 \oplus \cdots \oplus B_s$ of Λ as two-sided ideals, given by $B_i = e_i\Lambda$.

Now suppose that Λ is Artinian. Then we can write $\Lambda = B_1 \oplus \cdots \oplus B_s$ with the B_i indecomposable two-sided ideals.

LEMMA 1.8.2. *This decomposition is unique; i.e., if $\Lambda = B_1 \oplus \cdots \oplus B_s = B'_1 \oplus \cdots \oplus B'_t$ then $s = t$ and after renumbering if necessary, $B_i = B'_i$.*

PROOF. Write $1 = e_1 + \cdots + e_s = e'_1 + \cdots + e'_t$. Then $e_ie'_j$ is either a central idempotent or zero for each pair i, j. Thus $e_i = e_ie'_1 + \cdots + e_ie'_t$, so that for a unique j, $e_i = e_ie'_j = e'_j$. □

DEFINITION 1.8.3. The indecomposable two-sided ideals in this decomposition are called the **blocks** of Λ.

Now suppose that M is an indecomposable Λ-module. Then $M = e_1M \oplus \cdots \oplus e_sM$ shows that for some i, $e_iM = M$, and $e_jM = 0$ for $j \neq i$. We then say that M **belongs to** the block B_i. Thus the simple modules and projective indecomposables are classified into blocks. Clearly if an indecomposable module is in a certain block, then so are all its composition factors.

The following proposition states that the block decomposition is determined by what happens modulo the square of the radical. It first appears in this form in the literature in Külshammer **[144]**, although equivalent statements have been well known for a long time.

PROPOSITION 1.8.4. *Suppose that Λ is Artinian and I is a two sided ideal contained in $J^2(\Lambda)$. Then the natural map $\Lambda \to \Lambda/I$ induces a bijection between the set of idempotents in the centre $Z(\Lambda)$ and the set of idempotents in $Z(\Lambda/I)$.*

PROOF. If f is an idempotent in $Z(\Lambda)$ then clearly $\bar{f}$ is an idempotent in $Z(\Lambda/I)$. If $\bar{f} = \bar{f}'$ then $f - ff'$ is nilpotent and idempotent, hence zero, so $f = ff' = f'$.

Conversely if e is an idempotent in $Z(\Lambda/I)$ then by Theorem 1.7.3 there is an idempotent f in Λ with $\bar{f} = e$. So we must show that $f \in Z(\Lambda)$. Since $\bar{f} \in Z(\Lambda/I)$, we have $\bar{f}(\Lambda/I)(1-\bar{f}) = 0$ and so $f\Lambda(1-f) \subseteq I \subseteq J^2$. Since f and $1-f$ are idempotent it follows that $f\Lambda(1-f) = fJ^2(1-f)$. We show by induction on n that $f\Lambda(1-f) = fJ^n(1-f)$. Namely

$$\begin{aligned} f\Lambda(1-f) &= fJ^{n-1}J(1-f) \subseteq fJ^{n-1}fJ(1-f) + fJ^{n-1}(1-f)J(1-f) \\ &\subseteq fJ^{n-1}fJ^2(1-f) + fJ^n(1-f)J(1-f) \subseteq fJ^{n+1}(1-f). \end{aligned}$$

Since J is nilpotent we thus have $f\Lambda(1-f) = 0$, and so for $a \in \Lambda$ we have $fa = faf + fa(1-f) = faf$. Similarly $af = faf$ and so $fa = af$, so that $f \in Z(\Lambda)$. □

The following should be compared with the Wedderburn–Artin theorem 1.3.5.

PROPOSITION 1.8.5. *Suppose that M is a simple Λ-module which is both projective and injective. Then M is the unique simple module in a block B of Λ with $B \cong \mathrm{Mat}_n(\Delta)$. Here, Δ is the division ring $\mathrm{End}_\Lambda(M)^{\mathrm{op}}$ and $n = \dim_{\Delta^{\mathrm{op}}}(M)$.*

PROOF. Since M is both projective and injective, we can write ${}_\Lambda\Lambda = n.M \oplus P$, where P is a projective module which does not involve M. Hence by Lemma 1.3.3 $\Lambda = \mathrm{End}_\Lambda({}_\Lambda\Lambda)^{\mathrm{op}} \cong \mathrm{Mat}_n(\mathrm{End}_\Lambda(M))^{\mathrm{op}} \oplus \mathrm{End}_\Lambda(P)^{\mathrm{op}}$. □

EXERCISE. Show that every commutative Artinian ring is a direct sum of local rings.

1.9. Algebras over a complete domain

In order to compare representations in characteristic zero with representations in characteristic p, we use representations over the p-adic integers as an intermediary. This is easier than using the ordinary integers because, as we shall see, we have a Krull–Schmidt theorem. It is better than using the p-local integers (i.e., the integers with numbers coprime to p inverted) because of the idempotent refinement theorem, which enables us to lift projective indecomposables from characteristic p.

Since it is often convenient to deal with fields larger than the rationals, we also look at rings of $\mathfrak{p}$-adic integers for $\mathfrak{p}$ a prime ideal in a ring of algebraic integers. The most general set up of this sort is a complete rank one discrete valuation ring, but we shall be content with rings of $\mathfrak{p}$-adic integers. If $\mathcal{O}$ is the ring of integers in an algebraic extension K of $\mathbb{Q}$ and $\mathfrak{p}$ is a prime ideal in $\mathcal{O}$ lying above a rational prime p, we form the completion

$$\mathcal{O}_{\mathfrak{p}} = \varprojlim_{n} \mathcal{O}/\mathfrak{p}^n.$$

The natural map $\mathcal{O} \to \mathcal{O}_{\mathfrak{p}}$ is injective, and so K is a subfield of the field of fractions $K_{\mathfrak{p}}$ of $\mathcal{O}_{\mathfrak{p}}$. The ring $\mathcal{O}_{\mathfrak{p}}$ has a unique maximal ideal $\mathfrak{p}_{\mathfrak{p}}$, which is principal, $\mathfrak{p}_{\mathfrak{p}} = (\pi)$. In particular $\mathcal{O}_{\mathfrak{p}}$ is a principal ideal domain, so that finitely generated torsion-free modules are free. The quotient field

$$k = \mathcal{O}_{\mathfrak{p}}/\mathfrak{p}_{\mathfrak{p}} \cong \mathcal{O}/\mathfrak{p}$$

is a field of characteristic p. We say that $(K_{\mathfrak{p}}, \mathcal{O}_{\mathfrak{p}}, k)$ is a **p-modular system**. More generally, if $\mathcal{O}$ is a complete rank one discrete valuation ring with field of fractions K of characteristic zero, maximal ideal $\mathfrak{p} = (\pi)$, and quotient field $k = \mathcal{O}/\mathfrak{p}$ of characteristic p, we shall say that $(K, \mathcal{O}, k)$ is a p-modular system. For the remainder of this section, K, $\mathcal{O}$ and k will be of this form.

Let Λ be an algebra over $\mathcal{O}$ which as an $\mathcal{O}$-module is free of finite rank. Let $\hat{\Lambda} = K \otimes_{\mathcal{O}} \Lambda$ and $\bar{\Lambda} = k \otimes_{\mathcal{O}} \Lambda = \Lambda/\pi\Lambda$. By a **$\Lambda$-lattice** we mean a finitely generated $\mathcal{O}$-free Λ-module. If M is a Λ-lattice then we set $\hat{M} = K \otimes_{\mathcal{O}} M$ as a $\hat{\Lambda}$-module, and $\bar{M} = k \otimes_{\mathcal{O}} M = M/\pi M$ as a $\bar{\Lambda}$-module. If K is a splitting field for $\hat{\Lambda}$ and k is a splitting field for $\bar{\Lambda}$, we say that $(K, \mathcal{O}, k)$ is a **splitting p-modular system** for Λ.

We call $\hat{\Lambda}$-modules **ordinary representations**, Λ-lattices **integral representations** and $\bar{\Lambda}$-modules **modular representations**.

LEMMA 1.9.1. *If V is a $\hat{\Lambda}$-module then there is a Λ-lattice M with $\hat{M} \cong V$.*

PROOF. Choose a basis $v_1, \ldots, v_n$ for V as a vector space over K and let $M = \Lambda v_1 + \cdots + \Lambda v_n \subseteq V$. As an $\mathcal{O}$-module, M is finitely generated and torsion free, and hence free. Choose a free basis $x_1, \ldots, x_m$. Then the x_i span V and are K-independent, and hence $m = n$, and $V = K \otimes_{\mathcal{O}} M$. □

Such a Λ-lattice M is called an **$\mathcal{O}$-form** of V. In general a $\hat{\Lambda}$-module has many non-isomorphic $\mathcal{O}$-forms.

LEMMA 1.9.2 (Fitting's lemma, $\mathfrak{p}$-adic version). *Let M be a Λ-lattice and suppose that $f \in \mathrm{End}_\Lambda(M)$. Write $\mathrm{Im}(f^\infty) = \bigcap_{n=1}^\infty \mathrm{Im}(f^n)$ and $\mathrm{Ker}(f^\infty) = \{x \in M \mid \forall\, n \geq 0\ \exists\, m \geq 0 \text{ s.t. } f^m(x) \in J(\Lambda)^n M\}$. Then*

$$M = \mathrm{Im}(f^\infty) \oplus \mathrm{Ker}(f^\infty).$$

PROOF. This follows from the usual version of Fitting's lemma. □

THEOREM 1.9.3 (Krull–Schmidt theorem, $\mathfrak{p}$-adic version). (i) *If M is an indecomposable Λ-lattice then $\mathrm{End}_\Lambda(M)$ is a local ring.*

(ii) *The unique decomposition property holds for Λ-lattices.*

PROOF. The proof of (i) is the same as the proof of 1.4.5, and (ii) follows by Theorem 1.4.3. □

THEOREM 1.9.4 (Idempotent refinement). (i) *Let e be an idempotent in $\bar{\Lambda}$. Then there is an idempotent f in Λ with $e = \bar{f}$. If e_1 is conjugate to e_2 in $\bar{\Lambda}$, $\bar{f_1} = e_1$ and $\bar{f_2} = e_2$ then f_1 is conjugate to f_2 in Λ.*

(ii) *Let $1 = e_1 + \cdots + e_n$ with the e_i primitive orthogonal idempotents in $\bar{\Lambda}$. Then we can write $1 = f_1 + \cdots + f_n$ with the f_i primitive orthogonal idempotents in Λ and $\bar{f_i} = e_i$. If e_i is conjugate to e_j then f_i is conjugate to f_j.*

(iii) *Suppose that reduction modulo $\mathfrak{p}$ is a surjective map from the centre $Z(\Lambda)$ to $Z(\bar{\Lambda})$. Let $1 = e_1 + \cdots + e_n$ with the e_i primitive central idempotents in $\bar{\Lambda}$. Then we can write $1 = f_1 + \cdots + f_n$ with the f_i primitive central idempotents in Λ and $\bar{f_i} = e_i$.*

PROOF. (i) We may apply the idempotent refinement theorem 1.7.3 for nilpotent ideals to obtain idempotents $f_i \in \Lambda/\pi^i\Lambda$ whose image in $\Lambda/\pi^{i-1}\Lambda$ is f_{i-1}. These define an element of $\Lambda = \varprojlim_n \Lambda/\pi^n\Lambda$ which is easily seen to be idempotent.

The conjugacy statement is proved exactly as in 1.7.3.

(ii) Apply the same argument to Corollary 1.7.4.

(iii) Apply (ii) to the centre of Λ. □

REMARK. We shall see that the hypothesis in (iii) is satisfied by group algebras of finite groups.

It follows from the above theorem that the decomposition of the regular representation ${}_{\bar{\Lambda}}\bar{\Lambda}$ into projective indecomposable modules lifts to a decomposition of ${}_\Lambda\Lambda$. So given a simple $\bar{\Lambda}$-module S_j, it has a projective cover $P_j = \bar{Q}_j$ for some projective indecomposable Λ-module Q_j unique up to isomorphism.

DEFINITION 1.9.5. Suppose that $V_1, \ldots, V_t$ are representatives for the isomorphism classes of irreducible $\hat{\Lambda}$-modules, and $M_1, \ldots, M_t$ are $\mathcal{O}$-forms of them (see the above lemma). Then we define the **decomposition number** d_{ij} to be the multiplicity of S_j as a composition factor of $\bar{M}_i$.

The following proposition shows that the decomposition numbers are independent of the choices of $\mathcal{O}$-forms.

PROPOSITION 1.9.6. *Suppose that $(K, \mathcal{O}, k)$ is a splitting system for Λ, and that $\hat{\Lambda}$ is semisimple. Then d_{ij} is the multiplicity of V_i as a composition factor of $\hat{Q}_j$. In particular*

$$c_{ij} = \sum_k d_{ki} d_{kj}.$$

PROOF. We have

$$\begin{aligned} d_{ij} &= \dim_k \operatorname{Hom}_{\bar{\Lambda}}(P_j, \bar{M}_i) \quad \text{by 1.7.6} \\ &= \operatorname{rank}_{\mathcal{O}} \operatorname{Hom}_{\Lambda}(Q_j, M_i) \quad \text{since } Q_j \text{ is projective} \\ &= \dim_k \operatorname{Hom}_{\hat{\Lambda}}(\hat{Q}_j, V_i) \end{aligned}$$

which is equal to the multiplicity of V_i as a composition factor of $\hat{Q}_j$ since $\hat{\Lambda}$ is semisimple. □

REMARKS. (i) Note carefully what this proposition is saying. It is saying that the decomposition matrix can be read in two different ways. The rows give the modular composition factors of modular reductions of the ordinary irreducibles, while the columns give the ordinary composition factors of lifts of the modular projective indecomposables. It is thus clear that the decomposition matrix times its transpose gives the modular irreducible composition factors of the modular projective indecomposables, namely the Cartan matrix.

(ii) If Λ is a group ring, we shall see in Chapter 3 that $\hat{\Lambda}$ is semisimple, so that this proposition applies in this case.

(iii) This proposition makes it clear that the decomposition numbers d_{ij} are independent of the choice of $\mathcal{O}$-form M_i chosen for the V_i.

(iv) It also follows from this proposition that the Cartan matrix (c_{ij}) is symmetric in this case. This is not true for more general algebras, even over splitting systems.

(v) In case $(K, \mathcal{O}, k)$ is not a splitting system, a modification of the above proposition is true. Namely the multiplicity of V_i as a composition factor of $\hat{Q}_j$ is

$$d_{ij} . \dim_k \operatorname{End}_{\bar{\Lambda}}(S_j) / \dim_k \operatorname{End}_{\hat{\Lambda}}(V_i)$$

and so

$$c_{ij} = \sum_k d_{ki} d_{kj} . \dim_k \operatorname{End}_{\bar{\Lambda}}(S_j) / \dim_k \operatorname{End}_{\hat{\Lambda}}(V_k).$$

The proof is the same.

CHAPTER 2

Homological algebra

2.1. Categories and functors

We shall assume that the reader is familiar with the elementary notions of **category** and **functor** (covariant and contravariant) as explained in Mac Lane [**149**, Sections 1.7 and 1.8].

DEFINITION 2.1.1. If $F, F' : \mathcal{C} \to \mathcal{D}$ are covariant functors, a **natural transformation** $\phi : F \rightsquigarrow F'$ assigns to each object $X \in \mathcal{C}$ a map $\phi_X : F(X) \to F'(X)$ in such a way that the square

$$\begin{array}{ccc} F(X) & \xrightarrow{\phi_X} & F'(X) \\ {\scriptstyle F(\alpha)}\downarrow & & \downarrow{\scriptstyle F'(\alpha)} \\ F(Y) & \xrightarrow{\phi_Y} & F'(Y) \end{array}$$

commutes for each morphism $\alpha : X \to Y$ in $\mathcal{C}$. Similarly if F and F' are contravariant, we make the same definition, but with the vertical arrows in the above diagram reversed. We write $\mathrm{Nat}(F, F')$ for the set of natural transformations from F to F'. A natural transformation $\phi : F \rightsquigarrow F'$ is a **natural isomorphism** if ϕ_X is an isomorphism for each $X \in \mathcal{C}$.

An **equivalence of categories** is a pair of functors $F : \mathcal{C} \to \mathcal{D}$ and $F' : \mathcal{D} \to \mathcal{C}$ such that $F \circ F'$ and $F' \circ F$ are naturally isomorphic to the appropriate identity functors.

The following are examples of categories we shall be interested in during the course of this book:

(i) The category **Grp** of groups and homomorphisms.
(ii) The categories ${}_\Lambda\mathbf{Mod}$ of left Λ-modules and ${}_\Lambda\mathbf{mod}$ of finitely generated left Λ-modules, for a ring Λ.
(iii) The categories **Set** of sets, **Ab** of abelian groups and ${}_k\mathbf{Vec}$ of k-vector spaces.
(iv) The category of functors from ${}_\Lambda\mathbf{mod}$ to **Ab**, or from ${}_\Lambda\mathbf{mod}$ to ${}_k\mathbf{Vec}$ if Λ is a k-algebra. In this category the morphisms are the natural transformations.
(v) The category of topological spaces and (continuous) maps.
(vi) The category of CW-complexes and homotopy classes of maps.
(vii) The category of chain complexes and chain maps.

The correct setting for doing homological algebra is an **abelian category**. A typical example of an abelian category is a category of modules for a ring.

DEFINITION 2.1.2. An **abelian category** is a category with the following extra structure.

(i) For each pair of objects A and B the set of maps $\mathrm{Hom}(A, B)$ is given the structure of an abelian group.

(ii) There is a **zero object** 0 with the property that $\mathrm{Hom}(A, 0)$ and $\mathrm{Hom}(0, A)$ are the trivial group for all objects A.

(iii) Composition of maps is a bilinear map

$$\mathrm{Hom}(B, C) \times \mathrm{Hom}(A, B) \to \mathrm{Hom}(A, C).$$

(iv) Finite direct sums exist (with the usual universal definition).

(v) Every morphism $\phi : A \to B$ has a **kernel**, namely a map $\sigma : K \to A$ such that $\phi \circ \sigma = 0$, and such that whenever $\sigma' : K' \to A$ with $\phi \circ \sigma' = 0$ there is a unique map $\lambda : K' \to K$ with $\sigma' = \sigma \circ \lambda$.

(vi) Every morphism has a **cokernel** (definition dual to that of kernel).

(vii) Every **monomorphism** (map with zero kernel) is the kernel of its cokernel.

(viii) Every **epimorphism** (map with zero cokernel) is the cokernel of its kernel.

(ix) Every morphism is the composite of a monomorphism and an epimorphism.

An **additive functor** F from one abelian category to another is one which induces a homomorphism of abelian groups

$$\mathrm{Hom}(A, B) \to \mathrm{Hom}(F(A), F(B))$$

for each pair A and B.

Freyd [**108**] has shown that given any *small* abelian category $\mathcal{A}$ (i.e., one where the class of objects is small enough to be a set) there is a *full exact* embedding $F : \mathcal{A} \to {}_{\Lambda}\mathbf{Mod}$ for a suitable ring Λ. Here, *full* means that for $X, Y \in \mathcal{A}$, every map in ${}_{\Lambda}\mathbf{Mod}$ from $F(X)$ to $F(Y)$ is in the image of F. *Exact* means that F takes exact sequences to exact sequences. This has the effect that diagram chasing may be performed in an abelian category as though the objects had elements. Since we shall only be working with categories where this is obviously true, we shall write our proofs this way. It is a simple matter and a worthless exercise to translate such a proof into a proof using only the axioms.

Thus you should not memorise the definition of an abelian category, but rather remember the Freyd category embedding theorem, and look up the definitions whenever you need them.

Often in representation theory, it is more convenient to work not in a module category but in a **stable module category**. We write ${}_{\Lambda}\underline{\mathbf{mod}}$ for the category of finitely generated Λ-modules modulo projectives. Namely, the objects of ${}_\Lambda\underline{\mathbf{mod}}$ are the same as those of ${}_\Lambda\mathbf{mod}$, but two maps in ${}_\Lambda\mathbf{mod}$ are regarded as the same in ${}_\Lambda\underline{\mathbf{mod}}$ if their difference factors through a projective module. Thus for example the projective modules are isomorphic to the zero object in ${}_\Lambda\underline{\mathbf{mod}}$. We write $\underline{\mathrm{Hom}}_\Lambda(M,N)$ and $\underline{\mathrm{End}}_\Lambda(M)$ for the hom sets in ${}_\Lambda\mathbf{mod}$, namely homomorphisms modulo those factoring through a projective module.

If the Krull–Schmidt theorem holds in ${}_\Lambda\mathbf{mod}$ then the indecomposable objects in ${}_\Lambda\underline{\mathbf{mod}}$ correspond to the non-projective indecomposable objects in ${}_\Lambda\mathbf{mod}$.

Recall from Section 1.5 that if M is a Λ-module then $\tilde{\Omega}(M)$ is defined to be the kernel of some epimorphism $P \to M$ with P projective. Schanuel's lemma can be interpreted as saying that while $\tilde{\Omega}$ is not a functor on ${}_\Lambda\mathbf{mod}$, it passes down to a well defined functor

$$\Omega : {}_\Lambda\underline{\mathbf{mod}} \to {}_\Lambda\underline{\mathbf{mod}}.$$

Similarly we write ${}_\Lambda\overline{\mathbf{mod}}$ for the category of finitely generated Λ-modules modulo injectives, and $\overline{\mathrm{Hom}}_\Lambda(M,N)$ and $\overline{\mathrm{End}}_\Lambda(M)$ for the hom sets in ${}_\Lambda\overline{\mathbf{mod}}$. The functor $\tilde{\Omega}^{-1}$ passes down to a well defined functor

$$\Omega^{-1} : {}_\Lambda\overline{\mathbf{mod}} \to {}_\Lambda\overline{\mathbf{mod}}.$$

If Λ is self injective, so that finitely generated projective and injective modules coincide, then ${}_\Lambda\underline{\mathbf{mod}} = {}_\Lambda\overline{\mathbf{mod}}$ and the functors Ω and Ω^{-1} are inverse to each other.

Representable functors.

Definition 2.1.3. A covariant functor $F : \mathcal{C} \to \mathbf{Set}$ is said to be **representable** if it is naturally isomorphic to a functor of the form

$$(X,-) : Y \to \mathrm{Hom}(X,Y).$$

A contravariant functor is representable if it is naturally isomorphic to a functor of the form

$$(-,Y) : X \to \mathrm{Hom}(X,Y).$$

If Hom sets in $\mathcal{C}$ have natural structures as abelian groups or vector spaces, then we have the same definition of representability of functors $F : \mathcal{C} \to \mathbf{Ab}$ or $F : \mathcal{C} \to {}_k\mathbf{Vec}$.

One of the most useful elementary lemmas from category theory is Yoneda's lemma, which says that natural transformations from representable functors are representable.

Lemma 2.1.4 (Yoneda). (i) *If $F : \mathcal{C} \to$ **Set** is a covariant functor and $(X,-)$ is a representable functor then the set of natural transformations from $(X,-)$ to F is in natural bijection with $F(X)$ via the map*

$$\begin{aligned} \mathrm{Nat}((X,-),F) &\xrightarrow{\cong} F(X) \\ (\phi : (X,-) \rightsquigarrow F) &\mapsto \phi_X(\mathrm{id}_X). \end{aligned}$$

(ii) *If* $F : \mathcal{C} \to \mathbf{Set}$ *is a contravariant functor and* $(-, X)$ *is a representable functor then the set of natural transformations from* $(-, X)$ *to* F *is in natural bijection with* $F(X)$ *via the map*

$$\mathrm{Nat}((-,X),F) \xrightarrow{\cong} F(X)$$
$$(\phi : (-,X) \rightsquigarrow F) \mapsto \phi_X(\mathrm{id}_X).$$

PROOF. (i) It is easy to check that the map

$$\begin{aligned} F(X) &\to \mathrm{Nat}((X,-),F) \\ x \in F(X) &\mapsto (\ \phi : (X,-) \rightsquigarrow F \\ &\qquad \phi_Y(\alpha : X \to Y) = F(\alpha)(x) \in F(Y)\ ) \end{aligned}$$

is inverse to the given map. The proof of (ii) is similar. □

ADJOINT FUNCTORS.

DEFINITION 2.1.5. An **adjunction** between functors $F : \mathcal{C} \to \mathcal{D}$ and $G : \mathcal{D} \to \mathcal{C}$ consists of bijections

$$\mathrm{Hom}(FX, Y) \to \mathrm{Hom}(X, GY)$$

natural in each variable $X \in \mathcal{C}$ and $Y \in \mathcal{D}$. We say that F is the **left adjoint** and G is the **right adjoint**.

It is not hard to see that if a functor has a right (or left) adjoint, then it is unique up to natural isomorphism. Examples of adjunctions abound. The most familiar example is probably the adjunction

$$\mathrm{Hom}(X \times Y, Z) \cong \mathrm{Hom}(Y, \mathrm{Hom}(X, Z))$$

between the functors $X \times -$ and $\mathrm{Hom}(X, -)$ on **Set**. Similarly in ${}_k\mathbf{Vec}$ we have

$$\mathrm{Hom}(U \otimes V, W) \cong \mathrm{Hom}(V, \mathrm{Hom}(U, W)).$$

Another class of examples is given by free objects. For example if $F : \mathbf{Set} \to \mathbf{Grp}$ takes a set to the free group with that set as basis, then F is left adjoint to the forgetful functor $G : \mathbf{Grp} \to \mathbf{Set}$ which assigns to each group its underlying set of elements.

LEMMA 2.1.6. *Suppose that* $\mathcal{C}$ *and* $\mathcal{D}$ *are abelian categories and* $F : \mathcal{C} \to \mathcal{D}$ *has a right adjoint* $G : \mathcal{D} \to \mathcal{C}$. *Then* F *takes epimorphisms to epimorphisms and* G *takes monomorphisms to monomorphisms.*

PROOF. A map $X \to X'$ is an epimorphism if and only if for every $Z \in \mathcal{C}$, the map $\mathrm{Hom}(X', Z) \to \mathrm{Hom}(X, Z)$ is injective. In particular

$$\mathrm{Hom}(X', GY) \to \mathrm{Hom}(X, GY)$$

is injective so that

$$\mathrm{Hom}(FX', Y) \to \mathrm{Hom}(FX, Y)$$

is injective for every $Y \in \mathcal{D}$. Thus $FX \to FX'$ is an epimorphism. The other statement is proved dually. □

2.2. Morita theory

When are two module categories ${}_\Lambda\mathbf{Mod}$ and ${}_\Gamma\mathbf{Mod}$ equivalent as abelian categories? Let $F : {}_\Lambda\mathbf{Mod} \to {}_\Gamma\mathbf{Mod}$, $F' : {}_\Gamma\mathbf{Mod} \to {}_\Lambda\mathbf{Mod}$ be an equivalence. Since the definition of a projective module is purely categorical, F and F' induce an equivalence between the full subcategories ${}_\Lambda\mathbf{Proj}$ and ${}_\Gamma\mathbf{Proj}$ of projective modules. Among all projective modules, one can recognise the finitely generated ones as the projective modules P for which $\mathrm{Hom}_\Lambda(P, -)$ distributes over direct sums. So F and F' induce an equivalence between the full subcategories ${}_\Lambda\mathbf{proj}$ and ${}_\Gamma\mathbf{proj}$ of finitely generated projective modules.

The image of the regular representation $P = F'({}_\Gamma\Gamma) \in {}_\Lambda\mathbf{Mod}$ has the following properties:

(i) P is a finitely generated projective module.

(ii) Every Λ-module is a homomorphic image of a direct sum of copies of P.

(iii) $\Gamma \cong \mathrm{End}_\Lambda(P)^{\mathrm{op}}$.

Conversely, we shall see that if P is a Λ-module satisfying (i) and (ii) then letting $\Gamma = \mathrm{End}_\Lambda(P)^{\mathrm{op}}$, ${}_\Lambda\mathbf{Mod}$ is equivalent to ${}_\Gamma\mathbf{Mod}$. The proof goes via an intermediate characterisation of equivalent module categories, using bimodules.

DEFINITION 2.2.1. A Λ-module P satisfying conditions (i) and (ii) above is called a **progenerator** for ${}_\Lambda\mathbf{Mod}$.

If Λ is an Artinian ring with $\Lambda/J(\Lambda) \cong \bigoplus \mathrm{Mat}_{n_i}(\Delta_i)$ and corresponding projective indecomposables P_i, so that ${}_\Lambda\Lambda = \bigoplus n_i P_i$, then a finitely generated projective module $P = \bigoplus m_i P_i$ is a progenerator if and only if each $m_i > 0$. If $\Gamma = \mathrm{End}_\Lambda(P)^{\mathrm{op}}$ then $\Gamma/J(\Gamma) \cong \bigoplus \mathrm{Mat}_{m_i}(\Delta_i)$. Thus the simple modules have changed dimension from n_i to m_i, without changing any other aspect of the representation theory. The smallest possibility for Γ is to take each $m_i = 1$. In this case, we say that Γ is the **basic algebra** of Λ. Basic algebras are characterised by the property that every simple module is one dimensional over the corresponding division ring.

DEFINITION 2.2.2. Two rings Λ and Γ are said to be **Morita equivalent** if there are bimodules ${}_\Lambda P_\Gamma$ and ${}_\Gamma Q_\Lambda$ and surjective maps $\phi : P \otimes_\Gamma Q \to \Lambda$ of Λ-Λ-bimodules and $\psi : Q \otimes_\Lambda P \to \Gamma$ of Γ-Γ-bimodules satisfying the identities $x\psi(y \otimes z) = \phi(x \otimes y)z$ and $y\phi(z \otimes w) = \psi(y \otimes z)w$ for x and z in P and y and w in Q.

LEMMA 2.2.3. *If P is a progenerator for ${}_\Lambda\mathbf{Mod}$, and $\Gamma = \mathrm{End}_\Lambda(P)^{\mathrm{op}}$ then Λ and Γ are Morita equivalent.*

PROOF. The ring Γ acts on P on the right, making P into a Λ-Γ-bimodule. Let $Q = \mathrm{Hom}_\Lambda(P, \Lambda)$, as a Γ-Λ-bimodule. The map $\phi : P \otimes_\Gamma \mathrm{Hom}_\Lambda(P, \Lambda) \to \Lambda$ given by evaluation is surjective, since Λ is a homomorphic image of a sum of copies of P, while the map $\psi : \mathrm{Hom}_\Lambda(P, \Lambda) \otimes_\Lambda P \to$

$\mathrm{End}_\Lambda(P)^{\mathrm{op}}$ given by $\psi(f \otimes x)^{\mathrm{op}}(y) = f(y).x$ is surjective since P is a summand of a finite sum of copies of ${}_\Lambda\Lambda$, so that every endomorphism is a sum of endomorphisms factoring through ${}_\Lambda\Lambda$. The identities are easy to check. □

LEMMA 2.2.4. *If ${}_\Lambda P_\Gamma$ and ${}_\Gamma Q_\Lambda$ are bimodules as in the definition of Morita equivalence, then the maps $\phi : P \otimes_\Gamma Q \to \Lambda$ and $\psi : Q \otimes_\Lambda P \to \Gamma$ are isomorphisms.*

PROOF. We shall show that $\mathrm{Ker}(\phi) = 0$. Let $\phi(\sum_i x_i \otimes y_i) = 1 \in \Gamma$ and suppose that $\phi(\sum_j z_j \otimes w_j) = 0$. Then

$$\begin{aligned}\sum_j z_j \otimes w_j &= \sum_{i,j}(z_j \otimes w_j)\phi(x_i \otimes y_i) = \sum_{i,j} z_j \otimes \psi(w_j \otimes x_i)y_i \\ &= \sum_{i,j} z_j\psi(w_j \otimes x_i) \otimes y_i = \sum_{i,j} \phi(z_j \otimes w_j)(x_i \otimes y_i) = 0.\end{aligned}$$ □

PROPOSITION 2.2.5. *Suppose that Λ and Γ are Morita equivalent, with bimodules P and Q and maps $\phi : P \otimes_\Gamma Q \to \Lambda$ and $\psi : Q \otimes_\Lambda P \to \Gamma$ as in the above definition. Then the functors*

$$Q \otimes_\Lambda - : {}_\Lambda\mathbf{Mod} \to {}_\Gamma\mathbf{Mod}, \quad P \otimes_\Gamma - : {}_\Gamma\mathbf{Mod} \to {}_\Lambda\mathbf{Mod}$$

provide an equivalence of abelian categories between ${}_\Lambda\mathbf{Mod}$ and ${}_\Gamma\mathbf{Mod}$. They also induce equivelences between ${}_\Lambda\mathbf{mod}$ and ${}_\Gamma\mathbf{mod}$.

PROOF. This follows directly from the associativity of tensor product and the above lemma. □

THEOREM 2.2.6 (Morita). *Two module categories ${}_\Lambda\mathbf{Mod}$ and ${}_\Gamma\mathbf{Mod}$ are equivalent if and only if ${}_\Lambda\mathbf{mod}$ and ${}_\Gamma\mathbf{mod}$ are equivalent. This happens if and only if $\Gamma \cong \mathrm{End}_\Lambda(P)^{\mathrm{op}}$ for some progenerator P of ${}_\Lambda\mathbf{Mod}$.*

PROOF. This follows from Lemma 2.2.3 and Proposition 2.2.5. □

PROPOSITION 2.2.7. *If ${}_\Lambda\mathbf{Mod}$ is equivalent to ${}_\Gamma\mathbf{Mod}$ then the centres $Z(\Lambda)$ and $Z(\Gamma)$ are isomorphic rings.*

PROOF. If $\lambda \in Z(\Lambda)$, then multiplication by λ is a natural transformation from the identity functor on ${}_\Lambda\mathbf{Mod}$ to itself. Conversely, we claim that all such natural transformations are of this form. Given such a natural transformation ϕ, let λ be the value on the identity element of the regular representation, $\lambda = \phi_{{}_\Lambda\Lambda}(1) \in \Lambda$. Then for any Λ-module M and $m \in M$, we define $f : {}_\Lambda\Lambda \to M$ by $f(\lambda) = \lambda m$. By naturality we have

$$\phi_M(m) = \phi_M(f(1)) = f(\phi_{{}_\Lambda\Lambda}(1)) = f(\lambda) = \lambda m.$$

Thus ϕ is equal to multiplication by λ, which in particular implies that $\lambda \in Z(\Lambda)$.

It follows that the ring $Z(\Lambda)$ may be recovered from ${}_\Lambda\mathbf{Mod}$, so that $Z(\Lambda) \cong Z(\Gamma)$. □

EXERCISES. 1. If Λ and Γ are Morita equivalent, prove that Λ is semisimple Artinian if and only if Γ is.

2. If Λ and Γ are finite dimensional algebras over a field, prove that a Morita equivalence between Λ and Γ induces a bijection between the simple Λ-modules and the simple Γ-modules, and that corresponding projective modules have the same multiplicities of corresponding simple modules in each Loewy layer.

3. If Λ and Γ are Morita equivalent $\mathcal{O}$-algebras of the form described in Section 1.9, prove that Λ and Γ have the same decomposition matrices.

4. Show that if Λ and Γ are Morita equivalent finite dimensional algebras then Λ is self injective if and only if Γ is self injective, and that Λ is symmetric if and only if Γ is symmetric. Show that if Λ is Frobenius then Γ does not have to be Frobenius. Show that the basic algebra of a finite dimensional self injective algebra is always Frobenius.

5. Show that if ${}_\Lambda P_\Gamma$ and ${}_\Gamma Q_\Lambda$ are bimodules inducing a Morita equivalence between Λ and Γ then there are adjunctions

$$\mathrm{Hom}_\Lambda(P \otimes_\Gamma -, -) \cong \mathrm{Hom}_\Gamma(-, Q \otimes_\Lambda -)$$
$$\mathrm{Hom}_\Lambda(-, P \otimes_\Gamma -) \cong \mathrm{Hom}_\Gamma(Q \otimes_\Lambda -, -)$$

so that $P \otimes_\Gamma -$ is both left and right adjoint to $Q \otimes_\Lambda -$.

Use these adjunctions and the fact that

$$Z(\Lambda) = \mathrm{Hom}_{\Lambda\otimes\Lambda^{\mathrm{op}}}(P \otimes_\Gamma Q, \Lambda)$$

to give an alternative proof that $Z(\Lambda) = Z(\Gamma)$.

2.3. Chain complexes and homology

Homological and cohomological concepts can be associated to groups, to modules, to topological spaces, to posets, and so on. These concepts form a major part of the subject matter of this book. They are defined in terms of **chain complexes** and **cochain complexes**.

DEFINITION 2.3.1. Let $\mathcal{A}$ be an abelian category. A **chain complex** of objects in $\mathcal{A}$ (for example, a chain complex of abelian groups, or of vector spaces, or of modules) consists of a collection $\mathbf{C} = \{C_n \mid n \in \mathbb{Z}\}$ of objects $C_n \in \mathcal{A}$ indexed by the integers, together with maps $\partial_n : C_n \to C_{n-1}$ (called the **differentials**) satisfying $\partial_n \circ \partial_{n+1} = 0$.

A **cochain complex** of objects in $\mathcal{A}$ consists of a collection $\mathbf{C} = \{C^n \mid n \in \mathbb{Z}\}$ of objects $C^n \in \mathcal{A}$ indexed by the integers, together with maps $\delta^n : C^n \to C^{n+1}$ satisfying $\delta^n \circ \delta^{n-1} = 0$.

If $x \in C_n$ or C^n, we write $\deg(x) = n$ and say x has **degree** n.

REMARK. If $\{C_n, \partial_n\}$ is a chain complex then letting $C^n = C_{-n}$, $\delta^n = \partial_{-n}$, we obtain a cochain complex $\{C^n, \delta^n\}$, and vice-versa. Thus in some sense chain complexes and cochain complexes are the same thing. In the end, whether we regard something as a chain complex or a cochain complex usually depends on where it came from. It often happens, for example, that

$C_n = 0$ (resp. $C^n = 0$) for $n < 0$ or for $n < -1$. We say that a (co)chain complex $\mathbf{C}$ is **bounded below** if $C_n = 0$ (resp. $C^n = 0$) for all n sufficiently large negative, and **bounded above** if this holds for all n sufficiently large positive. $\mathbf{C}$ is **bounded** if it is bounded both below and above.

DEFINITION 2.3.2. The **homology** of a chain complex $\mathbf{C}$ is given by

$$H_n(\mathbf{C}) = H_n(\mathbf{C}, \partial_*) = \frac{\operatorname{Ker}(\partial_n : C_n \to C_{n-1})}{\operatorname{Im}(\partial_{n+1} : C_{n+1} \to C_n)} = \frac{Z_n(\mathbf{C})}{B_n(\mathbf{C})}.$$

The **cohomology** of a cochain complex $\mathbf{C}$ is given by

$$H^n(\mathbf{C}) = H^n(\mathbf{C}, \delta^*) = \frac{\operatorname{Ker}(\delta^n : C^n \to C^{n+1})}{\operatorname{Im}(\delta^{n-1} : C^{n-1} \to C^n)} = \frac{Z^n(\mathbf{C})}{B^n(\mathbf{C})}.$$

If $x \in C_n$ with $\partial_n(x) = 0$ (resp. $x \in C^n$ with $\delta^n(x) = 0$) then $x \in Z_n(\mathbf{C})$ is called a **cycle** (resp. $x \in Z^n(\mathbf{C})$ is a **cocycle**), and we write $[x]$ for the image of x in $H_n(\mathbf{C})$ (resp. $H^n(\mathbf{C})$). If $x = \partial_{n+1}(y)$ with $y \in C_{n+1}$ (resp. $x = \delta^{n-1}(y)$ with $y \in C^{n-1}$) then $x \in B_n(\mathbf{C})$ is called a **boundary** (resp. $x \in B^n(\mathbf{C})$ is a **coboundary**). Thus $H_n(\mathbf{C})$ (resp. $H^n(\mathbf{C})$) consists of cycles modulo boundaries (resp. cocycles modulo coboundaries).

DEFINITION 2.3.3. If $\mathbf{C}$ and $\mathbf{D}$ are chain complexes (resp. cochain complexes), a **chain map** (resp. **cochain map**) $f : \mathbf{C} \to \mathbf{D}$ consists of maps $f_n : C_n \to D_n$ (resp. $f_n : C^n \to D^n$), $n \in \mathbb{Z}$, such that the following diagram commutes.

$$\begin{array}{ccc} C_n & \xrightarrow{\partial_n} & C_{n-1} \\ \downarrow{\scriptstyle f_n} & & \downarrow{\scriptstyle f_{n-1}} \\ D_n & \xrightarrow{\partial_n} & D_{n-1} \end{array} \qquad \left(\text{resp.} \quad \begin{array}{ccc} C^n & \xrightarrow{\delta^n} & C^{n+1} \\ \downarrow{\scriptstyle f_n} & & \downarrow{\scriptstyle f_{n+1}} \\ D^n & \xrightarrow{\delta^n} & D^{n+1} \end{array} \right).$$

Clearly a (co)chain map $f : \mathbf{C} \to \mathbf{D}$ induces a well defined map $f_* : H_n(\mathbf{C}) \to H_n(\mathbf{D})$ (resp. $f^* : H^n(\mathbf{C}) \to H^n(\mathbf{D})$) defined by $f_*[x] = [f(x)]$ (resp. $f^*[x] = [f(x)]$) for $x \in Z_n$ (resp. Z^n).

From now on, we shall formulate concepts and theorems for chain complexes, and leave the reader to formulate them for cochain complexes.

DEFINITION 2.3.4. If $f, f' : \mathbf{C} \to \mathbf{D}$ are chain maps, we say f and f' are **chain homotopic** (written $f \simeq f'$) if there are maps $h_n : C_n \to D_{n+1}$, $n \in \mathbb{Z}$, such that

$$f_n - f'_n = \partial_{n+1} \circ h_n + h_{n-1} \circ \partial_n.$$

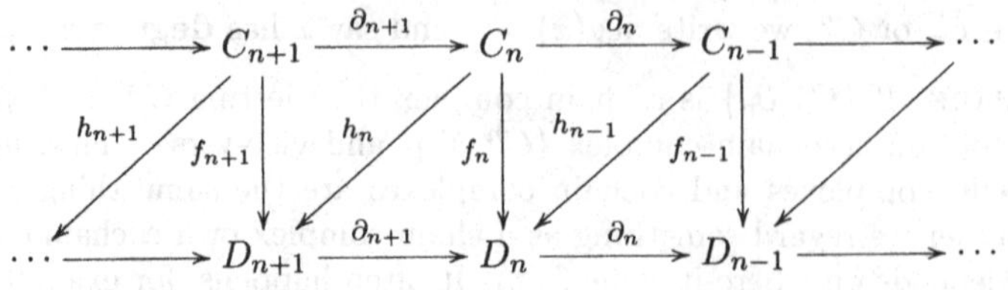

$$\begin{array}{ccccccccc} \cdots & \longrightarrow & C_{n+1} & \xrightarrow{\partial_{n+1}} & C_n & \xrightarrow{\partial_n} & C_{n-1} & \longrightarrow & \cdots \\ & \swarrow{\scriptstyle h_{n+1}} & \downarrow{\scriptstyle f_{n+1}} & \swarrow{\scriptstyle h_n} & \downarrow{\scriptstyle f_n} & \swarrow{\scriptstyle h_{n-1}} & \downarrow{\scriptstyle f_{n-1}} & \swarrow & \\ \cdots & \longrightarrow & D_{n+1} & \xrightarrow{\partial_{n+1}} & D_n & \xrightarrow{\partial_n} & D_{n-1} & \longrightarrow & \cdots \end{array}$$

We say $\mathbf{C}$ and $\mathbf{D}$ are **chain homotopy equivalent** (written $\mathbf{C} \simeq \mathbf{D}$) if there are chain maps $f : \mathbf{C} \to \mathbf{D}$ and $f' : \mathbf{D} \to \mathbf{C}$ such that the composites are chain homotopic to the identity maps $f \circ f' \simeq \mathrm{id}_{\mathbf{D}}$ and $f' \circ f \simeq \mathrm{id}_{\mathbf{C}}$.

We say $\mathbf{C}$ is **chain contractible** if it is chain homotopy equivalent to the zero complex. This is equivalent to the condition that there is a **chain contraction**, i.e., a collection of maps $s_n : C_n \to C_{n+1}$ with $\mathrm{id}_{C_n} = \partial_{n+1} \circ s_n + s_{n-1} \circ \partial_n$.

The reason for this definition is that homotopic maps between topological spaces (see Chapter 1 of Volume II) give rise to chain homotopic maps between their singular chain complexes. A contractible space will have a chain contractible *reduced* singular chain complex. See for example Spanier [**190**, Section 4.4]. Thus the following proposition is the algebraic counterpart of the fact that the singular homology groups of a topological space are homotopy invariants.

PROPOSITION 2.3.5. *If $f, f' : \mathbf{C} \to \mathbf{D}$ are chain homotopic then $f_* = f'_* : H_n(\mathbf{C}) \to H_n(\mathbf{D})$. Thus a homotopy equivalence $\mathbf{C} \simeq \mathbf{D}$ induces isomorphisms $H_n(\mathbf{C}) \cong H_n(\mathbf{D})$ for all $n \in \mathbb{Z}$.*

PROOF. If $x \in C_n$ with $\partial_n(x) = 0$ then

$$\begin{aligned} f_*[x] - f'_*[x] &= [f_n(x) - f'_n(x)] = [\partial_{n+1}(h_n(x)) + h_{n-1}(\partial_n(x))] \\ &= [\partial_{n+1}(h_n(x))] = 0. \end{aligned}$$

□

THE LONG EXACT SEQUENCE IN HOMOLOGY.

DEFINITION 2.3.6. A **short exact sequence** $0 \to \mathbf{C}' \to \mathbf{C} \to \mathbf{C}'' \to 0$ of chain complexes consists of maps of chain complexes $\mathbf{C}' \to \mathbf{C}$ and $\mathbf{C} \to \mathbf{C}''$ such that for each n, $0 \to C'_n \to C_n \to C''_n \to 0$ is a short exact sequence.

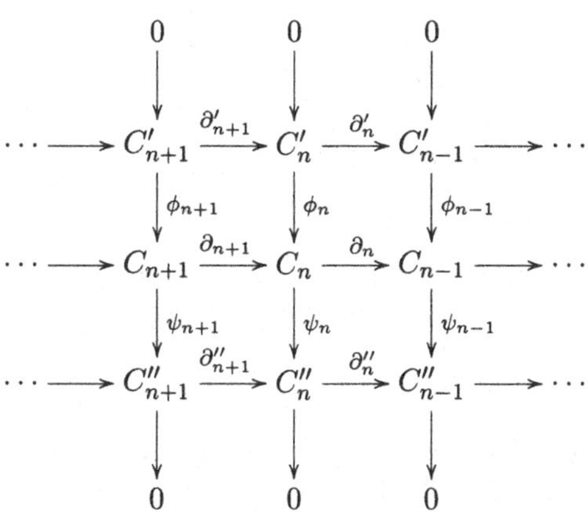

PROPOSITION 2.3.7. *A short exact sequence* $0 \to \mathbf{C}' \to \mathbf{C} \to \mathbf{C}'' \to 0$ *of chain complexes gives rise to a long exact sequence*

$$\cdots \to H_n(\mathbf{C}') \to H_n(\mathbf{C}) \to H_n(\mathbf{C}'') \to H_{n-1}(\mathbf{C}') \to H_{n-1}(\mathbf{C}) \to \cdots$$

PROOF. We define the **switchback map** or **connecting homomorphism**

$$\partial : H_n(\mathbf{C}'') \to H_{n-1}(\mathbf{C}')$$

as follows. If $x \in \mathbf{C}''_n$ with $\partial''_n(x) = 0$, so that $[x] \in H_n(\mathbf{C}'')$, choose $y \in C_n$ with $\psi_n(y) = x$. Then $\psi_{n-1}\partial_n(y) = \partial''_n\psi_n(y) = 0$ and so $\partial_n(y) = \phi_{n-1}(z)$ with $z \in C'_{n-1}$. We have $\phi_{n-2}\partial'_{n-1}(z) = \partial_{n-1}\phi_{n-1}(z) = \partial_{n-1}\partial_n(y) = 0$ and so $\partial'_{n-1}(z) = 0$. We define $\partial[x] = [z] \in H_{n-1}(\mathbf{C}')$.

If y' is another element of C_n with $\psi_n(y') = x$ and $z' \in C'_{n-1}$ with $\partial_n(y') = \phi_{n-1}(z')$, then $\psi_n(y-y') = 0$, and so $y-y' = \phi_n(u)$ for some $u \in C'_n$. We have $\phi_{n-1}\partial'_n(u) = \partial_n\phi_n(u) = \partial_n(y) - \partial_n(y') = \phi_{n-1}(z) - \phi_{n-1}(z')$ and so $\partial'_n(u) = z - z'$. Thus $[z] = [z'] \in H_{n-1}(\mathbf{C}')$. This shows that $\partial : H_n(\mathbf{C}'') \to H_{n-1}(\mathbf{C}')$ is well defined. Exactness of the sequence is not hard to check. □

We find it worthwhile to record the cohomological version of the above proposition.

PROPOSITION 2.3.8. *A short exact sequence of cochain complexes gives rise to a long exact sequence*

$$\cdots \to H^n(\mathbf{C}') \to H^n(\mathbf{C}) \to H^n(\mathbf{C}'') \to H^{n+1}(\mathbf{C}') \to H^{n+1}(\mathbf{C}) \to \cdots \qquad \square$$

A particular case of the above exact sequences is the following:

LEMMA 2.3.9 (Snake Lemma). *A commutative diagram of short exact sequences*

$$\begin{array}{ccccccccc} 0 & \longrightarrow & C'_1 & \longrightarrow & C_1 & \longrightarrow & C''_1 & \longrightarrow & 0 \\ & & \downarrow{\scriptstyle\alpha} & & \downarrow{\scriptstyle\beta} & & \downarrow{\scriptstyle\gamma} & & \\ 0 & \longrightarrow & C'_0 & \longrightarrow & C_0 & \longrightarrow & C''_0 & \longrightarrow & 0 \end{array}$$

gives rise to a six term exact sequence

$$0 \to \mathrm{Ker}\alpha \to \mathrm{Ker}\beta \to \mathrm{Ker}\gamma \to \mathrm{Coker}\alpha \to \mathrm{Coker}\beta \to \mathrm{Coker}\gamma \to 0.$$

PROOF. We regard the diagram as a short exact sequence of chain complexes of length two, and apply Proposition 2.3.7. □

2.4. Ext and Tor

Our first application of the theory of chain complexes and homology is to define functors Ext and Tor for modules over a ring. We shall interpret Ext in terms of extensions of modules.

DEFINITION 2.4.1. A **projective resolution** of a Λ-module M is a long exact sequence

$$\cdots \to P_2 \xrightarrow{\partial_2} P_1 \xrightarrow{\partial_1} P_0$$

of modules with the P_n projective and with $P_0/\mathrm{Im}(\partial_1) \cong M$. In other words, the sequence

$$\cdots \to P_2 \to P_1 \to P_0 \to M \to 0$$

is exact. Since every module is a homomorphic image of a free module, projective resolutions always exist.

We shall regard the sequences in the above definition as chain complexes. The module M appears in degree -1 in the second sequence.

THEOREM 2.4.2 (Comparison theorem). *Any map of modules $M \to M'$ can be extended to a map of projective resolutions*

$$\begin{array}{ccccccccc}
\cdots \longrightarrow & P_2 & \xrightarrow{\partial_2} & P_1 & \xrightarrow{\partial_1} & P_0 & \longrightarrow & M & \longrightarrow 0 \\
& \downarrow{\scriptstyle f_2} & & \downarrow{\scriptstyle f_1} & & \downarrow{\scriptstyle f_0} & & \downarrow & \\
\cdots \longrightarrow & Q_2 & \xrightarrow{\partial'_2} & Q_1 & \xrightarrow{\partial'_1} & Q_0 & \longrightarrow & M' & \longrightarrow 0
\end{array}$$

Given any two such maps $\{f_n\}$ and $\{f'_n\}$, there is a chain homotopy $h_n : P_n \to Q_{n+1}$, so that $f_n - f'_n = \partial'_{n+1} \circ h_n + h_{n-1} \circ \partial_n$.

PROOF. We construct the $f_n : P_n \to Q_n$ inductively as follows. Since

$$\partial'_{n-1} \circ f_{n-1} \circ \partial_n = f_{n-2} \circ \partial_{n-1} \circ \partial_n = 0,$$

we have

$$\begin{array}{ccccc}
& & P_n & & \\
& \swarrow & \downarrow{\scriptstyle f_{n-1}\circ\partial_n} & & \\
Q_n & \xrightarrow{\partial'_n} & \mathrm{Im}(\partial'_n) & \longrightarrow & 0
\end{array}$$

and so we can find a map $f_n : P_n \to Q_n$ with $\partial'_n \circ f_n = f_{n-1} \circ \partial_n$.

We also construct the h_n inductively. We have

$$\begin{aligned}
\partial'_n \circ (f_n - f'_n - h_{n-1} \circ \partial_n) &= (f_{n-1} - f'_{n-1} - \partial'_n \circ h_{n-1}) \circ \partial_n \\
&= h_{n-2} \circ \partial_{n-1} \circ \partial_n = 0
\end{aligned}$$

and so we may find a map $h_n : P_n \to Q_{n+1}$ with $f_n - f'_n - h_{n-1} \circ \partial_n = \partial'_{n+1} \circ h_n$. □

REMARK. The proof of the above theorem did not use all the hypotheses. It suffices for the upper complex to consist of projective modules but it need not be exact, and for the lower complex to be exact but not necessarily to consist of projective modules. We shall sometimes use this stronger form of the theorem.

If M' is a right Λ-module and

$$\cdots \to P_2 \xrightarrow{\partial_2} P_1 \xrightarrow{\partial_1} P_0$$

is a projective resolution of a left Λ-module M, we have a chain complex

$$\cdots M' \otimes_\Lambda P_2 \xrightarrow{1\otimes\partial_2} M' \otimes_\Lambda P_1 \xrightarrow{1\otimes\partial_1} M' \otimes_\Lambda P_0.$$

This complex is no longer necessarily exact, although it is clear that $(1 \otimes \partial_{n-1}) \circ (1 \otimes \partial_n) = 0$.

It follows from the above theorem that this complex is independent of choice of projective resolution, up to chain homotopy equivalence. Thus the homology groups are independent of this choice, and we define

$$\mathrm{Tor}_n^\Lambda(M', M) = H_n(M' \otimes \mathbf{P}, 1 \otimes \partial_*).$$

Similarly if M' is a left Λ-module and

$$\cdots \to P_2 \xrightarrow{\partial_2} P_1 \xrightarrow{\partial_1} P_0$$

is a projective resolution of a left Λ-module M, we have a cochain complex

$$\mathrm{Hom}_\Lambda(P_0, M') \xrightarrow{\delta^0} \mathrm{Hom}_\Lambda(P_1, M') \xrightarrow{\delta^1} \mathrm{Hom}_\Lambda(P_2, M') \to \cdots$$

where δ^n is given by composition with ∂_{n+1}. This complex is independent of choice of projective resolution, up to chain homotopy equivalence. Thus its cohomology groups are independent of this choice, and we define

$$\mathrm{Ext}_\Lambda^n(M, M') = H^n(\mathrm{Hom}_\Lambda(\mathbf{P}, M'), \delta^*).$$

Note that $\mathrm{Tor}_0^\Lambda(M', M) = M' \otimes_\Lambda M$ and $\mathrm{Ext}_\Lambda^0(M, M') = \mathrm{Hom}_\Lambda(M, M')$.

EXAMPLE. In case $\Lambda = \mathbb{Z}$, a Λ-module is the same as an abelian group. Since every subgroup of a free abelian group is again a free abelian group, it follows that every module has a projective resolution of length one (i.e., $P_n = 0$ for $n \geq 2$), and so $\mathrm{Tor}_n^{\mathbb{Z}}$ and $\mathrm{Ext}_{\mathbb{Z}}^n$ are zero for $n \geq 2$.

It was conjectured by J. H. C. Whitehead that if $\mathrm{Ext}_{\mathbb{Z}}^1(A, \mathbb{Z}) = 0$ then A is free as an abelian group. It is now known, thanks to the extraordinary work of S. Shelah [**187**] that the truth of this conjecture *depends on the set theory being used!*

REMARKS. (i) If M is projective, then $\mathbf{P}$ can be taken to be non-zero only in degree zero, and equal to M there, so that in this case $\mathrm{Ext}_\Lambda^n(M, M')$ and $\mathrm{Tor}_n^\Lambda(M, M')$ are zero for $n > 0$.

(ii) We write $\tilde{\Omega}^n(M)$ for $\mathrm{Ker}(\partial_{n-1})$ in a projective resolution of M. Note that by Schanuel's lemma if $\tilde{\Omega}^n(M)'$ is defined similarly using another projective resolution of M then there are projective modules P and P' with $\tilde{\Omega}^n(M) \oplus P' \cong \tilde{\Omega}^n(M)' \oplus P$. If M is finitely generated and the Krull–Schmidt theorem holds for finitely generated Λ-modules then there is a unique **minimal resolution** of M, and we write $\Omega^n(M)$ for $\mathrm{Ker}(\partial_{n-1})$ in this particular resolution.

Dually we write $\tilde{\Omega}^{-n}(M)$ for the nth cokernel in an injective resolution, and $\Omega^{-n}(M)$ if the resolution is minimal.

(iii) The discussion above of Ext and Tor is a particular case of the concept of **derived functors**. Suppose that $\mathcal{A}$ and $\mathcal{B}$ are abelian categories and that every object in $\mathcal{A}$ is a quotient of a projective object. If $F : \mathcal{A} \to \mathcal{B}$ is a covariant additive functor, and M is an object in $\mathcal{A}$, we form a projective resolution

$$\cdots \to P_2 \xrightarrow{\partial_2} P_1 \xrightarrow{\partial_1} P_0$$

of M, apply F to obtain a chain complex

$$\cdots \to F(P_2) \xrightarrow{F(\partial_2)} F(P_1) \xrightarrow{F(\partial_1)} F(P_0)$$

whose homology groups are the **left derived functors**

$$L_nF(M) = H_n(F(\mathbf{P}), F(\partial_*)).$$

Using the comparison theorem in the same way as before, we see that these are independent of the choice of resolution. If F is right exact then $L_0F(M) = F(M)$. Thus for example the left derived functors of $M' \otimes_\Lambda -$ are

$$L_n(M' \otimes_\Lambda -) = \mathrm{Tor}_n^\Lambda(M', -).$$

Similarly, the **right derived functors** of the covariant additive functor F are defined by applying F to an injective resolution

$$I_0 \xrightarrow{\delta^0} I_1 \xrightarrow{\delta^1} I_2 \to \cdots$$

of M. The right derived functors are then the cohomology groups

$$R^nF(M) = H^n(F(\mathbf{I}, F(\delta^*)).$$

If F is left exact then $R^0F(M) = F(M)$.

For contravariant functors, the left derived functors are defined using an injective resolution and the right derived functors are defined using a projective resolution. Thus for example the right derived functors of $\mathrm{Hom}_\Lambda(-, M')$ are $R^n\mathrm{Hom}_\Lambda(-, M') = \mathrm{Ext}_\Lambda^n(-, M')$.

The reader may wonder why we have not discussed $L_n(- \otimes_\Lambda M)$ and $R^n\mathrm{Hom}_\Lambda(M, -)$. This is because it turns out that we get nothing but Tor and Ext again, as we shall see in Proposition 2.5.5.

PROPOSITION 2.4.3. *Suppose that Λ is Artinian, and $P = \Lambda e$ and $P' = \Lambda e'$ are projective indecomposable Λ-modules, so that $P/\mathrm{Rad}(P) = S$ and $P'/\mathrm{Rad}(P') = S'$ are simple (by the idempotent refinement theorem). Then*

$$\mathrm{Ext}_\Lambda^1(S, S') \cong \mathrm{Hom}_\Lambda(\mathrm{Rad}(P)/\mathrm{Rad}^2(P), S').$$

As an $\mathrm{End}_\Lambda(S')$-$\mathrm{End}_\Lambda(S)$-bimodule this is dual to the $\mathrm{End}_\Lambda(S)$-$\mathrm{End}_\Lambda(S')$-bimodule $e'J(\Lambda)e/e'J^2(\Lambda)e$. In particular, if Λ is a finite dimensional algebra over a field k, then

$$\dim_k \mathrm{Ext}_\Lambda^1(S, S') = \dim_k(e'J(\Lambda)e/e'J^2(\Lambda)e).$$

PROOF. If $\operatorname{Rad}(P)/\operatorname{Rad}^2(P) = \bigoplus_i n_i S_i$ as a direct sum of simple modules, then letting $P_i/\operatorname{Rad}(P_i) = S_i$, the minimal projective resolution of S has the form

$$\cdots \to \bigoplus_i n_i P_i \to P$$

and so

$$\operatorname{Ext}^1_\Lambda(S, S') \cong \operatorname{Hom}_\Lambda(\bigoplus_i n_i P_i, S') \cong \operatorname{Hom}_\Lambda(\bigoplus_i n_i S_i, S')$$

which we can rewrite as $\operatorname{Hom}_\Lambda(\operatorname{Rad}(P)/\operatorname{Rad}^2(P), S')$. This is dual to

$$\begin{aligned}\operatorname{Hom}_\Lambda(S', \operatorname{Rad}(P)/\operatorname{Rad}^2(P)) &\cong \operatorname{Hom}_\Lambda(P', \operatorname{Rad}(P)/\operatorname{Rad}^2(P))\\ &\cong \operatorname{Hom}_\Lambda(P', \operatorname{Rad}(P))/\operatorname{Hom}_\Lambda(P', \operatorname{Rad}^2(P)) \cong e'J(\Lambda)e/e'J^2(\Lambda)e\end{aligned}$$

by Lemma 1.3.3. □

AUGMENTED ALGEBRAS.

DEFINITION 2.4.4. An **augmented algebra** Λ over a commutative ring of coefficients R is an algebra together with a surjective **augmentation map** $\varepsilon : \Lambda \to R$ of R-algebras.

If Λ is an augmented algebra, then R may be given the structure of a left Λ-module via $\lambda(x) = \varepsilon(\lambda)x$, and of a right Λ-module via $(x)\lambda = \varepsilon(\lambda)x$.

We define the **homology groups** of Λ with coefficients in a right Λ-module M to be

$$H_n(\Lambda, M) = \operatorname{Tor}^\Lambda_n(M, R)$$

and the **cohomology groups** of Λ with coefficients in a left Λ-module M to be

$$H^n(\Lambda, M) = \operatorname{Ext}^n_\Lambda(R, M).$$

The special case $M = R$ is of particular importance, since as we shall see in Section 2.6, there is a ring structure in this case.

REMARK. Suppose that $R \to R'$ is a homomorphism of coefficient rings, and Λ is projective as an R-module. Then tensoring with R' will take a projective resolution of R as a Λ-module to a projective resolution of R' as an $R' \otimes_R \Lambda$-module. Thus we have

$$H_n(\Lambda, R') \cong H_n(R' \otimes_R \Lambda, R'), \quad H^n(\Lambda, R') \cong H^n(R' \otimes_R \Lambda, R').$$

EXERCISE. Show that Ext and Tor are bilinear in the sense that there are natural isomorphisms

$$\begin{aligned}\operatorname{Ext}^n_\Lambda(M \oplus M', M'') &\cong \operatorname{Ext}^n_\Lambda(M, M'') \oplus \operatorname{Ext}^n_\Lambda(M', M'')\\ \operatorname{Ext}^n_\Lambda(M, M' \oplus M'') &\cong \operatorname{Ext}^n_\Lambda(M, M'') \oplus \operatorname{Ext}^n_\Lambda(M, M'')\end{aligned}$$

and similarly for $\operatorname{Tor}^\Lambda_n$.

2.5. Long exact sequences

LEMMA 2.5.1 (Horseshoe lemma). *If* $0 \to M' \to M \to M'' \to 0$ *is a short exact sequence of left* Λ*-modules, then given projective resolutions*

$$\cdots \to P_2' \to P_1' \to P_0', \qquad \cdots \to P_2'' \to P_1'' \to P_0''$$

of M' *and* M''*, we may complete to a short exact sequence of chain complexes*

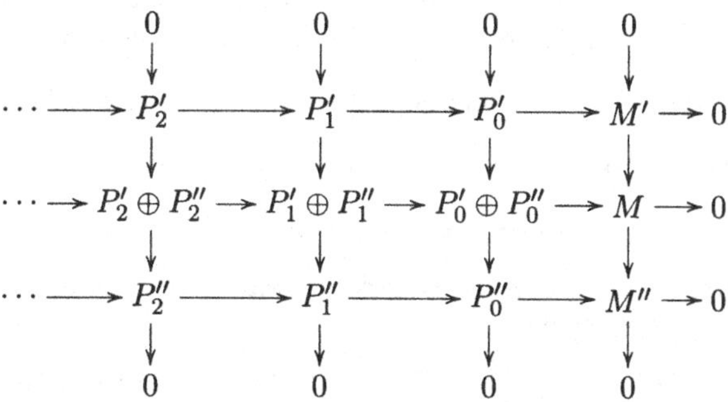

PROOF. It is easy to construct the required maps by induction, using the definition of a projective module. □

PROPOSITION 2.5.2. *Suppose that*

$$0 \to M' \to M \to M'' \to 0$$

is a short exact sequence of left Λ*-modules.*

(i) *If* M_0 *is a right* Λ*-module, there is a long exact sequence*

$$\cdots \to \mathrm{Tor}_n^\Lambda(M_0, M') \to \mathrm{Tor}_n^\Lambda(M_0, M) \to \mathrm{Tor}_n^\Lambda(M_0, M'') \to \mathrm{Tor}_{n-1}^\Lambda(M_0, M') \to \cdots$$
$$\cdots \to \mathrm{Tor}_1^\Lambda(M_0, M'') \to M_0 \otimes_\Lambda M' \to M_0 \otimes_\Lambda M \to M_0 \otimes_\Lambda M'' \to 0.$$

(ii) *If* M_0 *is a left* Λ*-module there is a long exact sequence*

$$0 \to \mathrm{Hom}_\Lambda(M'', M_0) \to \mathrm{Hom}_\Lambda(M, M_0) \to \mathrm{Hom}_\Lambda(M', M_0) \to \mathrm{Ext}_\Lambda^1(M'', M_0) \to \cdots$$
$$\cdots \to \mathrm{Ext}_\Lambda^n(M'', M_0) \to \mathrm{Ext}_\Lambda^n(M, M_0) \to \mathrm{Ext}_\Lambda^n(M', M_0) \to \mathrm{Ext}_\Lambda^{n+1}(M'', M_0) \to \cdots$$

PROOF. (i) Tensor M_0 with the diagram given in the lemma and use Proposition 2.3.7.

(ii) Take homs from the diagram given in the lemma to M_0 and use Proposition 2.3.8. □

Exactly the same proof shows in general that if $F : \mathcal{A} \to \mathcal{B}$ is a right exact covariant additive functor then there is a long exact sequence

$$\cdots \to L_nF(M') \to L_nF(M) \to L_nF(M'') \to L_{n-1}F(M') \to \cdots$$
$$\to L_1F(M'') \to F(M') \to F(M) \to F(M'') \to 0,$$

while if F is a left exact contravariant additive functor then there is a long exact sequence

$$\begin{aligned} 0 \to F(M'') \to F(M) \to F(M') \to R^1F(M'') \to \cdots \\ \to R^nF(M'') \to R^nF(M) \to R^nF(M') \to R^{n+1}F(M'') \to \cdots \end{aligned}$$

Of course, similar statements are true of the right derived functors of a left exact covariant functor and left derived functors of right exact contravariant functors. We leave the interested reader to formulate these cases.

We also obtain exact sequences in the other variable as follows.

PROPOSITION 2.5.3. (i) *Suppose that*

$$0 \to M_0 \to M_1 \to M_2 \to 0$$

is a short exact sequence of right Λ-modules, and M' is a left Λ-module. Then there is a long exact sequence

$$\begin{aligned} \cdots \to \mathrm{Tor}_n^\Lambda(M_0, M') \to \mathrm{Tor}_n^\Lambda(M_1, M') \to \mathrm{Tor}_n^\Lambda(M_2, M') \to \mathrm{Tor}_{n-1}^\Lambda(M_0, M') \to \cdots \\ \cdots \to \mathrm{Tor}_1^\Lambda(M_2, M') \to M_0 \otimes_\Lambda M' \to M_1 \otimes_\Lambda M' \to M_2 \otimes_\Lambda M' \to 0. \end{aligned}$$

(ii) *Suppose that $0 \to M_0 \to M_1 \to M_2 \to 0$ is a short exact sequence of left Λ-modules and M' is a left Λ-module. Then there is a long exact sequence*

$$\begin{aligned} 0 \to \mathrm{Hom}_\Lambda(M', M_0) \to \mathrm{Hom}_\Lambda(M', M_1) \to \mathrm{Hom}_\Lambda(M', M_2) \to \mathrm{Ext}_\Lambda^1(M', M_0) \to \cdots \\ \cdots \to \mathrm{Ext}_\Lambda^n(M', M_0) \to \mathrm{Ext}_\Lambda^n(M', M_1) \to \mathrm{Ext}_\Lambda^n(M', M_2) \to \mathrm{Ext}_\Lambda^{n+1}(M', M_0) \to \cdots \end{aligned}$$

PROOF. (i) Tensoring the short exact sequence with a resolution

$$\cdots \to P'_2 \to P'_1 \to P'_0$$

of M' as a left Λ-module gives a short exact sequence of chain complexes

$$\begin{array}{ccccc} & 0 & & 0 & & 0 \\ & \downarrow & & \downarrow & & \downarrow \\ \cdots \to & M_0 \otimes P'_2 & \to & M_0 \otimes P'_1 & \to & M_0 \otimes P'_0 \\ & \downarrow & & \downarrow & & \downarrow \\ \cdots \to & M_1 \otimes P'_2 & \to & M_1 \otimes P'_1 & \to & M_1 \otimes P'_0 \\ & \downarrow & & \downarrow & & \downarrow \\ \cdots \to & M_2 \otimes P'_2 & \to & M_2 \otimes P'_1 & \to & M_2 \otimes P'_0 \\ & \downarrow & & \downarrow & & \downarrow \\ & 0 & & 0 & & 0 \end{array}$$

Applying Proposition 2.3.7 yields the required long exact sequence.

(ii) If

$$\cdots \to P'_2 \to P'_1 \to P'_0$$

is a projective resolution of M' as a left Λ-module, then applying Proposition 2.3.8 to the short exact sequence of cochain complexes

$$\begin{array}{ccccccc}
0 & & 0 & & 0 & & \\
\downarrow & & \downarrow & & \downarrow & & \\
\mathrm{Hom}_\Lambda(P'_0, M_0) & \to & \mathrm{Hom}_\Lambda(P'_1, M_0) & \to & \mathrm{Hom}_\Lambda(P'_2, M_0) & \to & \cdots \\
\downarrow & & \downarrow & & \downarrow & & \\
\mathrm{Hom}_\Lambda(P'_0, M_1) & \to & \mathrm{Hom}_\Lambda(P'_1, M_1) & \to & \mathrm{Hom}_\Lambda(P'_2, M_1) & \to & \cdots \\
\downarrow & & \downarrow & & \downarrow & & \\
\mathrm{Hom}_\Lambda(P'_0, M_2) & \to & \mathrm{Hom}_\Lambda(P'_1, M_2) & \to & \mathrm{Hom}_\Lambda(P'_2, M_2) & \to & \cdots \\
\downarrow & & \downarrow & & \downarrow & & \\
0 & & 0 & & 0 & &
\end{array}$$

yields the required long exact sequence. □

COROLLARY 2.5.4. *If M is a module for an Artinian ring Λ and S is a simple Λ-module then*
(i) $\mathrm{Ext}^n_\Lambda(M, S) \cong \mathrm{Hom}_\Lambda(\Omega^n M, S)$
(ii) $\mathrm{Ext}^n_\Lambda(S, M) \cong \mathrm{Hom}_\Lambda(S, \Omega^{-n} M)$.

PROOF. (i) Let

$$\cdots \to P_2 \to P_1 \to P_0$$

be a minimal resolution of M. Then the complex

$$\mathrm{Hom}_\Lambda(P_0, S) \to \mathrm{Hom}_\Lambda(P_1, S) \to \mathrm{Hom}_\Lambda(P_2, S) \to \cdots$$

has zero differential, since if the composite $P_{n+1} \to P_n \to S$ is non-zero then P_n has a summand isomorphic to the projective cover of S and which is in the image of $P_{n+1} \to P_n$ and hence in the kernel of $P_n \to P_{n-1}$, contradicting the minimality of P_n. Hence

$$\begin{aligned}\mathrm{Ext}^n_\Lambda(M, S) &= \mathrm{Hom}_\Lambda(P_n, S) = \mathrm{Hom}_\Lambda(P_n/\mathrm{Im}(P_{n+1} \to P_n), S) \\ &= \mathrm{Hom}_\Lambda(\Omega^n M, S).\end{aligned}$$

(ii) is proved similarly, using part (ii) of the following proposition. □

PROPOSITION 2.5.5. (i) *Suppose that M' is a right Λ-module and*

$$\cdots \to P'_2 \xrightarrow{\partial'_2} P'_1 \xrightarrow{\partial'_1} P'_0 \to M' \to 0$$

is a resolution of M' by projective right Λ-modules. Then

$$\mathrm{Tor}^\Lambda_n(M', M) \cong H_n(\mathbf{P}' \otimes M, \partial'_* \otimes 1).$$

In particular if M' is projective then $\mathrm{Tor}^\Lambda_n(M', M) = 0$ for $n > 0$.
(ii) *Suppose that M' is a left Λ-module and*

$$0 \to M' \to I'_0 \xrightarrow{\delta^0} I'_1 \xrightarrow{\delta^1} I'_2 \to \cdots$$

is an injective resolution of M'. Then

$$\mathrm{Ext}^n_\Lambda(M, M') \cong H^n(\mathrm{Hom}_\Lambda(M, \mathbf{I}'), \delta^*).$$

In particular if M' is injective then $\mathrm{Ext}^n_\Lambda(M, M') = 0$ *for* $n > 0$.

PROOF. We shall prove (ii), since the proof of (i) is the dual of the same argument. The proof is an example of the inductive technique of **dimension shifting**. We denote by $\mathcal{EXT}^n_\Lambda(M, M')$ the groups $H^n(\mathrm{Hom}_\Lambda(M, \mathbf{I}'), \delta^*)$, and we wish to show that $\mathrm{Ext}^n_\Lambda(M, M') \cong \mathcal{EXT}^n_\Lambda(M, M')$.

Choose a short exact sequence

$$0 \to M_1 \to P \to M \to 0$$

with P projective. Then the long exact sequence

$$\cdots \to \mathrm{Ext}^{n-1}_\Lambda(P, M') \to \mathrm{Ext}^{n-1}_\Lambda(M_1, M') \to \\ \mathrm{Ext}^n_\Lambda(M, M') \to \mathrm{Ext}^n_\Lambda(P, M') \to \cdots$$

shows that $\mathrm{Ext}^n_\Lambda(M, M') \cong \mathrm{Ext}^{n-1}_\Lambda(M_1, M')$.

The functor $\mathcal{EXT}$ also clearly has long exact sequences in each variable by the same arguments as above, and so we obtain $\mathcal{EXT}^n_\Lambda(M, M') \cong \mathcal{EXT}^{n-1}_\Lambda(M_1, M')$.

We are now finished by induction, since the case $n = 1$ follows from the diagram

$$\begin{array}{ccccccccc} 0 & \to & \mathrm{Hom}_\Lambda(M,M') & \to & \mathrm{Hom}_\Lambda(P,M') & \to & \mathrm{Hom}_\Lambda(M_1,M') & \to & \mathrm{Ext}^1_\Lambda(M,M') & \to & 0 \\ & & \| & & \| & & \| & & & & \\ 0 & \to & \mathrm{Hom}_\Lambda(M,M') & \to & \mathrm{Hom}_\Lambda(P,M') & \to & \mathrm{Hom}_\Lambda(M_1,M') & \to & \mathcal{EXT}^1_\Lambda(M,M') & \to & 0. \end{array}$$ □

COROLLARY 2.5.6. *If either M or M' is flat then* $\mathrm{Tor}^\Lambda_n(M', M) = 0$ *for all* $n > 0$. □

The proof of the following may now be safely left to the reader.

PROPOSITION 2.5.7.
(i) $\mathrm{Ext}^n_\Lambda(M, M') \cong \mathrm{Ext}^{n-1}_\Lambda(\tilde{\Omega}(M), M') \cong \mathrm{Ext}^{n-1}_\Lambda(M, \tilde{\Omega}^{-1}(M'))$.
(ii) $\mathrm{Tor}^\Lambda_n(M, M') \cong \mathrm{Tor}^\Lambda_{n-1}(\tilde{\Omega}(M), M') \cong \mathrm{Tor}^\Lambda_{n-1}(M, \tilde{\Omega}(M'))$. □

EXERCISE. Formulate and prove a version of Proposition 2.5.5 for derived functors of functors of two variables with appropriate exactness properties.

2.6. Extensions

DEFINITION 2.6.1. If M and M' are left Λ-modules, an ***n*-fold extension** of M by M' is an exact sequence

$$0 \to M' \to M_{n-1} \to M_{n-2} \to \cdots \to M_0 \to M \to 0$$

beginning with M' and ending with M, and with n intermediate terms.

Two n-fold extensions are **equivalent** if there is a **map of n-fold extensions**

$$\begin{array}{ccccccccccccc}
0 & \longrightarrow & M' & \longrightarrow & M_{n-1} & \longrightarrow & \cdots & \longrightarrow & M_0 & \longrightarrow & M & \longrightarrow & 0 \\
 & & \| & & \downarrow & & & & \downarrow & & \| & & \\
0 & \longrightarrow & M' & \longrightarrow & M'_{n-1} & \longrightarrow & \cdots & \longrightarrow & M'_0 & \longrightarrow & M & \longrightarrow & 0.
\end{array}$$

We complete this to an equivalence relation by symmetry and transitivity in the usual way.

For $n = 1$, we simply call an n-fold extension an **extension**. Note that an equivalence of extensions is an isomorphism of short exact sequences.

LEMMA 2.6.2. *Suppose that Λ is an algebra over a field k, and*

$$0 \to M_1 \to M_2 \to M_3 \to 0$$

is a short exact sequence of Λ-modules of finite k-dimension. If $M_2 \cong M_1 \oplus M_3$ then the sequence splits; i.e., it represents the zero element of $\mathrm{Ext}^1_\Lambda(M_3, M_1)$.

PROOF. By dimension counting, the last map in the exact sequence

$$0 \to \mathrm{Hom}_\Lambda(M_3, M_1) \to \mathrm{Hom}_\Lambda(M_3, M_2) \to \mathrm{Hom}_\Lambda(M_3, M_3) \to \mathrm{Ext}^1_\Lambda(M_3, M_1)$$

is zero, so the previous map is surjective. A pre-image under this map of the identity homomorphism of M_3 is a splitting for the sequence. □

An n-fold extension of M by M' determines an element of $\mathrm{Ext}^n_\Lambda(M, M')$ by completing the diagram

$$\begin{array}{ccccccccccccccc}
\cdots & \longrightarrow & P_{n+1} & \longrightarrow & P_n & \xrightarrow{\partial_n} & P_{n-1} & \longrightarrow & \cdots & \longrightarrow & P_0 & \longrightarrow & M & \longrightarrow & 0 \\
 & & & & \downarrow{\scriptstyle\phi} & & \downarrow & & & & \downarrow & & \| & & \\
 & & 0 & \longrightarrow & M' & \longrightarrow & M_{n-1} & \longrightarrow & \cdots & \longrightarrow & M_0 & \longrightarrow & M & \longrightarrow & 0
\end{array}$$

using the remark after Theorem 2.4.2. By enlarging the projective resolution of M if necessary, we may assume that ϕ is surjective.

It is easy to see that equivalent n-fold extensions define the same element of $\mathrm{Ext}^n_\Lambda(M, M')$. Conversely, if two n-fold extensions define the same element of $\mathrm{Ext}^n_\Lambda(M, M')$ then we have a commutative diagram

$$\begin{array}{ccccccccccccccc}
0 & \to & M' & \longrightarrow & M_{n-1} & \longrightarrow & M_{n-2} & \to & \cdots & \to & M_0 & \to & M & \to & 0 \\
 & & \| & & \uparrow & & \uparrow & & & & \uparrow & & \| & & \\
0 & \to & M' & \to & P_{n-1}/\partial_n(\mathrm{Ker}\phi) & \to & P_{n-2} & \to & \cdots & \to & P_0 & \to & M & \to & 0 \\
 & & \| & & \downarrow & & \downarrow & & & & \downarrow & & \| & & \\
0 & \to & M' & \longrightarrow & M'_{n-1} & \longrightarrow & M'_{n-2} & \to & \cdots & \to & M'_0 & \to & M & \to & 0
\end{array}$$

and so they are equivalent. Thus we have interpreted $\mathrm{Ext}^n_\Lambda(M, M')$ as the set of equivalence classes of n-fold extensions of M by M'. In particular, $\mathrm{Ext}^1_\Lambda(M, M')$ is the set of equivalence classes of extensions $0 \to M' \to M_0 \to$

$M \to 0$, and in this case the equivalence relation reduces to isomorphism of short exact sequences.

If $\zeta \in \mathrm{Ext}^n_\Lambda(M, M')$, we write $\hat{\zeta}$ for the corresponding map $\tilde{\Omega}^n(M) \to M'$. By rechoosing the projective resolution of M if necessary, we may always assume that $\hat{\zeta}$ is an epimorphism, and we define L_ζ to be its kernel. Thus we have a commutative diagram

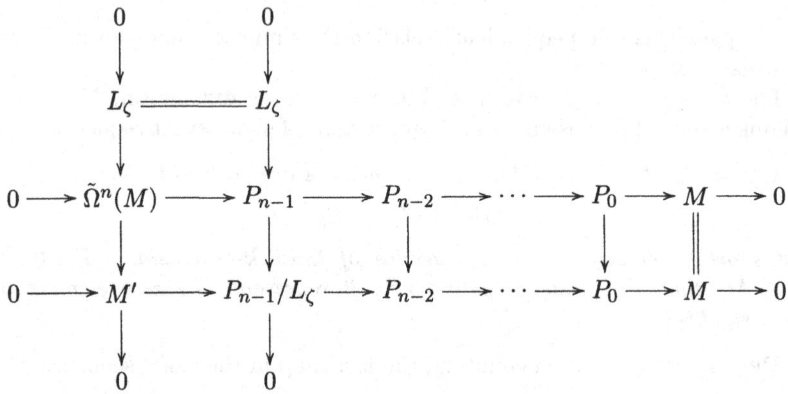

where the bottom row is an n-fold extension representing ζ.

Note that L_ζ is only determined up to addition of projective modules, in the sense that if L'_ζ is defined using another projective resolution of M then there are projective modules P and P' with $L_\zeta \oplus P \cong L'_\zeta \oplus P'$.

Yoneda composition. If

$$0 \to M' \to M_{n-1} \to \cdots \to M_0 \to M \to 0,$$
$$0 \to M'' \to M'_{m-1} \to \cdots \to M'_0 \to M' \to 0$$

represent elements $\zeta \in \mathrm{Ext}^n_\Lambda(M, M')$ and $\eta \in \mathrm{Ext}^m_\Lambda(M', M'')$, then we can form their **Yoneda splice** as follows.

$$\begin{array}{ccccccccccc}
0 \to M'' \to M'_{m-1} \to \cdots \to & M'_0 & & \longrightarrow & & M_{n-1} & \to \cdots \to M_0 \to M \to 0 \\
& & \searrow & & \nearrow & & \\
& & & M' & & & \\
& & \nearrow & & \searrow & & \\
& 0 & & & & 0 &
\end{array}$$

to obtain an element $\eta \circ \zeta$ of $\mathrm{Ext}^{n+m}_\Lambda(M, M'')$. This way we obtain a bilinear map

$$\mathrm{Ext}^m_\Lambda(M', M'') \times \mathrm{Ext}^n_\Lambda(M, M') \to \mathrm{Ext}^{n+m}_\Lambda(M, M'')$$

called **Yoneda composition**. If m or n is equal to zero (recall $\mathrm{Ext}^0 = \mathrm{Hom}$) this map is defined by pushing out or pulling back in the obvious way. This composition is clearly associative, and so it defines a ring structure on $\mathrm{Ext}^*_\Lambda(M, M)$, and $\mathrm{Ext}^*_\Lambda(M, M')$ is an $\mathrm{Ext}^*_\Lambda(M', M')$-$\mathrm{Ext}^*_\Lambda(M, M)$-bimodule. The reader should be warned that the ring $\mathrm{Ext}^*_\Lambda(M, M)$ is often far from being commutative, even when Λ is a group ring RG, although it does turn out that $H^*(RG, R) = \mathrm{Ext}^*_{RG}(R, R)$ is *graded commutative.* For further details see Section 3.2.

If $\zeta \in \mathrm{Ext}^n_\Lambda(M, M')$ is represented by a map $\hat{\zeta} : \tilde{\Omega}^n(M) \to M'$ and $\eta \in \mathrm{Ext}^m_\Lambda(M', M'')$ is represented by $\hat{\eta} : \tilde{\Omega}^m(M') \to M''$, we obtain a representative $\widehat{\eta \circ \zeta}$ of the Yoneda splice as follows. As in Section 1.5, we may form $\tilde{\Omega}^m(\hat{\zeta}) : \tilde{\Omega}^{n+m}(M) \to \tilde{\Omega}^m(M')$ and then set

$$\widehat{\eta \circ \zeta} = \hat{\eta} \circ \tilde{\Omega}^m(\hat{\zeta}) : \tilde{\Omega}^{n+m}(M) \to M''.$$

EXERCISES. 1. Show that a chain complex is chain contractible if and only if it is a Yoneda splice of split short exact sequences.

2. Show that the natural isomorphism

$$\mathrm{Ext}^n_\Lambda(M \oplus M', M'') \cong \mathrm{Ext}^n_\Lambda(M, M'') \oplus \mathrm{Ext}^n_\Lambda(M', M'')$$

corresponds to a pushout diagram

$$\begin{array}{ccccccccccccc} 0 \to & M''\oplus M'' & \to & M_{n-1}\oplus M'_{n-1} & \to & M_{n-2}\oplus M'_{n-2} & \to & \cdots & \to & M_0\oplus M'_0 & \to & M\oplus M' & \to 0 \\ & \downarrow & & \downarrow & & \| & & & & \| & & \| & \\ 0 \to & M'' & \to & X & \to & M_{n-2}\oplus M'_{n-2} & \to & \cdots & \to & M_0\oplus M'_0 & \to & M\oplus M' & \to 0. \end{array}$$

Similarly, write down a pullback diagram for the natural isomorphism

$$\mathrm{Ext}^n_\Lambda(M, M' \oplus M'') \cong \mathrm{Ext}^n_\Lambda(M, M') \oplus \mathrm{Ext}^n_\Lambda(M, M'').$$

Deduce that Yoneda composition is compatible with direct sum decomposition, so that elements of

$$\mathrm{Ext}^n_\Lambda(M_1 \oplus \cdots \oplus M_m, N_1 \oplus \cdots \oplus N_n)$$

may be written as matrices (ζ_{ij}) with $\zeta_{ij} \in \mathrm{Ext}^n_\Lambda(M_i, N_j)$, and Yoneda composition corresponds to matrix multiplication.

3. (Feit) If Λ is self injective, show that for $n > 0$ there is a natural isomorphism

$$\mathrm{Ext}^n_\Lambda(M, M') \cong \mathrm{Ext}^n_\Lambda(\Omega M, \Omega M')$$

compatible with Yoneda composition, so that in particular

$$\mathrm{Ext}^*_\Lambda(M, M) \cong \mathrm{Ext}^*_\Lambda(\Omega M, \Omega M)$$

as graded rings
(after quotienting out the elements of $\mathrm{Ext}^0_\Lambda(M, M) = \mathrm{Hom}_\Lambda(M, M)$ which factor through a projective module).

4. Suppose that Λ is a ring and

$$0 \to M_1 \to M_2 \to M_3 \to 0$$

is a short exact sequence of Λ-modules representing an element

$$\rho \in \mathrm{Ext}^1_\Lambda(M_3, M_1).$$

Show that for any Λ-module N, the connecting homomorphisms

$$\mathrm{Ext}^{n-1}_\Lambda(N, M_3) \to \mathrm{Ext}^n_\Lambda(N, M_1), \quad \mathrm{Ext}^{n-1}_\Lambda(M_1, N) \to \mathrm{Ext}^n_\Lambda(M_3, N)$$

are equal to Yoneda composition with ρ.

5. Suppose that Λ is an Artinian ring, and that P_M is the projective cover of a Λ-module M with kernel $\Omega(M)$. Show that the isomorphism between $\mathrm{Hom}_\Lambda(\Omega M, S)$ and $\mathrm{Ext}^1_\Lambda(M, S)$ given in Corollary 2.5.4 can be described as follows. If $0 \neq \phi \in \mathrm{Hom}_\Lambda(\Omega M, S)$ then applying the snake lemma to the diagram

$$\begin{array}{ccccccccc} 0 & \longrightarrow & \operatorname{Ker}\phi & \longrightarrow & \Omega M & \xrightarrow{\phi} & S & \longrightarrow & 0 \\ & & \downarrow & & \downarrow & & \downarrow & & \\ 0 & \longrightarrow & P_M & \longrightarrow & P_M & \longrightarrow & 0 & \longrightarrow & 0 \end{array}$$

we obtain a non-split extension of M by S.

2.7. Operations on chain complexes

In this section, we discuss tensors and homs, duality, Ext and Tor for chain complexes. The theory is parallel to the theory for modules, but the signs need some attention. We leave the reader to formulate the corresponding notions for cochain complexes.

We begin by discussing tensor products. If $\mathbf{C}$ and $\mathbf{D}$ are chain complexes of right, resp. left Λ-modules, we define

$$(\mathbf{C} \otimes_\Lambda \mathbf{D})_n = \bigoplus_{i+j=n} C_i \otimes_\Lambda D_j.$$

The differential

$$\partial_n : (\mathbf{C} \otimes_\Lambda \mathbf{D})_n \to (\mathbf{C}.\otimes_\Lambda \mathbf{D})_{n-1}$$

is given by

$$\partial_n(x \otimes y) = \partial_i(x) \otimes y + (-1)^i x \otimes \partial_j(y)$$

for $x \in C_i$, $y \in D_j$. The introduction of the signs $(-1)^i$ ensures that $\partial_n \circ \partial_{n+1} = 0$, as is easily checked.

The general convention about signs is that if we pass something of degree i through something of degree j, we should multiply by the sign $(-1)^{ij}$. We regard ∂ as being of degree -1, so that the above sign of $(-1)^i$ comes from passing the ∂ through the x. As long as one follows this convention carefully, the signs should take care of themselves.

This boundary formula shows that if x and y are cycles then so is $x \otimes y$, and if one is a cycle and the other is a boundary then $x \otimes y$ is a boundary. Thus we have a well defined product map

$$H_i(\mathbf{C}) \otimes_\Lambda H_j(\mathbf{D}) \to H_{i+j}(\mathbf{C} \otimes_\Lambda \mathbf{D}).$$

THEOREM 2.7.1 (Künneth theorem). *If the cycles $Z_n(\mathbf{C})$ and the boundaries $B_n(\mathbf{C})$ are flat Λ-modules for all n, then there is a short exact sequence*

$$0 \to \bigoplus_{i+j=n} H_i(\mathbf{C}) \otimes_\Lambda H_j(\mathbf{D}) \to H_n(\mathbf{C} \otimes_\Lambda \mathbf{D}) \to \bigoplus_{i+j=n-1} \operatorname{Tor}_1^\Lambda(H_i(\mathbf{C}), H_j(\mathbf{D})) \to 0.$$

PROOF. Since $Z(\mathbf{C})$ is flat, we have

$$(Z(\mathbf{C}) \otimes_\Lambda Z(\mathbf{D}))_n = \operatorname{Ker}(1 \otimes \partial : (Z(\mathbf{C}) \otimes_\Lambda \mathbf{D})_n \to (Z(\mathbf{C}) \otimes_\Lambda \mathbf{D})_{n-1})$$
$$(Z(\mathbf{C}) \otimes_\Lambda B(\mathbf{D}))_n = \operatorname{Im}(1 \otimes \partial : (Z(\mathbf{C}) \otimes_\Lambda \mathbf{D})_{n+1} \to (Z(\mathbf{C}) \otimes_\Lambda \mathbf{D})_n)$$

and so

$$H(Z(\mathbf{C}) \otimes_\Lambda \mathbf{D}) = Z(\mathbf{C}) \otimes_\Lambda H(\mathbf{D}).$$

Similarly

$$H(B(\mathbf{C}) \otimes_\Lambda \mathbf{D}) = B(\mathbf{C}) \otimes_\Lambda H(\mathbf{D}).$$

We now tensor the short exact sequence of complexes

$$0 \to B(\mathbf{C}) \xrightarrow{i} Z(\mathbf{C}) \to H(\mathbf{C}) \to 0$$

with $H(\mathbf{D})$. By Corollary 2.5.6, $\operatorname{Tor}_1^\Lambda(Z(\mathbf{C}), H(\mathbf{D})) = 0$ and so the long exact sequence of Section 2.5 becomes

$$0 \to \operatorname{Tor}_1^\Lambda(H(\mathbf{C}), H(\mathbf{D})) \to H(B(\mathbf{C}) \otimes_\Lambda \mathbf{D}) \xrightarrow{i_*} H(Z(\mathbf{C}) \otimes_\Lambda \mathbf{D}) \to H(\mathbf{C}) \otimes_\Lambda H(\mathbf{D}) \to 0. \quad \dagger$$

We next tensor the short exact sequence of complexes

$$0 \to Z(\mathbf{C}) \to \mathbf{C} \to B(\mathbf{C})[-1] \to 0$$

(where $[-1]$ denotes a shift of degree -1, so that $(B(\mathbf{C})[-1])_n = B_{n-1}(\mathbf{C})$) with $\mathbf{D}$. Since $\operatorname{Tor}_1^\Lambda(B(\mathbf{C}), \mathbf{D}) = 0$ we obtain a short exact sequence

$$0 \to Z(\mathbf{C}) \otimes_\Lambda \mathbf{D} \to \mathbf{C} \otimes_\Lambda \mathbf{D} \to (B(\mathbf{C}) \otimes_\Lambda \mathbf{D})[-1] \to 0.$$

Taking homology we obtain

$$\cdots \to H(B(\mathbf{C}) \otimes_\Lambda \mathbf{D}) \to H(Z(\mathbf{C}) \otimes_\Lambda \mathbf{D}) \to H(\mathbf{C} \otimes_\Lambda \mathbf{D}) \to H(B(\mathbf{C}) \otimes_\Lambda \mathbf{D})[-1] \to H(Z(\mathbf{C}) \otimes_\Lambda \mathbf{D})[-1] \to \cdots$$

It is not hard to show that the boundary map of this sequence is the map i_*, and so we have

$$0 \to \operatorname{Coker}(i_*) \to H(\mathbf{C} \otimes_\Lambda \mathbf{D}) \to \operatorname{Ker}(i_*)[-1] \to 0$$

which, using the exact sequence (†) above, becomes

$$0 \to H(\mathbf{C}) \otimes_\Lambda H(\mathbf{D}) \to H(\mathbf{C} \otimes_\Lambda \mathbf{D}) \to \operatorname{Tor}_1^\Lambda(H(\mathbf{C}), H(\mathbf{D}))[-1] \to 0. \qquad \square$$

COROLLARY 2.7.2. *If $Z_n(\mathbf{C})$ and $H_n(\mathbf{C})$ are projective Λ-modules for all n then*

$$H_n(\mathbf{C} \otimes_\Lambda \mathbf{D}) \cong \bigoplus_{i+j=n} H_i(\mathbf{C}) \otimes_\Lambda H_j(\mathbf{D}).$$

PROOF. The sequence

$$0 \to B_n(\mathbf{C}) \to Z_n(\mathbf{C}) \to H_n(\mathbf{C}) \to 0$$

splits since $H_n(\mathbf{C})$ is projective, and so $B_n(\mathbf{C})$ is also projective. Thus $Z_n(\mathbf{C})$ and $B_n(\mathbf{C})$ are flat, and also $\mathrm{Tor}_1^\Lambda(H_i(\mathbf{C}), H_j(\mathbf{D})) = 0$, so the result follows from the Künneth theorem. □

COROLLARY 2.7.3. *If $Z_n(\mathbf{C})$ and $H_n(\mathbf{C})$ are projective Λ-modules and either $\mathbf{C}$ or $\mathbf{D}$ is exact then so is $\mathbf{C} \otimes_\Lambda \mathbf{D}$.* □

REMARK. The case $\Lambda = R$ a commutative ring of coefficients is a useful special case of the above theorems. In particular, if R is **hereditary**, namely if every submodule of a projective module is projective (for example this happens if R is a Dedekind domain) then as long as the modules C_n are projective over R then the cycles $Z_n(\mathbf{C})$ and boundaries $B_n(\mathbf{C})$ are projective and hence flat, so that the hypothesis of Theorem 2.7.1 is satisfied.

We shall see in Section 3.6 of Volume II that the Künneth theorem is an especially simple case of the **Künneth spectral sequence** in which the hypotheses on $Z_n(\mathbf{C})$ and $B_n(\mathbf{C})$ are dropped. This special case is the one where the spectral sequence has only two non-vanishing columns, and no non-zero differentials.

Next we discuss homomorphisms. If $\mathbf{C}$ and $\mathbf{D}$ are chain complexes of left Λ-modules, we define a new chain complex

$$\mathbf{Hom}_\Lambda(\mathbf{C}, \mathbf{D})_n = \prod_{i+n=j} \mathrm{Hom}_\Lambda(C_i, D_j)$$

with differential

$$\partial_n : \mathbf{Hom}_\Lambda(\mathbf{C}, \mathbf{D})_n \to \mathbf{Hom}_\Lambda(\mathbf{C}, \mathbf{D})_{n-1}$$

defined so that

$$\partial_j(f(x)) = \partial_n(f)(x) + (-1)^n f(\partial_{i+1}(x))$$

for $f \in \mathrm{Hom}_\Lambda(C_i, D_j)$. In other words, ∂_n is defined by

$$(\partial_n f)(x) = \partial_j(f(x)) - (-1)^n f(\partial_{i+1}(x)),$$

i.e., $\partial f = [\partial, f]$, the graded commutator of ∂ and f.

If M is a left Λ-module, we write $M[n]$ for the chain complex consisting of M in degree $-n$ and zero elsewhere. If $\mathbf{C}$ is a chain complex of left Λ-modules, we write $\mathbf{C}[n]$ for $\Lambda[n] \otimes_\Lambda \mathbf{C}$. Namely we have $(\mathbf{C}[n])_i = (\mathbf{C})_{i+n}$. Note that the differential in $\mathbf{C}[n]$ is $(-1)^n$ times the differential in $\mathbf{C}$.

The **dual** of $\mathbf{C}$ is the chain complex of right Λ-modules $\mathbf{Hom}_\Lambda(\mathbf{C}, \Lambda[0])$. Note that the differential on the dual is given by

$$(\partial_n f)(x) = (-1)^{n-1} f(\partial_{-n+1}(x)).$$

Evaluation is a map of chain complexes

$$\mathbf{Hom}_\Lambda(\mathbf{C}, \Lambda[0]) \otimes_\Lambda \mathbf{C} \to \Lambda[0]$$

sending $f \otimes x$ to $f(x)$.

We now discuss Ext and Tor for chain complexes. This is sometimes also called **hypercohomology**.

DEFINITION 2.7.4. Suppose that $\mathbf{C}$ is a chain complex of left Λ-modules, bounded below. Then a **projective resolution** of $\mathbf{C}$ is a chain complex $\mathbf{P}$ of projective left Λ-modules, bounded below, together with a map of chain complexes $\mathbf{P} \to \mathbf{C}$ which is an isomorphism on homology.

Note that in the case $\mathbf{C} = M[0]$, this agrees with Definition 2.4.1. Existence of projective resolutions is easy to prove inductively using the definition of projective modules. The Comparison Theorem 2.4.2 also holds for projective resolutions of chain complexes, with exactly the same proof.

If $\mathbf{D}$ is a chain complex of right Λ-modules, we may thus define

$$\mathrm{Tor}_n^\Lambda(\mathbf{D}, \mathbf{C}) = H_n(\mathbf{D} \otimes_\Lambda \mathbf{P}),$$

and this will be independent of the choice of projective resolution $\mathbf{P}$. Note that if $\mathbf{D}$ is also bounded below, and $\mathbf{P}'$ is a projective resolution of $\mathbf{D}$, then $\mathbf{D} \otimes_\Lambda \mathbf{P} \leftarrow \mathbf{P}' \otimes_\Lambda \mathbf{P} \to \mathbf{P}' \otimes_\Lambda \mathbf{C}$ are homotopy equivalences, so that

$$\mathrm{Tor}_n^\Lambda(\mathbf{D}, \mathbf{C}) = H_n(\mathbf{P}' \otimes_\Lambda \mathbf{C}).$$

This gives an alternative proof of Proposition 2.5.5.

Similarly if $\mathbf{D}$ is a chain complex of left Λ-modules, we may define

$$\mathrm{Ext}_\Lambda^n(\mathbf{C}, \mathbf{D}) = H^n(\mathbf{Hom}_\Lambda(\mathbf{P}, \mathbf{D})).$$

(Note that H^n really means H_{-n}; we are regarding a chain complex as a cochain complex by negating the degrees.) If $\mathbf{D}$ is bounded above, and $\mathbf{I}$ is an injective resolution of $\mathbf{D}$ (defined in the obvious way), then

$$\mathbf{Hom}_\Lambda(\mathbf{P}, \mathbf{D}) \to \mathbf{Hom}_\Lambda(\mathbf{P}, \mathbf{I}) \leftarrow \mathbf{Hom}_\Lambda(\mathbf{C}, \mathbf{I})$$

are homotopy equivalences, and so

$$\mathrm{Ext}_\Lambda^n(\mathbf{C}, \mathbf{D}) = H^n(\mathbf{Hom}_\Lambda(\mathbf{C}, \mathbf{I})).$$

If Λ is an augmented algebra, we write $H_n(\Lambda, \mathbf{C})$ and $H^n(\Lambda, \mathbf{C})$ for the abelian groups $\mathrm{Tor}_n^\Lambda(\mathbf{C}, R)$ and $\mathrm{Ext}_\Lambda^n(R, \mathbf{C})$.

Composition of maps in **Hom** gives rise to Yoneda composition

$$\mathrm{Ext}_\Lambda^*(\mathbf{D}, \mathbf{E}) \otimes \mathrm{Ext}_\Lambda^*(\mathbf{C}, \mathbf{D}) \to \mathrm{Ext}_\Lambda^*(\mathbf{C}, \mathbf{E}),$$

which agrees with the usual Yoneda composition in case $\mathbf{C}$, $\mathbf{D}$ and $\mathbf{E}$ are modules concentrated in degree zero. Indeed, this was part of the original motivation for the definition of the derived category, which is really what we are using in disguise (see [**70**], Appendix 1).

EXERCISE. Show that an element $f \in \mathbf{Hom}_\Lambda(\mathbf{C},\mathbf{D})_0$ is a map of chain complexes if and only if f is a cycle ($\partial_0 f = 0$), and that f and g are homotopic if and only if $f - g$ is a boundary.

2.8. Induction and restriction

DEFINITION 2.8.1. If Γ is a subring of a ring Λ and M is a Λ-module then we may restrict the action to give the **restricted** Γ-module which we write as $M\downarrow_\Gamma$. If N is a Γ-module we define the **induced** Λ-module $N\uparrow^\Lambda = \Lambda \otimes_\Gamma N$, where Λ is regarded as a Λ-Γ-bimodule by left and right multiplication. We define the **co-induced** Λ-module $N\Uparrow^\Lambda = \mathrm{Hom}_\Gamma(\Lambda, N)$, where Λ is regarded as a Γ-Λ-bimodule by left and right multiplication.

LEMMA 2.8.2. *If Γ and Λ are rings, and we have a left Λ-module M, a Λ-Γ-bimodule A and a left Γ-module N, then there is a natural isomorphism*

$$\mathrm{Hom}_\Gamma(N, \mathrm{Hom}_\Lambda(A, M)) \cong \mathrm{Hom}_\Lambda(A \otimes_\Gamma N, M).$$

In other words, $A \otimes_\Gamma -$ is left adjoint to $\mathrm{Hom}_\Lambda(A, -)$.

PROOF. We have maps

$$\phi : \mathrm{Hom}_\Gamma(N, \mathrm{Hom}_\Lambda(A, M)) \to \mathrm{Hom}_\Lambda(A \otimes_\Gamma N, M)$$

given by $\phi(\alpha)(a \otimes n) = \alpha(n)(a)$ and

$$\psi : \mathrm{Hom}_\Lambda(A \otimes_\Gamma N, M) \to \mathrm{Hom}_\Gamma(N, \mathrm{Hom}_\Lambda(A, M))$$

given by $\psi(\beta)(n)(a) = \beta(a \otimes n)$. It is easy to check that these are inverse isomorphisms. □

PROPOSITION 2.8.3 (Nakayama relations). *If Γ is a subring of Λ, M is a Λ-module and N is a Γ-module then there are natural isomorphisms*
(i) $\mathrm{Hom}_\Gamma(N, M\downarrow_\Gamma) \cong \mathrm{Hom}_\Lambda(N\uparrow^\Lambda, M)$,
(ii) $\mathrm{Hom}_\Gamma(M\downarrow_\Gamma, N) \cong \mathrm{Hom}_\Lambda(M, N\Uparrow^\Lambda)$.
In other words, $\downarrow_\Gamma$ has a left adjoint $\uparrow^\Gamma$ and a right adjoint $\Uparrow^\Lambda$.

PROOF. (i) Put A equal to Λ as a Λ-Γ-bimodule in the lemma, and notice that $\mathrm{Hom}_\Lambda(\Lambda, M)$ is equal to $M\downarrow_\Gamma$ as a Γ-module.

(ii) Put A equal to Λ as a Γ-Λ-bimodule and swap the rôles of Λ and Γ, and of M and N in the lemma. Notice that $\Lambda \otimes_\Lambda M$ is equal to $M\downarrow_\Gamma$ as a Γ-module. □

It is worth pointing out that the natural maps $\eta : M \to M\downarrow_\Gamma\Uparrow^\Lambda$ and $\eta' : M\downarrow_\Gamma\uparrow^\Lambda \to M$ corresponding to the identity map on $M\downarrow_\Gamma$ under the Nakayama isomorphisms have an obvious interpretation. Namely $\eta(m) \in \mathrm{Hom}_\Gamma(\Lambda, M)$ is given by $\eta(m)(\lambda) = \lambda m$; and if $\lambda \otimes m \in \Lambda \otimes_\Gamma M$ then $\eta'(\lambda \otimes m) = \lambda m$.

COROLLARY 2.8.4 (Eckmann–Shapiro Lemma).

(sometimes this is known as Shapiro's lemma, but see **[93]***) Suppose that* Γ *is a subring of* Λ*, and that* Λ *is projective as a* Γ*-module. If* M *is a* Λ*-module and* N *is a* Γ*-module, then*

(i) $\mathrm{Ext}^n_\Gamma(N, M\downarrow_\Gamma) \cong \mathrm{Ext}^n_\Lambda(N\uparrow^\Lambda, M)$.

(ii) $\mathrm{Ext}^n_\Gamma(M\downarrow_\Gamma, N) \cong \mathrm{Ext}^n_\Lambda(M, N\Uparrow^\Lambda)$.

PROOF. Choose a projective resolution

$$\mathbf{P}: \qquad \cdots \to P_2 \to P_1 \to P_0$$

of N as a Γ-module. Since $({}_\Gamma\Gamma)\uparrow^\Lambda = {}_\Lambda\Lambda$ and induction preserves direct sums, induction takes projective Γ-modules to projective Λ-modules. Since Λ is projective as a Γ-module, induction takes exact sequences to exact sequences, and so

$$\mathbf{P}\uparrow^\Lambda: \qquad \cdots \to P_2\uparrow^\Lambda \to P_1\uparrow^\Lambda \to P_0\uparrow^\Lambda$$

is a projective resolution of $N\uparrow^\Lambda$ as a Λ-module. Hence

$$\begin{aligned}\mathrm{Ext}^n_\Gamma(N, M\downarrow_\Gamma) &= H^n(\mathrm{Hom}_\Gamma(\mathbf{P}, M\downarrow_\Gamma))\\ &\cong H^n(\mathrm{Hom}_\Lambda(\mathbf{P}\uparrow^\Lambda, M)) \cong \mathrm{Ext}^n_\Lambda(N\uparrow^\Lambda, M).\end{aligned}$$

The argument for co-induction is similar. □

Note that there is a natural surjective map from $M\downarrow_\Gamma\uparrow^\Lambda$ to M given by $\lambda \otimes m \mapsto \lambda m$, and injective map from N to $N\uparrow^\Lambda\downarrow_\Gamma$ given by $n \mapsto 1 \otimes n$.

In terms of n-fold extensions, the correspondence given by the Eckmann–Shapiro lemma is as follows. If

$$0 \to M\downarrow_\Gamma \to N_{n-1} \to \cdots \to N_0 \to N \to 0$$

represents an element of $\mathrm{Ext}^n_\Gamma(N, M\downarrow_\Gamma)$ then we can form the pushout

$$\begin{array}{ccc} M\downarrow_\Gamma\uparrow^\Lambda & \longrightarrow & N_{n-1}\uparrow^\Lambda \\ \downarrow & & \downarrow \\ M & \longrightarrow & X \end{array}$$

to obtain an n-fold extension

$$0 \to M \to X \to N_{n-2}\uparrow^\Lambda \to \cdots \to N_0\uparrow^\Lambda \to N\uparrow^\Lambda \to 0.$$

Conversely if

$$0 \to M \to M_{n-1} \to \cdots \to M_0 \to N\uparrow^\Lambda \to 0$$

represents an element of $\mathrm{Ext}^n_\Lambda(N\uparrow^\Lambda, M)$ then we form the pullback

$$\begin{array}{ccc} Y & \longrightarrow & N \\ \downarrow & & \downarrow \\ M_0\downarrow_\Gamma & \longrightarrow & N\uparrow^\Lambda\downarrow_\Gamma \end{array}$$

to obtain an n-fold extension

$$0 \to M\downarrow_\Gamma \to M_{n-1}\downarrow_\Gamma \to \cdots \to M_1\downarrow_\Gamma \to Y \to N \to 0.$$

Finally, the following is a dimension-shifted version of Lemma 2.8.2.

PROPOSITION 2.8.5. *If Γ and Λ are rings, and we have an injective left Λ-module I, a Λ-Γ-bimodule A and a left Γ-module N, then*

$$\mathrm{Ext}^n_\Gamma(N, \mathrm{Hom}_\Lambda(A, I)) \cong \mathrm{Hom}_\Lambda(\mathrm{Tor}^\Gamma_n(A, N), I).$$

PROOF. Choose a projective resolution

$$\mathbf{P}: \qquad \cdots \to P_2 \to P_1 \to P_0$$

of N as a Γ-module. Then

$$\begin{aligned}\mathrm{Ext}^n_\Gamma(N, \mathrm{Hom}_\Lambda(A, I)) &= H^n(\mathrm{Hom}_\Gamma(\mathbf{P}, \mathrm{Hom}_\Lambda(A, I))) \\ &\cong H^n(\mathrm{Hom}_\Lambda(A \otimes_\Gamma \mathbf{P}, I))\end{aligned}$$

Since I is injective $\mathrm{Hom}_\Lambda(-, I)$ takes exact sequences to exact sequences and so this is equal to

$$\mathrm{Hom}_\Lambda(H_n(A \otimes_\Gamma \mathbf{P}), I) = \mathrm{Hom}_\Lambda(\mathrm{Tor}^\Gamma_n(A, N), I). \qquad \square$$

EXERCISE. Suppose that Γ is a subring of Λ, and that Λ is projective as a Γ-module. If M and N are Λ-modules, denote by $\eta : M \to M\downarrow_\Gamma\Uparrow^\Lambda$ and $\eta' : M\downarrow_\Gamma\uparrow^\Lambda \to M$ the natural maps discussed above. Show that the following diagrams commute.

$$\begin{array}{ccc} \mathrm{Ext}^n_\Lambda(N, M) & \xrightarrow{\eta_*} & \mathrm{Ext}^n_\Lambda(N, M\downarrow_\Gamma\Uparrow^\Lambda) \\ & {\scriptstyle \mathrm{res}_{\Lambda,\Gamma}} \searrow & \downarrow \cong \\ & & \mathrm{Ext}^n_\Gamma(N\downarrow_\Gamma, M\downarrow_\Gamma) \end{array} \qquad \begin{array}{ccc} \mathrm{Ext}^n_\Lambda(M, N) & \xrightarrow{(\eta')^*} & \mathrm{Ext}^n_\Lambda(M\downarrow_\Gamma\uparrow^\Lambda, N) \\ & {\scriptstyle \mathrm{res}_{\Lambda,\Gamma}} \searrow & \downarrow \cong \\ & & \mathrm{Ext}^n_\Gamma(M\downarrow_\Gamma, N\downarrow_\Gamma) \end{array}$$

Here, the vertical maps are the Eckmann–Shapiro isomorphisms.

CHAPTER 3

Modules for group algebras

In Chapter 1 we gave a brief summary of some standard material on rings and modules. In this chapter we investigate what more we can say if the ring is the group algebra RG of a finite group G over a ring of coefficients R. The major new feature we find here is that we may give the tensor product over R of two RG-modules the structure of an RG-module. Of course, we also try to relate the subgroup structure of the group with the representation theory.

Throughout this chapter, R will denote a commutative ring of coefficients, and k will denote a field of coefficients. All RG-modules and kG-modules will be *finitely generated.*

3.1. Operations on RG-modules

DEFINITION 3.1.1. If G is a finite group and R is a commutative ring, we may form the **group ring** RG whose elements are the formal linear combinations $\sum_i r_i g_i$ with $r_i \in R$ and $g_i \in G$. Addition and multiplication are given by

$$\sum_i r_i g_i + \sum_j r'_j g_j = \sum_i (r_i + r'_i) g_i, \quad \sum_i r_i g_i \cdot \sum_j r'_j g_j = \sum_{i,j} r_i r'_j (g_i g_j).$$

Thus RG is an R-algebra, which as an R-module is free of rank $|G|$.

Of course, this definition also makes sense for infinite groups, provided we restrict our attention to finite sums.

The group ring RG is an **augmented algebra** with augmentation $\varepsilon : RG \to R$ given by

$$\varepsilon(\sum_i r_i g_i) = \sum_i r_i$$

(cf. Section 2.4). Thus it makes sense to talk of the **trivial** RG-module R. We write $H_n(G, M)$ and $H^n(G, M)$ for the homology and cohomology groups with coefficients in M, namely the groups $H_n(RG, M)$ and $H^n(RG, M)$ defined in Section 2.6. Note that in the former case we should regard the left RG-module M as a right module via $mg = g^{-1}m$. This is a standard device which we shall make explicit in this section. Similarly we write $H_n(G, \mathbf{C})$ and $H^n(G, \mathbf{C})$ for the hyperhomology and hypercohomology groups discussed in Section 2.7.

We shall see in Section 3.4 that a particular resolution called the **standard resolution** or **bar resolution** may be used to write down explicit cocycles and coboundaries. This approach is only really useful for low degree cohomology, and in Section 3.7 we provide group theoretical interpretations of degree one and degree two cohomology.

In case $R = k$ is a field, the group ring kG is a **symmetric algebra** by the following proposition:

PROPOSITION 3.1.2. *The linear map $\lambda : kG \to k$ sending a linear combination of group elements to the coefficient of the identity element satisfies the conditions of Definition 1.6.1. In particular, every projective kG-module is injective, and if P is a projective indecomposable kG-module then* $\mathrm{Soc}(P) \cong P/\mathrm{Rad}(P)$.

PROOF. Suppose I is a non-zero left ideal of kG. If $\sum_i r_i g_i$ is in I with $r_j \neq 0$, then $g_j^{-1}(\sum_i r_i g_i)$ is also an element of I and is not contained in $\mathrm{Ker}(\lambda)$. The same applies to right ideals and so condition (i) is satisfied. Condition (ii) is clearly satisfied. The remaining statements now follow by applying Proposition 1.6.2 and Theorem 1.6.3. □

DEFINITION 3.1.3. A **representation** of G over R is a homomorphism $G \to GL_n(R)$, the group of non-singular $n \times n$ matrices over R, for some n. Two representations are equivalent if one may be transformed into the other by a change of basis in R^n. A representation is called an **ordinary representation** if R is a field of characteristic zero (or more generally of characteristic not dividing $|G|$), a **modular representation** if R is a field of characteristic p dividing $|G|$, and an **integral representation** if R is a ring of algebraic integers or one of its localisations or completions at a prime ideal.

Note that a representation $\phi : G \to GL_n(R)$ makes R^n into an RG-module via

$$(\sum_i r_i g_i)x = \sum_i r_i \phi(g_i)(x).$$

This gives a one–one correspondence between equivalence classes of representations and isomorphism classes of finitely generated R-free RG-modules.

One is often interested in a slightly wider class of modules. An **RG-lattice** is a finitely generated R-projective RG-module. We demand that a map of RG-lattices, when restricted to R, is a composite of a split epi and a split mono.

TENSORS AND HOMS. If M and N are RG-modules, then we make $M \otimes_R N$ and $\mathrm{Hom}_R(M, N)$ into RG-modules via

$$(\sum_i r_i g_i)(m \otimes n) = \sum_i r_i (g_i(m) \otimes g_i(n)),$$

$$(\sum_i r_i g_i)(\phi)(m) = \sum_i r_i g_i(\phi(g_i^{-1} m)).$$

Note that only the group elements, and not general elements of the group algebra, act "diagonally" in this definition. Thus we need to know where the group is, inside the group algebra. This notion is captured in the following definition:

DEFINITION 3.1.4. A **bialgebra** consists of an algebra Λ over a commutative ring R together with maps of algebras

$$\Delta : \Lambda \to \Lambda \otimes_R \Lambda$$

called the **comultiplication** and

$$\varepsilon : \Lambda \to R$$

called the **co-unit**, such that the following diagrams commute:

$$\begin{array}{ccc} \Lambda & \xrightarrow{\Delta} & \Lambda \otimes_R \Lambda \\ {\scriptstyle\Delta}\downarrow & & \downarrow{\scriptstyle\Delta\otimes 1} \\ \Lambda \otimes_R \Lambda & \xrightarrow{1\otimes\Delta} & \Lambda \otimes \Lambda \otimes_R \Lambda \end{array}$$

(co-associativity) and

$$\begin{array}{ccccc} & \swarrow^{\cong} & \Lambda & \searrow^{\cong} & \\ & & \downarrow{\scriptstyle\Delta} & & \\ R \otimes_R \Lambda & \xleftarrow{\varepsilon\otimes 1} & \Lambda \otimes_R \Lambda & \xrightarrow{1\otimes\varepsilon} & \Lambda \otimes_R R \end{array}$$

(co-unitary property)

Let $\tau : \Lambda \otimes_R \Lambda \to \Lambda \otimes_R \Lambda$ be the twist map $\tau(\mu \otimes \nu) = \nu \otimes \mu$. We say that Λ is **cocommutative** if the following diagram commutes:

$$\begin{array}{ccc} \Lambda & & \\ {\scriptstyle\Delta}\downarrow & \searrow^{\Delta} & \\ \Lambda \otimes_R \Lambda & \xrightarrow{\tau} & \Lambda \otimes_R \Lambda \end{array}$$

A **Hopf algebra** is a bialgebra Λ together with an R-linear map

$$\eta : \Lambda \to \Lambda$$

called the **antipode** such that if $\Delta(\lambda) = \sum_i \mu_i \otimes \nu_i$ then

$$\sum_i \mu_i \eta(\nu_i) = \sum_i \eta(\mu_i)\nu_i = \varepsilon(\lambda).1 \in \Lambda.$$

We are mostly interested in the case where Λ is either commutative or cocommutative. In this case, the antipode is unique, it is an anti-automorphism of Λ (i.e., $\eta(\lambda\mu) = \eta(\mu)\eta(\lambda)$), and its square is the identity map. See for example Sweedler [**194**, Chapter 4] for more details.

If Λ is a bialgebra over R and M, N are (left) Λ-modules, we make $M \otimes_R N$ into a Λ-module as follows. If

$$\Delta(\lambda) = \sum_i \mu_i \otimes \nu_i \in \Lambda \otimes_R \Lambda$$

then

$$\lambda(m \otimes n) = \sum_i \mu_i(m) \otimes \nu_i(n).$$

We also make R into a Λ-module via $\lambda(r) = \varepsilon(\lambda)r$. This module is called the **trivial module.**

If Λ is a Hopf algebra over R and M and N are (left) Λ-modules, then we make $\mathrm{Hom}_R(M,N)$ into a Λ-module as follows. If $\Delta(\lambda) = \sum_i \mu_i \otimes \nu_i \in \Lambda \otimes_R \Lambda$ and $\phi \in \mathrm{Hom}_R(M,N)$ then

$$\lambda(\phi)(m) = \sum_i \mu_i(\phi(\eta(\nu_i)(m))).$$

We shall write M^* for the **dual module** to M, namely the module $\mathrm{Hom}_R(M,R)$. Note that we are viewing the dual of a *left* Λ-module as a *left* Λ-module. What is happening here is that because of the antipode, we can regard right Λ-modules as left Λ-modules via $\lambda m = m\eta(\lambda)$ and vice-versa.

Suppose that $R = k$ is a field and Λ is cocommutative. If M and N are finite dimensional as k-vector spaces, then the natural vector space isomorphisms

$$\mathrm{Hom}_k(M,N) \cong M^* \otimes_k N, \qquad M^{**} \cong M$$

etc. are Λ-module isomorphisms.

For example, a group algebra RG is a cocommutative Hopf algebra, with

$$\Delta\left(\sum_i r_i g_i\right) = \sum_i r_i g_i \otimes g_i, \quad \varepsilon\left(\sum_i r_i g_i\right) = \sum_i r_i,$$

$$\eta\left(\sum_i r_i g_i\right) = \sum_i r_i g_i^{-1}.$$

With these definitions, the action of RG on a tensor product $M \otimes_R N$ is given by

$$g(m \otimes n) = gm \otimes gn.$$

The action on $\mathrm{Hom}_R(M,N)$ is given by

$$(g\phi)(m) = g(\phi(g^{-1}m)).$$

If R is an integral domain, we can recover the group G from the Hopf algebra RG as the set of **grouplike elements**, i.e., the set of non-zero elements $\lambda \in RG$ satisfying $\Delta(\lambda) = \lambda \otimes \lambda$. Thus while many groups may have the same group algebra as an algebra, the same is not true as a Hopf algebra.

PROPOSITION 3.1.5. *Suppose P is a projective module and M is an R-free module for a Hopf algebra Λ over R. Then $P \otimes_R M$ is projective.*

PROOF. Since projective modules are the same as direct summands of free modules, it suffices to prove that $\Lambda \otimes_R M$ is free. Notice that the action of Λ on $\Lambda \otimes_R M$ is the diagonal one defined through the map Δ, rather than the action given in the definition (2.8.1) of an induced module. For the induced module definition it is clear that $\Lambda \otimes_R M$ is free, because if $\{m_\alpha\}$ is a free R-basis of M then $\{1 \otimes m_\alpha\}$ is a free Λ-basis of $\Lambda \otimes_R M$. So we define an R-linear map

$$\Lambda \otimes_R M \to \Lambda \otimes_R M$$

as follows. If $\Delta(\lambda) = \sum_i \mu_i \otimes \nu_i$ then $\lambda \otimes m \mapsto \sum_i \mu_i \otimes \nu_i m$. This is an isomorphism of R-modules with inverse given by $\lambda \otimes m \mapsto \sum_i \mu_i \otimes \eta(\nu_i)m$. Furthermore, it defines a Λ-module isomorphism from $\Lambda \otimes_R M$ with the structure of induced module to $\Lambda \otimes_R M$ with the required diagonal action. □

COROLLARY 3.1.6. *Suppose M and N are modules for a Hopf algebra Λ, and N is R-free. Then*

$$\tilde{\Omega}(M) \otimes_R N \oplus (\text{projective}) \cong \tilde{\Omega}(M \otimes_R N) \oplus (\text{projective}).$$

If Λ is finite dimensional over a field k, then

$$\Omega(M) \otimes_k N \cong \Omega(M \otimes_k N) \oplus (\text{projective}).$$

PROOF. Tensor the short exact sequence

$$0 \to \tilde{\Omega}(M) \to P \to M \to 0$$

with N and use the proposition. □

For group algebras we have another identity which is not obvious in the more general case of a Hopf algebra.

PROPOSITION 3.1.7. *If M and N are RG-modules then*

$$\mathrm{Hom}_{RG}(R, \mathrm{Hom}_R(M, N)) \cong \mathrm{Hom}_{RG}(M, N).$$

PROOF. We can regard the left hand side of this as an R-submodule of $\mathrm{Hom}_R(M, N)$ by evaluating at the identity. Then it consists of the elements ϕ of $\mathrm{Hom}_R(M, N)$ such that for all $g \in G$, $g\phi(g^{-1}m) = \phi(m)$. This is equivalent to the statement that ϕ is an RG-module homomorphism. □

PROPOSITION 3.1.8. *Suppose M_1, M_2 and M_3 are modules for a group algebra RG. Then there are natural isomorphisms*

(i) $\mathrm{Hom}_{RG}(M_1 \otimes_R M_2, M_3) \cong \mathrm{Hom}_{RG}(M_2, \mathrm{Hom}_R(M_1, M_3))$

(ii) *If M_1 is R-projective, then*

$$\mathrm{Ext}^n_{RG}(M_1 \otimes_R M_2, M_3) \cong \mathrm{Ext}^n_{RG}(M_2, \mathrm{Hom}_R(M_1, M_3))$$

(iii) *If M_1 is R-projective, then*

$$\mathrm{Ext}^n_{RG}(M_1, M_3) \cong H^n(G, \mathrm{Hom}_R(M_1, M_3)).$$

PROOF. (i) It is easy to check that the isomorphism

$$\mathrm{Hom}_R(M_1 \otimes_R M_2, M_3) \cong \mathrm{Hom}_R(M_2, \mathrm{Hom}_R(M_1, M_3))$$

of Lemma 2.8.2 is an isomorphism of RG-modules. Now apply $\mathrm{Hom}_{RG}(R, -)$ to both sides.

(ii) This follows from (i) by the usual dimension-shifting argument.

(iii) This is the special case of (ii) in which $M_2 = R$. □

In general, tensor products of modules are hard to decompose. However, under suitable conditions we can always tell when the trivial module is a direct summand of a tensor product. The following theorem is proved in [**20**]:

THEOREM 3.1.9. *Suppose $R = k$ is an algebraically closed field of characteristic p (possibly $p = 0$), and G is a group. If M and N are finite dimensional indecomposable kG-modules, then $M \otimes_k N$ has the trivial module k as a direct summand if and only if the following two conditions are satisfied.*

(i) $M \cong N^*$

(ii) $p \nmid \dim(N)$.

(Note that if $p = 0$ then (ii) *is automatically satisfied.)*

Moreover if k is a direct summand of $N^* \otimes_k N$ then it has multiplicity one (i.e., $k \oplus k$ is not a summand).

PROOF. The trivial kG-module k is a direct summand of $M \otimes_k N$ if and only if we can find homomorphisms

$$k \to M \otimes_k N \to k$$

whose composite is non-zero.

Using the isomorphism

$$\mathrm{Hom}_{kG}(k, M \otimes_k N) \cong \mathrm{Hom}_{kG}(N^*, M)$$

we see that the sum of the submodules of $M \otimes_k N$ isomorphic to k is $\mathrm{Hom}_{kG}(N^*, M)$. Thus the trivial module is a summand if and only if the composite map

$$\mathrm{Hom}_{kG}(N^*, M) \overset{i}{\hookrightarrow} M \otimes_k N \overset{p}{\twoheadrightarrow} (\mathrm{Hom}_{kG}(M, N^*))^*$$

is non-zero.

Associated to $p \circ i$, there is a map

$$\eta : \mathrm{Hom}_{kG}(N^*, M) \otimes_k \mathrm{Hom}_{kG}(M, N^*) \to k$$

with the property that $p \circ i \neq 0$ if and only if $\eta \neq 0$. This map is given as follows. Choose a basis $n_1, \ldots, n_r$ for N and let $n'_1, \ldots, n'_r$ be the dual basis for N^*. Let $\alpha \in \mathrm{Hom}_{kG}(N^*, M)$ and $\beta \in \mathrm{Hom}_{kG}(M, N^*)$. Then

$$i(\alpha) = \sum_{j=1}^{r} \alpha(n'_j) \otimes n_j, \qquad p(m \otimes n)(\beta) = \beta(m)(n),$$

and so by definition

$$\eta(\alpha \otimes \beta) = ((p \circ i)(\alpha))(\beta) = p\left(\sum_{j=1}^{r} \alpha(n'_j) \otimes n_j\right)(\beta)$$
$$= \sum_{j=1}^{r} p(\alpha(n'_j) \otimes n_j)(\beta) = \sum_{j=1}^{r} \beta(\alpha(n'_j))(n_j) = \mathrm{tr}(\beta \circ \alpha).$$

Hence we may factor η as composition followed by trace

$$\mathrm{Hom}_{kG}(N^*, M) \otimes_k \mathrm{Hom}_{kG}(M, N^*) \xrightarrow{\circ} \mathrm{End}_{kG}(N^*) \xrightarrow{\mathrm{tr}} k.$$

Since N^* is indecomposable, $\mathrm{End}_{kG}(N^*)$ is a local ring (Lemma 1.4.5), and so since k is algebraically closed, every kG-module endomorphism of N^* is of the form $\lambda I + n$ with $\lambda \in k$ and n nilpotent. Now we have $\mathrm{tr}(n) = 0$, and $\mathrm{tr}(I) = \dim(N^*) = \dim(N)$. Thus for k to be a direct summand of $M \otimes_k N$ (i.e., for η to be non-zero) p cannot divide $\dim(N)$. Moreover, we must have elements $\alpha \in \mathrm{Hom}_{kG}(N^*, M)$ and $\beta \in \mathrm{Hom}_{kG}(M, N^*)$ such that $\mathrm{tr}(\beta \circ \alpha) \neq 0$; namely, such that $\beta \circ \alpha$ is an isomorphism. Since M is indecomposable this means we must have $M \cong N^*$. Moreover, in the case where $p \nmid \dim(N)$ and $M \cong N^*$, it is clear that $\eta(\beta \circ \alpha) \neq 0$ for any isomorphisms α and β.

Finally, suppose k is a direct summand of $N^* \otimes_k N$ with multiplicity greater than one. Then the image of $p \circ i$ has dimension greater than one. This means that there are subspaces of $\mathrm{Hom}_{kG}(N^*, M)$ and $\mathrm{Hom}_{kG}(M, N^*)$ of dimension greater than one, on which η is a non-singular pairing. Thus there is a subspace of $\mathrm{Hom}_{kG}(N^*, M)$ of dimension greater than one all of whose non-zero elements are isomorphisms, and this we know to be impossible. □

See also Auslander and Carlson [**9**] for a discussion of more general coefficient rings R. The following is also useful:

PROPOSITION 3.1.10. *If M is a finite dimensional module for a cocommutative Hopf algebra Λ over a field k, then M is a summand of $M \otimes_k M^* \otimes_k M$. Thus the following are equivalent:*
(i) M is projective
(ii) $M \otimes_k M^*$ is projective
(iii) $M \otimes_k M$ is projective
(iv) M^* is projective
(v) M is injective.

PROOF. Choose dual bases $m_1, \ldots, m_r$ of M and $m'_1, \ldots, m'_r$ of M^*. We have maps

$$\begin{array}{ccccccccc} M & \cong & k \otimes_k M \longrightarrow (M \otimes_k M^*) \otimes_k M & \cong & M \otimes_k (M^* \otimes_k M) \longrightarrow M \otimes_k k & \cong & M \\ m & = & 1 \otimes m \longmapsto (\sum m_i \otimes m'_i) \otimes m & = & \sum m_i \otimes (m'_i \otimes m) \longmapsto \sum m_i \otimes m'_i(m) & = & m. \end{array}$$

Thus M is a summand of $M \otimes_k M^* \otimes_k M$. □

COROLLARY 3.1.11. *If Λ is a finite dimensional cocommutative Hopf algebra over a field k, then Λ is self injective.* □

In fact, this corollary is still true without the assumption of cocommutativity, but the argument is harder.

3.2. Cup products

Suppose Λ is a Hopf algebra over R which is projective as an R-module. If M, M', N and N' are Λ-modules which are projective as R-modules, then we define the **cup product**

$$\cup : \mathrm{Ext}^m_\Lambda(M, M') \times \mathrm{Ext}^n_\Lambda(N, N') \to \mathrm{Ext}^{m+n}_\Lambda(M \otimes_R N, M' \otimes_R N')$$

as follows. Choose exact sequences

$$0 \to M' \to M_{m-1} \to \cdots \to M_0 \to M \to 0$$
$$0 \to N' \to N_{n-1} \to \cdots \to N_0 \to N \to 0$$

representing the given elements $\zeta \in \mathrm{Ext}^m_\Lambda(M, M')$ and $\eta \in \mathrm{Ext}^n_\Lambda(N, N')$. We can make sure the M_i and N_j are R-projective, for example by choosing the sequences

$$0 \to M' \to P_{m-1}/L_\zeta \to \cdots \to P_0 \to M \to 0$$
$$0 \to N' \to Q_{n-1}/L_\eta \to \cdots \to Q_0 \to N \to 0$$

of Section 2.4. By Corollary 2.7.3 of the Künneth theorem, the tensor product over R of the complexes

$$0 \to M' \to M_{m-1} \to \cdots \to M_0$$
$$0 \to N' \to N_{n-1} \to \cdots \to N_0$$

is a sequence

$$0 \to M' \otimes_R N' \to (M_{m-1} \otimes_R N') \oplus (M' \otimes_R N_{n-1}) \to \cdots \to M_0 \otimes_R N_0$$

whose homology is $M \otimes_R N$ in degree zero and is exact elsewhere. Thus we may complete to an exact sequence

$$0 \to M' \otimes_R N' \to (M_{m-1} \otimes_R N') \oplus (M' \otimes_R N_{n-1}) \to \cdots \to M_0 \otimes_R N_0 \to M \otimes_R N \to 0$$

representing an element of $\mathrm{Ext}^{m+n}_\Lambda(M \otimes_R N, M' \otimes_R N')$. Maps of m-fold and n-fold extensions, as in Definition 2.6.1, tensor together to give a map of $(m+n)$-fold extensions, and so the resulting element $\zeta \cup \eta$ of $\mathrm{Ext}^{m+n}_\Lambda(M \otimes_R N, M' \otimes_R N')$ depends only on ζ and η.

Since tensor products of chain complexes are associative, cup product is also associative.

If Λ is cocommutative, then tensor products of complexes are **graded commutative** in the sense that given complexes $\mathbf{C}$ and $\mathbf{D}$ of Λ-modules, there is an isomorphism of complexes

$$\mathbf{C} \otimes_R \mathbf{D} \to \mathbf{D} \otimes_R \mathbf{C}$$
$$x \otimes y \mapsto (-1)^{\deg(x)\deg(y)} y \otimes x$$

It follows that the cup product is graded commutative in the same sense. In particular the cup product makes $\mathrm{Ext}^*_\Lambda(R, R)$ a graded commutative ring with unit.

In terms of projective resolutions, cup product can be viewed as follows. If

$$\mathbf{P}: \quad \cdots \to P_2 \to P_1 \to P_0$$
$$\mathbf{Q}: \quad \cdots \to Q_2 \to Q_1 \to Q_0$$

are resolutions of M and N respectively then again by the Künneth theorem the tensor product over R of these complexes is a projective resolution of $M \otimes_R N$. If $\zeta \in \mathrm{Ext}^m_\Lambda(M, M')$ and $\eta \in \mathrm{Ext}^n_\Lambda(N, N')$ are represented by maps $\hat{\zeta} : P_m \to M'$ and $\hat{\eta} : Q_n \to N'$ then $\hat{\zeta} \otimes \hat{\eta} : P_m \otimes_R Q_n \to M' \otimes_R N'$ can be extended by zero to give a map $(\mathbf{P} \otimes_R \mathbf{Q})_{m+n} \to M' \otimes_R N'$. Since

$$\delta(\hat{\zeta} \otimes \hat{\eta}) = \delta(\hat{\zeta}) \otimes \hat{\eta} + (-1)^{\deg(\zeta)} \hat{\zeta} \otimes \delta(\hat{\eta})$$

the product of two cocycles is a cocycle, and the product of a cocycle with a coboundary or a coboundary with a cocycle is a coboundary. This product therefore passes down to a well defined product on cohomology which may easily be checked to agree with the map defined above.

In particular, if $M = N = R$ then we may take $\mathbf{P} = \mathbf{Q}$. Thus the extension of $\hat{\zeta} \otimes \hat{\eta}$ to a cocycle $\mathbf{P} \otimes_R \mathbf{P} \to M' \otimes_R N'$ may be composed with a chain map $\mathbf{P} \to \mathbf{P} \otimes_R \mathbf{P}$ extending the obvious isomorphism $R \otimes_R R \to R$ to give a cocycle $\mathbf{P} \to M' \otimes_R N'$. Such a chain map $\Delta : \mathbf{P} \to \mathbf{P} \otimes \mathbf{P}$ exists and is unique up to homotopy by the Comparison Theorem 2.4.2. It also follows from the comparison theorem that such a map is homotopy cocommutative and co-associative, in the sense that the composite of Δ with the map $\mathbf{P} \otimes_R \mathbf{P} \to \mathbf{P} \otimes_R \mathbf{P}$ sending $x \otimes y \mapsto (-1)^{\deg(x)\deg(y)} y \otimes x$ is homotopic to Δ, and $(1 \otimes \Delta) \circ \Delta$ is homotopic to $(\Delta \otimes 1) \circ \Delta$. Such a map Δ is called a **diagonal approximation**. In fact we shall see that using the standard resolution, which we shall investigate soon, the diagonal approximation may be chosen to be co-associative (not just up to homotopy). The same is not true of cocommutativity, and this is really the reason for the existence of Steenrod operations, which we shall discuss in Chapter 4 of Volume II.

We now have two different definitions of products on $\mathrm{Ext}^*_\Lambda(R, R)$, namely Yoneda composition (Section 2.6) and cup product. The following proposition shows that these agree, and that more generally every cup product may be viewed as a Yoneda composition. The converse is not true, because for example Yoneda composition is in general not graded commutative on $\mathrm{Ext}^*_{kG}(M, M)$.

PROPOSITION 3.2.1. *If M, M', N and N' are modules for a Hopf algebra Λ, and $\zeta \in \operatorname{Ext}^m_\Lambda(M, M')$, $\eta \in \operatorname{Ext}^n_\Lambda(N, N')$, then the cup product*

$$\zeta \cup \eta \in \operatorname{Ext}^{m+n}_\Lambda(M \otimes_R N, M' \otimes_R N')$$

is equal to the Yoneda composite of

$$\zeta \otimes \mathrm{id}_{N'} \in \operatorname{Ext}^m_\Lambda(M \otimes_R N', M' \otimes_R N')$$

and

$$\mathrm{id}_M \otimes \eta \in \operatorname{Ext}^n_\Lambda(M \otimes_R N, M \otimes_R N').$$

PROOF. Denote by $\mathbf{M}$ and $\mathbf{N}$ complexes

$$0 \to M' \to M_{m-1} \to \cdots \to M_0$$
$$0 \to N' \to N_{n-1} \to \cdots \to N_0$$

which together with maps $\varepsilon : M_0 \to M$ and $\varepsilon' : N_0 \to N$ correspond to elements $\zeta \in \operatorname{Ext}^m_\Lambda(M, M')$ and $\eta \in \operatorname{Ext}^n_\Lambda(N, N')$. Then we have a map of $(m+n)$-fold extensions

$$\begin{array}{ccccccccccccc} 0 \to M' \otimes_R N' & \to & (\mathbf{M} \otimes_R \mathbf{N})_{m+n-1} & \to \cdots & & & \cdots \to & (\mathbf{M} \otimes_R \mathbf{N})_0 & \to & M \otimes_R N \to 0 \\ \| & & \downarrow \phi_{m+n-1} & & & & & \downarrow \phi_0 & & \| \\ 0 \to M' \otimes_R N' & \to & M_{m-1} \cdots \to M_0 & \to & N_{n-1} & \cdots \to & & N_0 & \to & M \otimes_R N \to 0 \end{array}$$

given by the composite maps

$$\phi_i : \begin{cases} (\mathbf{M} \otimes_R \mathbf{N})_i \to M_0 \otimes_R N_i \xrightarrow{\varepsilon} M \otimes_R N_i & 0 \le i \le n-1 \\ (\mathbf{M} \otimes_R \mathbf{N})_i \to M_{i-n} \otimes_R N' & n \le i \le m+n-1. \end{cases}$$

It is easy to check that this diagram commutes, which proves the proposition. □

COROLLARY 3.2.2. *Suppose Λ is a cocommutative Hopf algebra (for example a group algebra). If $\zeta \in H^*(\Lambda, R) = \operatorname{Ext}^*_\Lambda(R, R)$, denote by ζ_M the image of ζ under the natural map*

$$\operatorname{Ext}^*_\Lambda(R, R) \xrightarrow{\otimes_R M} \operatorname{Ext}^*_\Lambda(M, M)$$

given by tensoring exact sequences with M. Then for any $\xi \in \operatorname{Ext}^*_\Lambda(M, M)$ we have

$$\zeta_M \circ \xi = (-1)^{\deg(\zeta)\deg(\xi)} \xi \circ \zeta_M. \qquad \square$$

In fact, the above discussion becomes clearer at the level of hypercohomology (see Section 2.7). The tensor product map

$$\mathbf{Hom}_\Lambda(\mathbf{C}, \mathbf{D}) \otimes_R \mathbf{Hom}_\Lambda(\mathbf{C}', \mathbf{D}') \to \mathbf{Hom}_\Lambda(\mathbf{C} \otimes_R \mathbf{C}', \mathbf{D} \otimes_R \mathbf{D}')$$

gives rise to a cup product map

$$\begin{array}{rcl} \operatorname{Ext}^*_\Lambda(\mathbf{C}, \mathbf{D}) \otimes_R \operatorname{Ext}^*_\Lambda(\mathbf{C}', \mathbf{D}') & \to & \operatorname{Ext}^*_\Lambda(\mathbf{C} \otimes_R \mathbf{C}', \mathbf{D} \otimes_R \mathbf{D}') \\ \zeta \otimes \eta & \mapsto & \zeta \cup \eta. \end{array}$$

LEMMA 3.2.3. *If* $\zeta \in \mathrm{Ext}^m_{RG}(\mathbf{C},\mathbf{D})$, $\eta \in \mathrm{Ext}^n_{RG}(\mathbf{C}',\mathbf{D}')$ *then the cup product*

$$\zeta \cup \eta \in \mathrm{Ext}^{m+n}_{RG}(\mathbf{C}\otimes_R \mathbf{C}', \mathbf{D}\otimes_R \mathbf{D}')$$

is equal to the Yoneda composite of

$$\zeta \otimes \mathrm{id}_{\mathbf{D}'} \in \mathrm{Ext}^m_{RG}(\mathbf{C}\otimes_R \mathbf{D}', \mathbf{D}\otimes_R \mathbf{D}')$$

and

$$\mathrm{id}_{\mathbf{C}} \otimes \eta \in \mathrm{Ext}^n_{RG}(\mathbf{C}\otimes_R \mathbf{C}', \mathbf{C}\otimes_R \mathbf{D}').$$

PROOF. This follows easily from the corresponding (obvious) statement at the level of **Hom**. □

Since tensor products are graded commutative, so are cup products, in the sense that the following diagram commutes:

$$\begin{array}{ccc}
\mathrm{Ext}^m_\Lambda(\mathbf{C},\mathbf{D})\otimes_R \mathrm{Ext}^n_\Lambda(\mathbf{C}',\mathbf{D}') & \longrightarrow & \mathrm{Ext}^{m+n}_\Lambda(\mathbf{C}\otimes_R \mathbf{C}', \mathbf{D}\otimes_R \mathbf{D}') \\
\downarrow \cong & & \downarrow \cong \\
\mathrm{Ext}^n_\Lambda(\mathbf{C}',\mathbf{D}')\otimes_R \mathrm{Ext}^m_\Lambda(\mathbf{C},\mathbf{D}) & \longrightarrow & \mathrm{Ext}^{m+n}_\Lambda(\mathbf{C}'\otimes_R \mathbf{C}, \mathbf{D}'\otimes_R \mathbf{D})
\end{array}$$

But in general Yoneda products are not graded commutative. So $\mathrm{Ext}^*_\Lambda(\mathbf{C},\mathbf{C})$ is in general a non-commutative graded ring and a module over the graded commutative ring $H^*(\Lambda, R) = \mathrm{Ext}^*_\Lambda(R,R)$.

EXERCISES. 1. (Massey products) Suppose $\mathbf{P}$ is a projective resolution of the trivial module R over a Hopf algebra Λ, and $\Delta : \mathbf{P} \to \mathbf{P}\otimes_R \mathbf{P}$ is a diagonal approximation. Choose a chain homotopy $\theta : \mathbf{P} \to \mathbf{P}\otimes_R \mathbf{P}\otimes_R \mathbf{P}$ from $(\Delta\otimes 1)\circ\Delta$ to $(1\otimes\Delta)\circ\Delta$. If α, β and γ are cocycles $\mathbf{P}\to R$ representing cohomology classes a, b and c, and with cup products $ab = 0$ and $bc = 0$, choose cochains u and v with $(\alpha\otimes\beta)\circ\Delta = \delta u$ and $(\beta\otimes\gamma)\circ\Delta = \delta v$. Show that the cochain

$$(u\otimes\gamma)\circ\Delta - (-1)^{\deg(\alpha)}(\alpha\otimes v)\circ\Delta - (-1)^{\deg(\alpha)+\deg(\beta)+\deg(\gamma)}(\alpha\otimes\beta\otimes\gamma)\circ\theta$$

is a cocycle, and its cohomology class is well defined modulo the ideal generated by a and c, and independent of the choices of $\mathbf{P}$, Δ, θ, u and v. This cohomology class is called the Massey product $\langle a,b,c\rangle$.

Note that there are two extreme cases in the above set-up. In case Δ is strictly co-associative (for example for the bar resolution, Section 3.4), the formula simplifies since we may take $\theta = 0$. In the other extreme, if we are in a situation where we may take for $\mathbf{P}$ the minimal resolution (for example if Λ is finite dimensional over a field) then we have $u = v = 0$ and so the other two terms drop out. Thus non-trivial Massey products may be viewed as obstructions to finding a strictly co-associative comultiplication on the minimal resolution.

2. Give a definition of Massey product with respect to Yoneda composition. Show that in the circumstances where the Yoneda composition is

expressible as a cup product as above, your definition agrees with the above definition.

3.3. Induction and restriction

Let G be a finite group. If $H \leq G$, then RH is a subring of RG, and so by Section 2.8 if M is an RG-module we can restrict it to an RH-module which we write as $M\downarrow_H$. If N is an RH-module we have the induced RG-module $N\uparrow^G = RG \otimes_{RH} N$. Since RG is free as a right RH-module, of rank $|G:H|$, we can write $N\uparrow^G = \bigoplus_{g_i \in G/H} g_i \otimes N$ as a sum of R-modules, where the sum runs over a set of coset representatives of H in G. The action of G is given by $g(g_i \otimes n) = g_j \otimes hn$, where g_j is the coset representative with $gg_i = g_jh$ and $h \in H$.

Since $RG \cong (RG)^* = \mathrm{Hom}_R(RG, R)$ as RG-RG-bimodules, induction and restriction behave well under duality in the sense that we have $(M^*)\downarrow_H \cong (M\downarrow_H)^*$ and $(N^*)\uparrow^G \cong (N\uparrow^G)^*$. It also follows that induction and co-induction coincide since

$$N\uparrow^G = RG \otimes_{RH} N \cong (RG)^* \otimes_{RH} N \cong \mathrm{Hom}_{RH}(RG, N) = N\Uparrow^G .$$

PROPOSITION 3.3.1 (Nakayama relations).
(i) $\mathrm{Hom}_{RH}(N, M\downarrow_H) \cong \mathrm{Hom}_{RG}(N\uparrow^G, M)$
(ii) $\mathrm{Hom}_{RH}(M\downarrow_H, N) \cong \mathrm{Hom}_{RG}(M, N\uparrow^G)$.

PROOF. This was proved in Section 2.8. This version for groups is also called **Frobenius reciprocity**. □

EXERCISE. Show that the Nakayama isomorphism

$$\mathrm{Hom}_{RH}(M\downarrow_H, N) \cong \mathrm{Hom}_{RG}(M, N\uparrow^G)$$

is given by the **exterior trace map** $\widehat{\mathrm{Tr}}_{H,G}$ defined by

$$\widehat{\mathrm{Tr}}_{H,G}(\phi)(m) = \sum_{g \in G/H} g \otimes \phi(g^{-1}m).$$

Note that the expression $g \otimes \phi(g^{-1}m)$ only depends on the left coset gH of H in G.

COROLLARY 3.3.2 (Eckmann–Shapiro Lemma).

$$\mathrm{Ext}^n_{RH}(N, M\downarrow_H) \cong \mathrm{Ext}^n_{RG}(N\uparrow^G, M)$$

PROOF. This was proved in Section 2.8, since RG is projective (even free) as an RH-module. □

PROPOSITION 3.3.3. (i) $(N \otimes_R M\downarrow_H)\uparrow^G \cong N\uparrow^G \otimes_R M$
(ii) $\mathrm{Hom}_R(N, M\downarrow_H)\uparrow^G \cong \mathrm{Hom}_R(N\uparrow^G, M)$.

PROOF. (i) By the associativity of tensor product we have

$$\begin{aligned}(N \otimes_R M\downarrow_H)\uparrow^G &= RG \otimes_{RH} (N \otimes_R M\downarrow_H) \\ &\cong (RG \otimes_{RH} N) \otimes_R M = N\uparrow^G \otimes_R M.\end{aligned}$$

(ii) By Lemma 2.8.2 we have

$$\begin{aligned}\mathrm{Hom}_R(N, M\downarrow_H)\uparrow^G &\cong \mathrm{Hom}_R(N, M\downarrow_H)\Uparrow^G = \mathrm{Hom}_{RH}(RG, \mathrm{Hom}_R(N, M\downarrow_H))\\ &\cong \mathrm{Hom}_R(RG\otimes_{RH} N, M) = \mathrm{Hom}_R(N\uparrow^G, M). \qquad \square\end{aligned}$$

Now if H and K are subgroups of G, we wish to describe what happens if we induce a module from K to G and then restrict to H. Thus we need to examine RG as an RH-RK-bimodule, with left and right action by multiplication. We have

$$_{RH}RG_{RK} \cong \bigoplus_{HgK} R(HgK)$$

where $R(HgK)$ denotes a free R-module on the elements of the double coset HgK, with left RH-action and right RK-action given by multiplication as before.

If M is a module for K, we write $g\otimes M$ or ${}^gM$ for the ${}^gK$-module with

$$(gkg^{-1})(g\otimes m) = g\otimes km.$$

With this notation we have

$$\begin{aligned}R(HgK) &\cong RH\otimes_{R(H\cap {}^gK)}(g\otimes RK)\\ hgk &\longleftrightarrow h\otimes g\otimes k.\end{aligned}$$

THEOREM 3.3.4 (Mackey Decomposition Theorem).
If M is an RK-module then

$$M\uparrow^G\downarrow_H \cong \bigoplus_{HgK} ({}^gM)\downarrow_{H\cap {}^gK}\uparrow^H.$$

PROOF. As RH-modules we have

$$\begin{aligned}M\uparrow^G\downarrow_H &\cong {}_{RH}RG_{RK}\otimes_{RK} M \cong \bigoplus_{HgK} R(HgK)\otimes_{RK} M\\ &\cong \bigoplus_{HgK} RH\otimes_{R(H\cap {}^gK)}(g\otimes RK)\otimes_{RK} M\\ &\cong \bigoplus_{HgK} RH\otimes_{R(H\cap {}^gK)}(g\otimes M)\downarrow_{H\cap {}^gK} \cong \bigoplus_{HgK} ({}^gM)\downarrow_{H\cap {}^gK}\uparrow^H. \qquad \square\end{aligned}$$

REMARK. For a more traditional proof without this notational overload, see for example Curtis and Reiner [**64**], §44.

COROLLARY 3.3.5. *If M is an RK-module and N is an RH-module then we have*

(i) $M\uparrow^G \otimes_R N\uparrow^G \cong \bigoplus_{HgK} ({}^gM\downarrow_{H\cap {}^gK} \otimes_R N\downarrow_{H\cap {}^gK})\uparrow^G$

(ii) $\mathrm{Hom}_R(M\uparrow^G, N\uparrow^G) \cong \bigoplus_{HgK} (\mathrm{Hom}_R({}^gM\downarrow_{H\cap {}^gK}, N\downarrow_{H\cap {}^gK}))\uparrow^G$

(iii) $M\uparrow^G \otimes_{RG} N\uparrow^G \cong \bigoplus_{HgK} ({}^gM_{H\cap {}^gK} \otimes_{R(H\cap {}^gK)} N\downarrow_{H\cap {}^gK})$

(iv) $\mathrm{Hom}_{RG}(M\!\uparrow^G, N\!\uparrow^G) \cong \bigoplus_{HgK} (\mathrm{Hom}_{R(H\cap {}^gK)}({}^gM\!\downarrow_{H\cap {}^gK}, N\!\downarrow_{H\cap {}^gK}))$

(v) $\mathrm{Tor}_n^{RG}(M\!\uparrow^G, N\!\uparrow^G) \cong \bigoplus_{HgK} (\mathrm{Tor}_n^{R(H\cap {}^gK)}({}^gM\!\downarrow_{H\cap {}^gK}, N\!\downarrow_{H\cap {}^gK}))$

(vi) $\mathrm{Ext}^n_{RG}(M\!\uparrow^G, N\!\uparrow^G) \cong \bigoplus_{HgK} (\mathrm{Ext}^n_{R(H\cap {}^gK)}({}^gM\!\downarrow_{H\cap {}^gK}, N\!\downarrow_{H\cap {}^gK}))$.

PROOF. (i) Using Proposition 3.3.3 and the Mackey decomposition theorem we have

$$M\!\uparrow^G \otimes_R N\!\uparrow^G \cong (M\!\uparrow^G\downarrow_H \otimes_R N)\!\uparrow^G \cong ((\bigoplus_{HgK} {}^gM\!\downarrow_{H\cap {}^gK}\uparrow^H) \otimes_R N)\!\uparrow^G$$
$$\cong (((\bigoplus_{HgK} {}^gM\!\downarrow_{H\cap {}^gK}) \otimes_R N\!\downarrow_{H\cap {}^gK})\!\uparrow^H)\!\uparrow^G \cong \bigoplus_{HgK} ({}^gM\!\downarrow_{H\cap {}^gK} \otimes_R N\!\downarrow_{H\cap {}^gK})\!\uparrow^G.$$

The proofs of (ii), (iii) and (iv) are similar, and (v) and (vi) follow from (iii) and (iv) by dimension shifting. □

In terms of the above descriptions, composition of maps is given as follows. Suppose M_1, M_2 and M_3 are RH, RK and RL-modules respectively, where H, K and L are three subgroups of G. If double coset multiplication is given by

$$(HgK)(Kg'L) = \sum_i n_i(Hgk_ig'L)$$

and

$$\alpha \in \mathrm{Hom}_{RG}(M_2\!\uparrow^G, M_1\!\uparrow^G), \quad \beta \in \mathrm{Hom}_{RG}(M_3\!\uparrow^G, M_2\!\uparrow^G)$$

are elements corresponding to

$$\bar\alpha \in \mathrm{Hom}_{R(H\cap {}^gK)}({}^gM_2\!\downarrow_{H\cap {}^gK}, M_1\!\downarrow_{H\cap {}^gK}),$$
$$\bar\beta \in \mathrm{Hom}_{R(K\cap {}^{g'}L)}({}^{g'}M_3\!\downarrow_{K\cap {}^{g'}L}, M_2\!\downarrow_{K\cap {}^{g'}L})$$

then $\alpha \circ \beta$ is the sum of the elements corresponding to $n_i\bar\alpha \circ ({}^{gk_i}\bar\beta)$, where

$$\bar\alpha \circ ({}^{gk_i}\bar\beta) \in \mathrm{Hom}_{R(H\cap {}^{gk_ig'}L)}({}^{gk_ig'}M_3, M_1)$$

is the composite map

$${}^{gk_ig'}M_3 \xrightarrow{{}^{gk_i}\bar\beta} {}^{gk_i}M_2 = {}^gM_2 \xrightarrow{\bar\alpha} M_1.$$

Applying dimension shifting as usual, we obtain a similar formula for Yoneda composition of elements

$$\alpha \in \mathrm{Ext}^m_{RG}(M_2\!\uparrow^G, M_1\!\uparrow^G), \quad \beta \in \mathrm{Ext}^n_{RG}(M_3\!\uparrow^G, M_2\!\uparrow^G)$$

as the sum of the elements corresponding to the Yoneda composites $n_i\bar\alpha \circ ({}^{gk_i}\bar\beta)$.

3.4. Standard resolutions

Suppose G is a group and R is a commutative ring of coefficients. The **standard resolution** of R as an RG-module is the resolution

$$\cdots \longrightarrow F_2 \xrightarrow{\partial_2} F_1 \xrightarrow{\partial_1} F_0$$

where F_n is the free R-module on $(n+1)$-tuples $(g_0, \dots, g_n)$ of elements of G, with G-action given by

$$g(g_0, \dots, g_n) = (gg_0, \dots, gg_n).$$

The boundary map ∂_n is given by

$$\partial_n(g_0, \dots, g_n) = \sum_{i=0}^{n} (-1)^i (g_0, \dots \overset{i}{\uparrow} \dots, g_n)$$

where the vertical arrow denotes that g_i is missing from the list. This definition should be compared with the usual boundary map in simplicial homology. See for example Spanier [**190**, Chapter 4]. It is easy to check that $\partial_n \circ \partial_{n+1} = 0$.

The map $h_n : F_n \to F_{n+1}$ given by

$$h_n(g_0, \dots, g_n) = (1, g_0, \dots, g_n)$$

satisfies

$$\mathrm{id}_{F_n} = \partial_{n+1} \circ h_n + h_{n-1} \circ \partial_n$$

and is hence a chain contraction (as a complex of R-modules, not of RG-modules) of the augmented complex

$$\cdots \longrightarrow F_2 \xrightarrow{\partial_2} F_1 \xrightarrow{\partial_1} F_0 \xrightarrow{\varepsilon} R \to 0$$

where $\varepsilon(g) = 1$ for all $g \in G$. Thus the complex is acyclic, i.e., $\mathrm{Im}(\partial_{n+1}) = \mathrm{Ker}(\partial_n)$, see Section 2.3.

To see that F_n is a free RG-module, we rewrite it in the **bar notation**

$$\begin{aligned}
(g_0, \dots, g_n) &= g_0[g_0^{-1}g_1 | g_1^{-1}g_2 | \cdots | g_{n-1}^{-1}g_n] \\
[g_1 | \cdots | g_n] &= (1, g_1, g_1g_2, \dots, g_1 \cdots g_n) \\
\partial_n[g_1 | \cdots | g_n] &= g_1[g_2 | \cdots | g_n] + \sum_{i=1}^{n-1} (-1)^i [g_1 | \cdots | g_ig_{i+1} | \cdots | g_n] \\
&\quad + (-1)^n [g_1 | \cdots | g_{n-1}].
\end{aligned}$$

It is easy to see that F_n is the free RG-module on basis elements given by the symbols $[g_1 | \cdots | g_n]$. Thus for example F_0 is the free module on the single symbol $[\,]$. Because of this notation, the standard resolution is also known as the **bar resolution**.

So we see that the $\{F_n, \partial_n\}$ form a **free resolution** of $\mathrm{Coker}(\partial_1) = R$ by RG-modules.

Now if M is an RG-module we may form complexes

$$\cdots \longrightarrow M \otimes_{RG} F_2 \xrightarrow{1 \otimes \partial_2} M \otimes_{RG} F_1 \xrightarrow{1 \otimes \partial_1} M \otimes_{RG} F_0$$

and

$$\mathrm{Hom}_{RG}(F_0, M) \xrightarrow{\delta^0} \mathrm{Hom}_{RG}(F_1, M) \xrightarrow{\delta^1} \mathrm{Hom}_{RG}(F_2, M) \to \cdots$$

where δ^n is given by composition with ∂_{n+1}. Since the $\{F_n, \partial_n\}$ form a free resolution of R as an RG-module we have

$$H_n(G, M) = \mathrm{Tor}_n^{RG}(M, R) = H_n(M \otimes_{RG} \mathbf{F}, 1 \otimes \partial_*)$$
$$H^n(G, M) = \mathrm{Ext}^n_{RG}(R, M) = H^n(\mathrm{Hom}_{RG}(\mathbf{F}, M), \delta^*),$$

see Section 2.3. Some people start with this definition in terms of the standard resolution as their basic definition of group homology and cohomology.

Notice how explicit the above description is. Since $\mathrm{Hom}_{RG}(F_n, M) \cong \mathrm{Hom}_R(RG^n, M)$, an n-cocycle is a function from G^n to M, and the coboundary formula above is given by

$$\delta^n \alpha(g_1, \dots, g_{n+1}) = g_1 \alpha(g_2, \dots, g_{n+1}) +$$
$$\sum_{i=1}^{n} (-1)^i \alpha(g_1, \dots, g_i g_{i+1}, \dots, g_{n+1}) + (-1)^{n+1} \alpha(g_1, \dots, g_n).$$

For the standard resolution, there is a strictly co-associative diagonal approximation called the **Alexander–Whitney map** given by

$$\Delta(g_0, \dots, g_n) = \sum_{j=0}^{n} (g_0, \dots, g_j) \otimes (g_j, \dots, g_n)$$

or in bar notation,

$$\Delta[g_1|\cdots|g_n] = \sum_{j=0}^{n} [g_1|\cdots|g_j] \otimes g_1 \dots g_j [g_{j+1}|\cdots|g_n].$$

INFLATION. If N is a normal subgroup of G and M is an $R(G/N)$-module then we have an **inflation map**

$$\mathrm{inf}_{G/N,G} : H^n(G/N, M) \to H^n(G, M)$$

given as follows. If $\mathbf{P}$ is a projective resolution of M as an $R(G/N)$-module and $\mathbf{Q}$ is a projective resolution of M as an RG-module, then by the Comparison Theorem 2.4.2 and the remark after it, the identity map of M lifts to a chain map $\rho : \mathbf{Q} \to \mathbf{P}$ uniquely up to homotopy. Thus if $\hat{\zeta} : P_n \to M$ is a cocycle representing $\zeta \in H^n(G/N, M)$ then $\hat{\zeta} \circ \rho : Q_n \to M$ is also a cocycle and represents $\mathrm{inf}_{G,N}(\zeta) \in H^n(G, M)$. It is easy to check that any other cocycle representing ζ will give rise to the same element of $H^n(G, M)$.

If N does not act trivially on M, then we still write $\mathrm{inf}_{G/N,G}$ for the composite of the inflation from $H^n(G/N, M^N)$ to $H^n(G, M^N)$ with the map $H^n(G, M^N) \to H^N(G, M)$ induced by the inclusion.

In terms of the standard resolution, we may take ρ to be the obvious map sending group elements to their coset representatives, so that $\hat{\zeta} \circ \rho$ is constant on cosets.

EXERCISE. Using the standard resolutions for G and G/N, show that if M is an RG-module then there is a five-term exact sequence

$$0 \to H^1(G/N, M^N) \xrightarrow{\inf} H^1(G,M) \to H^1(N,M)^{G/N} \\ \to H^2(G/N, M^N) \xrightarrow{\inf} H^2(G,M).$$

This sequence does not continue. We shall derive this sequence in Chapter 3 of Volume II as a consequence of the spectral sequence of the group extension.

Note that by using the resolution for G to calculate $H^*(N,M)$, it is clear that G/N acts, and that the image of restriction $H^*(G,M) \to H^*(N,M)$ lies in the fixed points of this action.

3.5. Cyclic and abelian groups

The standard resolution constructed above is of exponential growth, whereas we shall see in Chapter 5 of Volume II that for finite groups there always exist resolutions of polynomial growth. The problem is that resolutions of polynomial growth are in general much harder to write down. One case where there is an easily describable **periodic** resolution is the case of a cyclic group.

Let $\mathbb{Z}/n = \langle x \mid x^n = 1\rangle$ be a cyclic group of order n. We have two maps

$$N : R(\mathbb{Z}/n) \to R(\mathbb{Z}/n) \qquad\qquad T : R(\mathbb{Z}/n) \to R(\mathbb{Z}/n)$$
$$x \mapsto 1 + x + \cdots + x^{n-1} \qquad\qquad x \mapsto 1 - x.$$

PROPOSITION 3.5.1. *The sequence*

$$\cdots \xrightarrow{N} R(\mathbb{Z}/n) \xrightarrow{T} R(\mathbb{Z}/n) \xrightarrow{N} R(\mathbb{Z}/n) \xrightarrow{T} R(\mathbb{Z}/n)$$

is a free resolution of R as an $R(\mathbb{Z}/n)$-module.

PROOF. This is easy to check. □

COROLLARY 3.5.2. *If M is an $R(\mathbb{Z}/n)$-module then*

$$H^0(\mathbb{Z}/n, M) = M^{\mathbb{Z}/n} = \{m \in M \mid (1-x)m = 0\}$$
$$H^{2r}(\mathbb{Z}/n, M) = M^{\mathbb{Z}/n}/(1 + x + \cdots + x^{n-1})M \quad (r > 0)$$
$$H^{2r+1}(\mathbb{Z}/n, M) = \{m \in M \mid (1 + x + \cdots + x^{n-1})m = 0\}/(1-x)M.$$ □

COROLLARY 3.5.3. *Suppose M is an RG-module, and H is a normal subgroup of G with G/H cyclic. If $M{\downarrow_H}$ is projective then*

$$M \oplus (\text{projective}) \cong \tilde{\Omega}^2(M) \oplus (\text{projective}).$$

PROOF. Since $M{\downarrow_H}$ is projective, so is

$$M{\downarrow_H}{\uparrow^G} \cong M \otimes_R R{\downarrow_H}{\uparrow^G} = M \otimes_R R(G/H).$$

Thus tensoring M with the sequence

$$0 \to R \to R(G/H) \xrightarrow{T} R(G/H) \to R \to 0$$

we obtain a sequence of the form

$$0 \to M \to (\text{projective}) \to (\text{projective}) \to M \to 0. \qquad \square$$

The proof of the above corollary also shows the following.

COROLLARY 3.5.4. *Suppose $G = \mathbb{Z}/n$ is cyclic of order n. Then there is an element $\sigma \in H^2(G, R)$ such that for all RG-modules M, cup product with σ induces an isomorphism*

$$H^r(G, M) \xrightarrow{\cong} H^{r+2}(G, M)$$

for all $r > 0$. In particular, if n is not invertible in R then σ is not nilpotent.

PROOF. We take $\sigma \in \operatorname{Ext}^2_{RG}(R, R)$ to correspond to the 2-fold extension

$$0 \to R \to RG \xrightarrow{T} RG \to R \to 0. \qquad \square$$

We now calculate the ring structure of $H^*(\mathbb{Z}/n, k)$, where k is a field of characteristic p. The above calculation shows that $H^r(\mathbb{Z}/n, k)$ is one dimensional over k for each r. Choose a non-zero element $x \in H^1(\mathbb{Z}/n, k)$. This is represented, as in Section 2.6, by a map $\hat{x} : \Omega(k) \to k$. Now the projective cover of k is $k(\mathbb{Z}/p^s)$, where p^s is the exact power of p dividing n. In other words, the generator for the cyclic group is represented by a Jordan block of size p^s with eigenvalue 1. The remaining projective modules are obtained by tensoring this one with the simple modules for $k(\mathbb{Z}/n)$ (if k is a splitting field, these are one dimensional and there are n/p^s of them). Thus $\Omega(k)$ is a uniserial module of length $p^s - 1$ with all its composition factors trivial. The map $\hat{x}$ takes the unique top composition factor of $\Omega(k)$ onto the trivial module.

Now choose a non-zero element $y \in H^2(\mathbb{Z}/n, k)$. Since $\Omega^2(k) \cong k$, $\hat{y} : \Omega^2(k) \to k$ is an isomorphism. We calculate products by the method at the end of Section 2.6. Namely, y^2 is represented by the map

$$\hat{y} \circ \Omega^2(\hat{y}) : \Omega^4(k) \to k$$

which is again an isomorphism, so that y^2 is a non-zero element of $H^4(\mathbb{Z}/n, k)$. Continuing this way, the powers y^r are non-zero elements in $H^{2r}(\mathbb{Z}/n, k)$. Similarly xy^r is represented by the surjective map

$$\hat{x} \circ \Omega(\hat{y}) \circ \Omega^3(\hat{y}) \circ \cdots \circ \Omega^{2r-1}(\hat{y}) : \Omega^{2r+1}(k) \cong \Omega(k) \to k$$

and is hence a non-zero element of $H^{2r+1}(\mathbb{Z}/n, k)$.

The only thing left to do now is to calculate x^2. This is represented by the map

$$\hat{x} \circ \Omega(\hat{x}) : \Omega^2(k) \cong k \to \Omega(k) \to k.$$

If $p^s - 1 = 1$ this composite is a non-zero multiple of y. If $p^s - 1 > 1$ then the composite is zero and so $x^2 = 0$.

We summarise the results of this calculation in the following proposition.

PROPOSITION 3.5.5. *Suppose p^s is the p-part of n, and k is a field of characteristic p. Then the cohomology ring $H^*(\mathbb{Z}/n, k)$ has the following structure.*

(i) *If $p^s = 2$ then $H^*(\mathbb{Z}/n, k) \cong k[x]$ with $\deg(x) = 1$.*

(ii) *If $p^s > 2$ then $H^*(\mathbb{Z}/n, k) \cong k[x, y]/(x^2)$ with $\deg(x) = 1$ and $\deg(y) = 2$.* □

An alternative proof is given in Cartan and Eilenberg [**53**, Chap. XII §7], where more general coefficient rings are discussed. They use the explicit diagonal approximation $\Delta_{rs} : R(\mathbb{Z}/n) \to R(\mathbb{Z}/n) \otimes_R R(\mathbb{Z}/n)$ given by

$$\Delta_{rs}(1) = \begin{cases} 1 \otimes 1 & r \text{ even} \\ 1 \otimes x & r \text{ odd, } s \text{ even} \\ \sum_{0 \leq i \leq j \leq n-1} x^i \otimes x^j & r \text{ odd, } s \text{ odd.} \end{cases}$$

To calculate the cohomology of a general finite abelian group, we need a version of the Künneth formula.

THEOREM 3.5.6. *Suppose R is a hereditary ring of coefficients (see the remark at the end of Section 2.7). If G_1 and G_2 are groups, M_1 is an RG_1-module and M_2 is an RG_2-module, then there is a short exact sequence*

$$0 \to \bigoplus_{i+j=n} H^i(G_1, M_1) \otimes_R H^j(G_2, M_2) \to H^n(G_1 \times G_2, M_1 \otimes_R M_2)$$
$$\to \bigoplus_{i+j=n-1} \mathrm{Tor}_1^R(H^i(G_1, M_1), H^j(G_2, M_2)) \to 0.$$

Here, $M_1 \otimes_R M_2$ is regarded as an $R(G_1 \times G_2)$-module via $(g_1, g_2)(m_1 \otimes m_2) = g_1 m_1 \otimes g_2 m_2$.

PROOF. If $\mathbf{P}$ and $\mathbf{Q}$ are projective resolutions of R as RG_1 and RG_2-modules respectively, then $\mathbf{P} \otimes_R \mathbf{Q}$ is a projective resolution of R as an $R(G_1 \times G_2)$-module (exactness follows from the Künneth theorem 2.7.1 since R is hereditary). The theorem now follows by applying Theorem 2.7.1 to the complex

$$\mathrm{Hom}_{R(G_1 \times G_2)}(\mathbf{P} \otimes_R \mathbf{Q}, M_1 \otimes_R M_2) \cong \mathrm{Hom}_{RG_1}(\mathbf{P}, M_1) \otimes_R \mathrm{Hom}_{RG_2}(\mathbf{Q}, M_2).$$
□

REMARK. If $R = k$ is a field then $H^*(G_1 \times G_2, k) \cong H^*(G_1, k) \otimes_k H^*(G_2, k)$ as graded rings; i.e., $(x \otimes y)(x' \otimes y') = (-1)^{\deg(y)\deg(x')} xx' \otimes yy'$. This follows from the cup product description of the ring structure and the corresponding commutativity formula for tensor products of chain complexes.

COROLLARY 3.5.7. *If $G = (\mathbb{Z}/p)^s$ is an elementary abelian group of order p^s and k is a field of characteristic p then the cohomology ring $H^*(G, k)$ has the following structure.*

(i) *If $p = 2$ then $H^*(G, k) \cong k[x_1, \ldots, x_s]$ is a polynomial algebra with $\deg(x_i) = 1$.*

(ii) *If* $p > 2$ *then* $H^*(G,k) \cong \Lambda(x_1, \dots, x_s) \otimes_k k[y_1, \dots, y_s]$ *is a tensor product of an exterior algebra and a polynomial algebra with* $deg(x_i) = 1$, $\deg(y_i) = 2$. □

3.6. Relative projectivity and transfer

DEFINITION 3.6.1. Suppose H is a subgroup of G. An RG-module M is said to be **projective relative to** H or **relatively** H**-projective** if whenever we are given modules M_1 and M_2, a map $\lambda : M \to M_1$ and an epimorphism $\mu : M_2 \to M_1$ such that there exists a map of RH-modules $\nu : M\downarrow_H \to M_2\downarrow_H$ with $\lambda = \mu \circ \nu$, then there exists a map of RG-modules $\nu' : M \to M_2$ with $\lambda = \mu \circ \nu'$.

$$\begin{array}{ccccc} & & M & & \\ & \swarrow^{\nu} & \downarrow \lambda & & \\ M_2 & \xrightarrow{\mu} & M_1 & \longrightarrow & 0 \end{array}$$

If $H = 1$ and R is a field, this agrees with the definition of projective RG-module given in Section 1.5.

An RG-module M is said to be **injective relative to** H or **relatively** H**-injective** if whenever we are given modules M_1 and M_2, a map $\lambda : M_1 \to M$ and a monomorphism $\mu : M_1 \to M_2$ such that there exists a map of RH-modules $\nu : M_1\downarrow_H \to M\downarrow_H$ with $\lambda = \nu \circ \mu$, then there exists a map of RG-modules $\nu' : M_1\downarrow_H \to M\downarrow_H$ with $\lambda = \nu' \circ \mu$.

$$\begin{array}{ccccc} 0 & \longrightarrow & M_1 & \xrightarrow{\mu} & M_2 \\ & & \lambda \downarrow & \swarrow_{\nu} & \\ & & M & & \end{array}$$

A short exact sequence of RG-modules is H**-split** if it splits on restriction to H.

The concept of relative projectivity was studied by D. G. Higman, who related it to the transfer map as follows:

DEFINITION 3.6.2. If H is a subgroup of G, and M and M' are RG-modules, we define the **transfer** or **trace map**

$$\mathrm{Tr}_{H,G} : \mathrm{Hom}_{RH}(M'\downarrow_H, M\downarrow_H) \to \mathrm{Hom}_{RG}(M', M)$$

as follows. Choose a set $\{g_i,\ i \in I\}$ of left coset representatives of H in G and set

$$\mathrm{Tr}_{H,G}(\phi)(m) = \sum_{i \in I} g_i \phi(g_i^{-1} m).$$

Since ϕ is an RH-module homomorphism, $g_i\phi(g_i^{-1}m)$ only depends on the left coset g_iH and so this map is independent of choice of coset representatives of H in G.

In other words, $\mathrm{Tr}_{H,G}$ is the composite of the exterior trace map

$$\widehat{\mathrm{Tr}}_{H,G} : \mathrm{Hom}_{RH}(M'\downarrow_H, M\downarrow_H) \xrightarrow{\cong} \mathrm{Hom}_{RG}(M', M\downarrow_H\uparrow^G)$$

defined in Section 3.3 and the map induced by the natural surjection $M\downarrow_H\uparrow^G \to M$.

We write $(M',M)^G$ for $\mathrm{Hom}_{RG}(M',M)$, we write $(M',M)^G_H$ for the image of $\mathrm{Tr}_{H,G}$, and we set

$$(M',M)^{G,H} = (M',M)^G/(M',M)^G_H$$

as an R-module. Similarly if $\mathcal{X}$ is a collection of subgroups of G we write $(M',M)^G_{\mathcal{X}}$ for the linear span of the $(M',M)^G_H$, $H \in \mathcal{X}$, and $(M',M)^{G,\mathcal{X}}$ for $(M',M)^G/(M',M)^G_{\mathcal{X}}$.

In case $M' = R$, $(R,M)^G$ is just the space of fixed points M^G. In this case we write M^G_H and $M^{G,H}$ for the image and cokernel of $\mathrm{Tr}_{H,G} : M^H \to M^G$.

LEMMA 3.6.3. (i) *If* $\alpha \in \mathrm{Hom}_{RH}(M_1, M_2)$ *and* $\beta \in \mathrm{Hom}_{RG}(M_2, M_3)$ *then*

$$\beta \circ \mathrm{Tr}_{H,G}(\alpha) = \mathrm{Tr}_{H,G}(\beta \circ \alpha).$$

(ii) *If* $\alpha \in \mathrm{Hom}_{RG}(M_1, M_2)$ *and* $\beta \in \mathrm{Hom}_{RH}(M_2, M_3)$ *then*

$$\mathrm{Tr}_{H,G}(\beta) \circ \alpha = \mathrm{Tr}_{H,G}(\beta \circ \alpha).$$

(iii) *If* $H \le K \le G$ *then* $\mathrm{Tr}_{K,G}\mathrm{Tr}_{H,K}(\alpha) = \mathrm{Tr}_{H,G}(\alpha)$.

(iv) *If* H *and* K *are two subgroups of* G*, then for* $\alpha \in \mathrm{Hom}_{RK}(M_1, M_2)$, $\mathrm{Tr}_{K,G}(\alpha) = \sum_{HgK} \mathrm{Tr}_{H\cap {}^gK,H}(g\alpha)$. *Recall that* $(g\alpha)(m) = g(\alpha(g^{-1}m))$.

(v) *If* $\alpha \in \mathrm{End}_{RH}(M)$ *and* $\beta \in \mathrm{End}_{RK}(M)$*, then*

$$\mathrm{Tr}_{H,G}(\alpha)\mathrm{Tr}_{K,G}(\beta) = \sum_{HgK} \mathrm{Tr}_{H\cap {}^gK,G}(\alpha g \beta).$$

(vi) *Suppose* $K \le H \le G$*,* M_1 *is an* RG*-module,* M_2 *is an* RH*-module and* $i : M_2 \to M_2\uparrow^G\downarrow_H$ *is the natural map taking* m *to* $1 \otimes m$*. Then*

$$\widehat{\mathrm{Tr}}_{H,G}\mathrm{Tr}_{K,H}(\alpha)(m) = \mathrm{Tr}_{K,G}(i \circ \alpha)(m).$$

PROOF. (i), (ii) and (iii) are clear from the definition.

(iv) For each double coset HgK, let $\lambda(g)$ be a set of left coset representatives of $H \cap {}^gK$ in H. Then $\bigcup_{HgK}\{hg \mid h \in \lambda(g)\}$ is a set of left coset representatives of K in G.

(v) We have

$$\begin{aligned}\mathrm{Tr}_{H,G}(\alpha)\mathrm{Tr}_{K,G}(\beta) &= \mathrm{Tr}_{H,G}(\alpha\mathrm{Tr}_{K,G}(\beta)) = \mathrm{Tr}_{H,G}(\sum_{HgK} \alpha\mathrm{Tr}_{H\cap {}^gK,H}(g\beta)) \\ &= \mathrm{Tr}_{H,G}(\sum_{HgK} \mathrm{Tr}_{H\cap {}^gK,H}(\alpha g\beta)) = \sum_{HgK} \mathrm{Tr}_{H\cap {}^gK,G}(\alpha g\beta).\end{aligned}$$

(vi) We have

$$\widehat{\mathrm{Tr}}_{H,G}\mathrm{Tr}_{K,H}(\alpha)(m) = \sum_{g\in G/H}\sum_{h\in H/K} g\otimes h\alpha h^{-1}g^{-1}(m)$$
$$= \sum_{g\in G/K} g\otimes \alpha g^{-1}(m) = \mathrm{Tr}_{K,G}(i\circ\alpha)(m). \qquad \square$$

Setting $M = M_1 = M_2 = M_3$ in parts (i) and (ii) of the above lemma, it follows that $(M,M)^G_H$ is a two-sided ideal in $\mathrm{End}_{RG}(M)$, and so the quotient $(M,M)^{G,H}$ inherits a ring structure. Similarly, $(M,M)^G_{\mathcal{X}}$ is a two-sided ideal and $(M,M)^{G,\mathcal{X}}$ is a ring.

The following two propositions explain the relationship between relative projectivity and transfer:

PROPOSITION 3.6.4 (D. G. Higman). *Let M be an RG-module and H a subgroup of G. Then the following are equivalent:*

(i) *M is projective relative to H.*

(ii) *Every H-split epimorphism of RG-modules $\lambda : M' \to M$ (i.e., one which splits as a map of RH-modules) splits.*

(iii) *M is injective relative to H.*

(iv) *Every H-split monomorphism of RG-modules splits.*

(v) *M is a direct summand of $M\downarrow_H\uparrow^G$.*

(vi) *M is a direct summand of some module induced from H.*

(vii) *(Higman's criterion) The identity map on M is in the image of* $\mathrm{Tr}_{H,G}$.

PROOF. The implications (i) $\Rightarrow$ (ii) $\Rightarrow$ (v) $\Rightarrow$ (vi) and (iii) $\Rightarrow$ (iv) $\Rightarrow$ (v) are clear, using the natural H-split maps $M \hookrightarrow M\downarrow_H\uparrow^G$ and $M\downarrow_H\uparrow^G \twoheadrightarrow M$.

(vi) $\Rightarrow$ (vii) : If M is a direct summand of $N\uparrow^G$ for an RH-module N then we denote by ρ the RH-module endomorphism $N\uparrow^G\downarrow_H \to N\uparrow^G\downarrow_H$ given by

$$\rho(g\otimes n) = \begin{cases} g\otimes n & \text{if } g\in H \\ 0 & \text{otherwise} \end{cases}$$

i.e., the projection onto N as a summand of $N\uparrow^G\downarrow_H$ (cf. the Mackey decomposition theorem). Then we have $\mathrm{Tr}_{H,G}(\rho) = 1$, the identity endomorphism of $N\uparrow^G$.

Now we denote by θ the RH-module endomorphism of M given by the composite

$$\theta : M\downarrow_H \hookrightarrow N\uparrow^G\downarrow_H \xrightarrow{\rho} N\uparrow^G\downarrow_H \twoheadrightarrow M\downarrow_H .$$

By Lemma 3.6.3 (i) and (ii), we have $\mathrm{Tr}_{H,G}(\theta) = 1$.

(vii) $\Rightarrow$ (i) : Let

$$\begin{array}{ccccc} & & M & & \\ & \overset{\nu}{\swarrow} & \downarrow \lambda & & \\ M_2 & \xrightarrow{\mu} & M_1 & \longrightarrow & 0 \end{array}$$

be as in the definition of relative projectivity. If $\theta \in \mathrm{End}_{RH}(M_1)$ with $\mathrm{Tr}_{H,G}(\theta) = 1$ then we let $\nu' = \mathrm{Tr}_{H,G}(\nu \circ \theta)$. By Lemma 3.6.3 (i) and (ii) we have

$$\begin{aligned} \mu \circ \nu' = \mu \circ \mathrm{Tr}_{H,G}(\nu \circ \theta) &= \mathrm{Tr}_{H,G}(\mu \circ \nu \circ \theta) \\ = \mathrm{Tr}_{H,G}(\lambda \circ \theta) &= \lambda \circ \mathrm{Tr}_{H,G}(\theta) = \lambda. \end{aligned}$$

(vii) $\Rightarrow$ (iii) is proved dually. □

COROLLARY 3.6.5. *Suppose an RG-module M is projective as an R-module and has a finite projective resolution as an RG-module. Then M is projective as an RG-module.*

PROOF. Let

$$0 \to P_n \to P_{n-1} \to \cdots \to P_0 \to M \to 0$$

be a projective resolution of M with n minimal. Since M is R-projective, this sequence is R-split. By Proposition 3.6.4, P_n is injective relative to the trivial subgroup, and so the map $P_n \to P_{n-1}$ splits, contradicting the minimality of n unless $n = 0$. □

PROPOSITION 3.6.6. *Suppose M_1 and M_2 are RG-modules, H is a subgroup of G and $\alpha \in \mathrm{Hom}_{RG}(M_1, M_2)$. Then the following are equivalent.*

(i) $\alpha \in (M_1, M_2)^G_H$.

(ii) *α factors as $M_1 \to M_2{\downarrow_H}{\uparrow^G} \to M_2$ where the latter map is the natural surjection $M_2{\downarrow_H}{\uparrow^G} \to M_2$.*

(iii) *α factors as $M_1 \xrightarrow{\alpha_1} M \xrightarrow{\alpha_2} M_2$ for some M projective relative to H.*

PROOF. (i) $\Rightarrow$ (ii) : This follows from the fact that $\mathrm{Tr}_{H,G}$ is the composite of $\widehat{\mathrm{Tr}}_{H,G}$ and the map induced by $M_2{\downarrow_H}{\uparrow^G} \to M_2$.

(ii) $\Rightarrow$ (iii) is clear.

(iii) $\Rightarrow$ (i) : Write $\mathrm{id}_M = \mathrm{Tr}_{H,G}(\theta)$ as in the previous proposition. If $\alpha = \alpha_2 \circ \alpha_1$ then

$$\alpha = \alpha_2 \circ \mathrm{Tr}_{H,G}(\theta) \circ \alpha_1 = \mathrm{Tr}_{H,G}(\alpha_2 \circ \theta \circ \alpha_1) \in (M_1, M_2)^G_H$$

by Lemma 3.6.3 (i) and (ii). □

COROLLARY 3.6.7. *If M and N are RG-modules with M relatively H-projective, then $M \otimes_R N$ is also relatively H-projective.*

PROOF. This follows from the Proposition 3.6.4, using the identity

$$M{\downarrow_H}{\uparrow^G} \otimes_R N \cong (M{\downarrow_H} \otimes_R N{\downarrow_H}){\uparrow^G}$$

(Proposition 3.3.3). □

COROLLARY 3.6.8. *If M and N are RG-modules with M relatively H-projective and N relatively K-projective, then $M \otimes_R N$ is a summand of a sum of modules which are projective relative to subgroups of the form $H \cap {}^gK$.*

PROOF. This follows in the same way, using Corollary 3.3.5 (i). □

COROLLARY 3.6.9. *Suppose $|G : H|$ is invertible in R. Then every RG-module M is projective relative to H.*

PROOF. $\mathrm{Tr}_{H,G}(\mathrm{id}_M/|G; H|) = \mathrm{id}_M$. □

COROLLARY 3.6.10. *Suppose R is a field of characteristic p or a commutative local ring whose residue class field has characteristic p. Suppose H contains a Sylow p-subgroup of G. Then every RG-module is projective relative to H.* □

COROLLARY 3.6.11. *Suppose $|G|$ is invertible in R. Then every RG-module which is projective as an R-module is projective as an RG-module. In particular every short exact sequence of such modules splits.* □

COROLLARY 3.6.12 (Maschke's theorem). *Suppose k is a field of characteristic coprime to $|G|$. Then every short exact sequence of kG-modules splits. In particular kG is semisimple.* □

REMARK. The usual proof of Maschke's theorem is as follows. If M_1 is a submodule of a kG-module M and if ρ is any linear projection of M onto M_1 then $\rho' = \sum_{g \in G} g\rho g^{-1}$ is a G-invariant projection and so $M = M_1 \oplus \mathrm{Ker}(\rho')$. This is in a sense the prototype for the proof of Higman's criterion.

DEFINITION 3.6.13. If X is a permutation representation of G, we write RX for the corresponding matrix representation whose basis elements are the elements of X, and we say a short exact sequence $0 \to M_1 \to M_2 \to M_3 \to 0$ is X-split if the sequence

$$0 \to RX \otimes_R M_1 \to RX \otimes_R M_2 \to RX \otimes_R M_3 \to 0$$

splits.

We say that M is projective relative to X, or relatively X-projective, if every X-split epimorphism $M' \to M \to 0$ splits.

The following two lemmas show that these concepts are equivalent to those introduced in Definition 3.6.1, but better behaved on restriction to subgroups.

LEMMA 3.6.14. *Suppose X is a transitive permutation representation of G with point stabiliser H. Then a short exact sequence of RG-modules is X-split if and only if it is H-split. An RG-module M is relatively X-projective if and only if it is relatively H-projective.* □

LEMMA 3.6.15. *Suppose X is a permutation representation of G and H is a subgroup of G.*

(i) *If a short exact sequence of RG-modules is X-split then its restriction to H is also X-split.*

(ii) *If an RG-module M is relatively X-projective then* $M\downarrow_H$ *is also relatively X-projective.* □

TRANSFER IN COHOMOLOGY. We can extend the notion of transfer from Hom to Ext as follows. First note that projective RG-modules are still projective as RH-modules for $H \le G$. So a projective resolution of M as an RG-module gives us a projective resolution of M as an RH-module. Thus we obtain a restriction map

$$\mathrm{res}_{G,H} : \mathrm{Ext}^n_{RG}(M, M') \to \mathrm{Ext}^n_{RH}(M, M').$$

If

$$\cdots \to P_2 \xrightarrow{\partial_2} P_1 \xrightarrow{\partial_1} P_0$$

is a projective resolution of an RG-module M, and M' is another RG-module then for any $\alpha \in \mathrm{Hom}_{RH}(P_n, M')$ we have $\mathrm{Tr}_{H,G}(\alpha)\circ\partial_{n+1} = \mathrm{Tr}_{H,G}(\alpha\circ\partial_{n+1})$ and $\partial_n \circ \mathrm{Tr}_{H,G}(\alpha) = \mathrm{Tr}_{H,G}(\partial_n \circ \alpha)$ by Lemma 3.6.3. Thus $\mathrm{Tr}_{H,G}$ induces a well defined map

$$\mathrm{Tr}_{H,G} : \mathrm{Ext}^n_{RH}(M, M') \to \mathrm{Ext}^n_{RG}(M, M').$$

It is easy to check that this is independent of the choice of the resolution.

LEMMA 3.6.16. (i) *If* $\alpha \in \mathrm{Ext}^m_{RH}(M_1, M_2)$ *and* $\beta \in \mathrm{Ext}^n_{RG}(M_2, M_3)$ *then*

$$\beta \circ \mathrm{Tr}_{H,G}(\alpha) = \mathrm{Tr}_{H,G}(\mathrm{res}_{G,H}(\beta)\circ\alpha) \in \mathrm{Ext}^{m+n}_{RG}(M_1, M_3).$$

(ii) *If* $\alpha \in \mathrm{Ext}^m_{RG}(M_1, M_2)$ *and* $\beta \in \mathrm{Ext}^n_{RH}(M_2, M_3)$ *then*

$$\mathrm{Tr}_{H,G}(\beta)\circ\alpha = \mathrm{Tr}_{H,G}(\beta \circ \mathrm{res}_{G,H}(\alpha)).$$

(iii) *If* $H \le K \le G$ *then* $\mathrm{Tr}_{K,G}\mathrm{Tr}_{H,K}(\alpha) = \mathrm{Tr}_{H,G}(\alpha)$.

(iv) *(Mackey formula) If* H, $K \le G$, *then for* $\alpha \in \mathrm{Ext}^n_{RK}(M_1, M_2)$ *we have*

$$\mathrm{res}_{G,H}\mathrm{Tr}_{K,G}(\alpha) = \sum_{HgK} \mathrm{Tr}_{H\cap {}^gK,H}\,\mathrm{res}_{K,H\cap {}^gK}(g\alpha).$$

(v) *If* $\alpha \in \mathrm{Ext}^m_{RH}(M, M)$ *and* $\beta \in \mathrm{Ext}^n_{RK}(M, M)$, *then*

$$\mathrm{Tr}_{H,G}(\alpha)\mathrm{Tr}_{K,G}(\beta) = \sum_{HgK} \mathrm{Tr}_{H\cap {}^gK,G}(\alpha g\beta).$$

PROOF. This is clear from Lemma 3.6.3. □

PROPOSITION 3.6.17. *If* $\alpha \in \mathrm{Ext}^n_{RG}(M, M')$ *then* $\mathrm{Tr}_{H,G}\,\mathrm{res}_{G,H}(\alpha) = |G : H|.\alpha$. *In particular for any* $\alpha \in \mathrm{Ext}^n_{RG}(M, M')$ *with* $n > 0$, *we have* $|G|.\alpha = 0$. □

COROLLARY 3.6.18. *If* $|G : H|$ *is invertible in R then the restriction map*

$$\mathrm{res}_{G,H} : \mathrm{Ext}^n_{RG}(M, M') \to \mathrm{Ext}^n_{RH}(M, M')$$

is injective. In particular, if $|G|$ *is invertible in R and M is R-projective, then* $\mathrm{Ext}^n_{RG}(M, M') = 0$ *for* $n > 0$. □

COROLLARY 3.6.19. *Suppose a Sylow p-subgroup P of G is a T.I. set (i.e., for all $g \in G$ we have either ${}^gP = P$ or ${}^gP \cap P = \{1\}$), and suppose $|G : P|$ is invertible in R. Then the restriction map*

$$\mathrm{res}_{G,N_G(P)} : \mathrm{Ext}^n_{RG}(M, M') \to \mathrm{Ext}^n_{RN_G(P)}(M, M')$$

is an isomorphism, and multiplication by $|G : N_G(P)| - 1$ annihilates every element of $\mathrm{Ext}^n_{RG}(M, M')$, for $n > 0$. Moreover, $\mathrm{Ext}^n_{RN_G(P)}(M, M')$ is equal to the invariants of $N_G(P)/P$ acting on $\mathrm{Ext}^n_{RP}(M, M')$.

PROOF. As above, $\mathrm{Tr}_{N_G(P),G}\mathrm{res}_{G,N_G(P)}(\alpha) = |G : N_G(P)|\alpha$, so that $\mathrm{res}_{G,N_G(P)}$ is injective. But on the other hand, the Mackey formula 3.6.16 (iv) shows that

$$\mathrm{res}_{G,N_G(P)}\mathrm{Tr}_{N_G(P),G}(\beta) = \beta$$

since the intersection of any two distinct conjugates of $N_G(P)$ is a group of order dividing $|G : P|$. Hence $\mathrm{res}_{G,N_G(P)}$ and $\mathrm{Tr}_{N_G(P),G}$ are inverse isomorphisms, and $|G : N_G(P)|\alpha = \alpha$.

Finally, the Mackey formula shows that if $\alpha \in \mathrm{Ext}^n_{RP}(M, M')$ then

$$\mathrm{res}_{N_G(P),P}\mathrm{Tr}_{P,N_G(P)}(\alpha) = \sum_{g \in N_G(P)/P} g\alpha.$$

Since $|N_G(P)/P|$ is invertible in R, the image of $\sum_{g\in N_G(P)/P} g$ is equal to the invariants. □

EXAMPLE. Suppose $G = \mathcal{S}_p$, the symmetric group on p letters, for p odd, and k is a field of characteristic p. A Sylow p-subgroup P of G is a cyclic group of order p, whose normaliser is a Frobenius group of order $p(p-1)$

$$N_G(P) = \langle g, h \mid g^p = 1, h^{p-1} = 1, hgh^{-1} = g^s\rangle$$

where s is a primitive root modulo p. Now by Proposition 3.5.5, we have

$$H^*(P, k) = k[x, y]/(x^2)$$

with $\deg(x) = 1$ and $\deg(y) = 2$. It is easy to check that the action of $N_G(P)$ on $H^*(P, k)$ is given by $h : x \mapsto sx$, $h : y \mapsto sy$ (for example, use the explicit description of degree one and degree two cohomology given in Section 3.7).

Now by Corollary 3.6.19, it follows that

$$H^*(\mathcal{S}_p, k) \cong H^*(N_G(P), k) \cong H^*(P, k)^{N_G(P)/P}.$$

Since $h : y^n \mapsto s^n y^n$ and $h : xy^{n-1} \mapsto s^n xy^{n-1}$, it follows that the invariants have a basis consisting of the elements of the form $y^{k(p-1)}$ and $xy^{k(p-1)-1}$. These elements therefore form a basis for $H^*(\mathcal{S}_p, k)$.

Similarly, if ε denotes the sign representation of $\mathcal{S}_p$, then $\varepsilon\downarrow_P = k\downarrow_P$, so $H^*(P, \varepsilon)$ is as before. However, as modules for $N_G(P)/P$ we have $H^*(P, \varepsilon) = H^*(P, k) \otimes_k \varepsilon$. Since h is an odd element of $\mathcal{S}_p$, it acts as $-1 = s^{(p-1)/2}$ on ε. Thus we have $h : y^n \mapsto s^{(p-1)/2+n}y^n$ and $h : xy^{n-1} \mapsto s^{(p-1)/2+n}xy^{n-1}$, and so

$$H^*(\mathcal{S}_p, \varepsilon) \cong H^*(N_G(P), \varepsilon) \cong [H^*(P, k) \otimes_k \varepsilon]^{N_G(P)/P}$$

has a basis consisting of the elements of the form

$$y^{(2k+1)(p-1)/2} \quad \text{and} \quad xy^{(2k+1)(p-1)/2-1}.$$

3.7. Low degree cohomology

We first give a group theoretic interpretation of

$$H^n(G, M) \cong \mathrm{Ext}^n_{RG}(R, M) \cong \mathrm{Ext}^n_{\mathbb{Z}G}(\mathbb{Z}, M)$$

for an RG-module M, for $n = 1$ and 2. Such group theoretic interpretations have been given for all n, but soon become contrived. The real home for cohomology of groups is in module theory.

DEFINITION 3.7.1. An **extension** of a group G by an RG-module M is a short exact sequence of groups

$$1 \to M \to \hat{G} \to G \to 1$$

where M is regarded as an abelian group (now written multiplicatively) and the action of G on M by conjugation is the same as the module action. A **central extension** of G is as above but with trivial G-action on M. The extension **splits** if there is a copy of G complementary to M in $\hat{G}$. In this case the group $\hat{G}$ may be written as matrices of the form $\left(\begin{smallmatrix} 1 & 0 \\ m & g \end{smallmatrix}\right)$.

An **isomorphism of extensions** must be the identity on both M and G.

PROPOSITION 3.7.2. *The cohomology group $H^1(G, M)$ parametrises the set of conjugacy classes of complements to the subgroup M in the split extension $\hat{G}$.*

PROOF. We write elements of $\hat{G}$ as matrices as above. Then a complementary copy of G in $\hat{G}$ is given by matrices $\left(\begin{smallmatrix} 1 & 0 \\ \beta[g] & g \end{smallmatrix}\right)$. Since

$$\begin{pmatrix} 1 & 0 \\ \beta[g_1] & g_1 \end{pmatrix}\begin{pmatrix} 1 & 0 \\ \beta[g_2] & g_2 \end{pmatrix} = \begin{pmatrix} 1 & 0 \\ \beta[g_1g_2] & g_1g_2 \end{pmatrix}$$

we have $\beta[g_1] + g_1\beta[g_2] = \beta[g_1g_2]$ and so $\beta : G \to M$ is a 1-cocycle for the bar resolution. Conjugating by $\left(\begin{smallmatrix} 1 & 0 \\ m & g' \end{smallmatrix}\right)$ replaces β by β' where

$$\beta'[g] - \beta[g] = (1 - g)(m - \beta[g'])$$

which is the coboundary of the 0-cochain $[\,] \mapsto m - \beta[g']$. □

In terms of extensions, there is an obvious action of $\hat{G}$ on $\hat{M} = R \oplus M$ given by

$$\begin{pmatrix} 1 & 0 \\ m_1 & g \end{pmatrix}\begin{pmatrix} n \\ m_2 \end{pmatrix} = \begin{pmatrix} n \\ nm_1 + g(m_2) \end{pmatrix}.$$

The above calculation shows that two complementary copies of G in $\hat{G}$ are conjugate if and only if the restrictions of the short exact sequence $0 \to M \to \hat{M} \to R \to 0$ correspond to the same element of $\mathrm{Ext}^1_{RG}(R, M)$.

DEGREE TWO COHOMOLOGY.

PROPOSITION 3.7.3. *The cohomology group $H^2(G,M)$ parametrises the set of isomorphism classes of extensions of G by the RG-module M.*

PROOF. If $1 \to M \to \hat{G} \to G \to 1$ is an extension, choose a set of coset representatives of M in $\hat{G}$, and write $\hat{g}$ for the representative corresponding to an element $g \in G$. Define a function $\alpha : G \times G \to M$ by

$$\hat{g}_1\hat{g}_2 = \alpha[g_1|g_2]\widehat{g_1g_2}.$$

Because of the associative law in $\hat{G}$, we have

$$\alpha[g_1|g_2] + \alpha[g_1g_2|g_3] = g_1\alpha[g_2|g_3] + \alpha[g_1|g_2g_3]$$

(recall that addition in M is really multiplication in $\hat{G}$, and the action of G on M corresponds to conjugation in $\hat{G}$) so that $\delta\alpha[g_1|g_2|g_3] = 0$, and α is a 2-cocycle for the bar resolution. Choosing new coset representatives $\tilde{g} = \beta[g]\hat{g}$ replaces $\alpha[g_1|g_2]$ by

$$\alpha[g_1|g_2] + \beta[g_1] + g_1\beta[g_2] - \beta[g_1g_2].$$

In other words it changes α by the coboundary of the 1-cochain β.

Conversely given a 2-cocycle α, we can define an extension $\hat{G}$ by putting a multiplication on the set of pairs $m \in M$, $g \in G$ as follows :

$$(m_1, g_1)(m_2, g_2) = (m_1 + g_1(m_2) + \alpha[g_1|g_2], g_1g_2).$$

As above, the associative law corresponds to the equation $\delta\alpha = 0$. □

If you like thinking in terms of matrices, suppose

$$0 \to M \to A \to B \to R \to 0$$

represents an element of $H^2(G,M) \cong \mathrm{Ext}^2_{RG}(R,M)$. Letting

$$M' = \mathrm{Coker}(M \to A) \cong \mathrm{Ker}(B \to R)$$

we can think of A and B as matrices of the form

$$A : \begin{pmatrix} \phi_{M'}(g) & 0 \\ a(g) & \phi_M(g) \end{pmatrix} \qquad B : \begin{pmatrix} 1 & 0 \\ b(g) & \phi_{M'}(g) \end{pmatrix}.$$

The group $\hat{G}$ then consists of all matrices of the form

$$\begin{pmatrix} 1 & 0 & 0 \\ b(g) & \phi_{M'}(g) & 0 \\ * & a(g) & \phi_M(g) \end{pmatrix}$$

where $*$ takes all possible values in $M = \mathrm{Hom}_R(R,M)$. The matrices

$$\begin{pmatrix} 1 & 0 & 0 \\ 0 & 1 & 0 \\ * & 0 & 1 \end{pmatrix}$$

form a normal subgroup isomorphic to M, and the quotient is isomorphic to G.

The splitting of the extension corresponds to being able to choose consistent values $c(g)$ for the entry $*$, preserved by multiplication. This is the same as the Yoneda product of $0 \to M \to A \to M' \to 0$ and $0 \to M' \to B \to R \to 0$ being the zero element of $\mathrm{Ext}^2_{RG}(R, M)$. The choice $c(g) + c'(g)$ also works if and only if

$$\begin{pmatrix} 1 & 0 \\ c'(g) & \phi_M(g) \end{pmatrix}$$

is an extension in $\mathrm{Ext}^1_{RG}(R, M) = H^1(G, M)$.

COROLLARY 3.7.4. *If $|G|$ is invertible in R and M is an RG-module then every extension of G by M splits.*

PROOF. This follows from Proposition 3.7.3, Corollary 3.6.18 and the fact that

$$H^2(G, M) = \mathrm{Ext}^2_{RG}(R, M). \qquad \square$$

PROPOSITION 3.7.5. *Let $\tilde{G}$ be a central extension of a finite group G by the multiplicative group $k^\times$ of a field k in which every element has a $|G|$th root. Then there is a finite subgroup $\hat{G}$ of $\tilde{G}$ with $\tilde{G} = \hat{G}.k^\times$, and $\hat{G} \cap k^\times$ a cyclic group of order prime to $p = \mathrm{char}\, k$. In particular, if G is a p-group and k is a perfect field, then the extension splits and $\tilde{G} = G \times k^\times$.*

PROOF. Since every element of k has a $|G|$th root, the map $k^\times \to k^\times$ given by raising to the $|G|$th power is onto, with kernel of order dividing $|G|_{p'}$. The short exact sequence

$$0 \to \mathbb{Z}/|G|_{p'} \to k^\times \to k^\times \to 1$$

(where 1 is the zero group written multiplicatively and $\mathbb{Z}/|G|_{p'}$ is the $|G|$th roots of unity written additively) gives rise to an exact sequence

$$\cdots \to H^2(G, \mathbb{Z}/|G|_{p'}) \to H^2(G, k^\times) \xrightarrow{0} H^2(G, k^\times) \to \cdots$$

where the second map is multiplication by $|G|$, and is hence zero by Proposition 3.6.17. Thus we may choose representatives $\tilde{g}_i$ in $\tilde{G}$ of the elements g_i of G so that letting $\tilde{g}_i\tilde{g}_j = \alpha[g_i|g_j]\widetilde{g_ig_j}$, the 2-cocycle α takes values in the $|G|$th roots of unity. We take $\hat{G}$ to be the group generated by these $\tilde{g}_i$, so that $\hat{G} \cap k^\times$ is contained in the finite cyclic group of $|G|$th roots of unity. $\square$

REMARKS. (i) The example $k = \mathbb{Q}$, $G = \mathbb{Z}/2$, $\tilde{G} = \langle \mathbb{Q}, \sqrt{2} \rangle$ shows that the hypothesis that every element has a $|G|$th root is necessary.

(ii) If $\alpha : G \times G \to k^\times$ is a 2-cocycle on G, then the **twisted group ring** $[kG]_\alpha$ has basis elements $\tilde{g}_i$ for $g_i \in G$ and multiplication $\tilde{g}_i\tilde{g}_j = \alpha[g_i|g_j]\widetilde{g_ig_j}$. According to the above proposition, if k is algebraically closed there is a finite cyclic p'-central extension

$$1 \to Z \to \hat{G} \to G \to 1$$

and an injective map $Z \hookrightarrow k^\times$ such that $[kG]_\alpha$ is the summand of $k\hat{G}$ on which Z acts as scalars via this map, so that as a $k\hat{G}$-module we have $[kG]_\alpha \cong \varepsilon_Z\uparrow^{\hat{G}}$.

COROLLARY 3.7.6. *Suppose that a finite group G acts as algebra automorphisms on* $\mathrm{End}_k(V)$, *where V is a finite dimensional vector space over an algebraically closed field k of characteristic p. Then a finite cyclic p'-central extension $\hat{G}$ of G acts on V in such a way that the induced action on* $V \otimes_k V^* = \mathrm{End}_k(V)$ *is the given one.*

PROOF. Every automorphism of $\mathrm{End}_k(V)$ is inner by Proposition 1.3.6, and so we obtain a map $G \to PGL(V) = GL(V)/k^\times$. This gives rise to a central extension $\tilde{G}$ of G by $k^\times$ and a map $\tilde{G} \to GL(V)$. By the proposition, we may pass down to a finite extension $\hat{G}$ of G and a map $\hat{G} \to GL(V)$, as required. □

EXERCISE. Show that $\mathrm{Ext}^1_{RG}(R, R)$ is isomorphic to the set $\mathrm{Hom}(G, R^+)$ of group homomorphisms from G to the additive group of R. Note that every such homomorphism has the derived group of G in its kernel, and that $\mathrm{Hom}(G, R^+)$ is an abelian group in an obvious way.

LIFTING THEOREMS. We now give an application of the above interpretation of degree two cohomology to the problem of lifting representations.

THEOREM 3.7.7 (Green). *Suppose $(K, \mathcal{O}, k)$ is a p-modular system (see Section 1.9) with $k = \mathcal{O}/\mathfrak{p}$, and M is a finitely generated kG-module such that* $\mathrm{Ext}^2_{kG}(M, M) = 0$. *Then there is an $\mathcal{O}G$-lattice $\hat{M}$ with $\hat{M}/\mathfrak{p} \cong M$.*

PROOF. We prove by induction that $M = M_1$ lifts to an $(\mathcal{O}/\mathfrak{p}^i)G$-lattice M_i with $M_i/\mathfrak{p}^{i-1} \cong M_{i-1}$. If such a lift M_i has been found, then finding M_{i+1} is the same as finding the diagonal map in the following diagram.

$$\begin{array}{ccccccccc} & & & & & & G & & \\ & & & & & \swarrow & \downarrow & & \\ 1 & \to & \mathrm{End}_k(M) & \to & GL_n(\mathcal{O}/\mathfrak{p}^{i+1}) & \to & GL_n(\mathcal{O}/\mathfrak{p}^i) & \to & 1 \end{array}$$

The obstruction to finding such a lift lies in

$$H^2(G, \mathrm{End}_k(M)) = \mathrm{Ext}^2_{kG}(M, M) = 0,$$

and so the diagonal map exists. Such a map gives a representation of G as matrices over $\mathcal{O}/\mathfrak{p}^{i+1}$. By completeness of $\mathcal{O}$, we now obtain an $\mathcal{O}G$-lattice $\hat{M}$ with the desired properties. □

In general, a representation may lift to $\mathcal{O}/\mathfrak{p}^i$ for i arbitrarily large, and then fail to lift further. For example, the trivial representation of the cyclic group of order two over $\mathbb{F}_2$ lifts to the matrix $(2^{i-1} + 1)$ as a representation over $\mathbb{Z}/2^i$, which then fails to lift to a representation over $\mathbb{Z}/2^{i+1}$. But in a sense, this lift only failed because we chose the wrong lift from $\mathbb{Z}/2^{i-1}$ to $\mathbb{Z}/2^i$. The following theorem shows that if we can lift as far as $\mathbb{Z}/p^{2r+1}$, where p^r is the order of the Sylow subgroups, then the reduction to $\mathbb{Z}/p^{r+1}$ lifts to the p-adics.

The following theorem is essentially due to Maranda **[150]**.

THEOREM 3.7.8. *Suppose G is a finite group of order $p^r q$ with q coprime to p, and $(K, \mathcal{O}, k)$ is a p-modular system. If M is a finitely generated kG-module which lifts to a finitely generated $\mathcal{O}/p^a$-free $(\mathcal{O}/p^a)G$-module $\tilde{M}$ with $a \geq 2r+1$, then there is a finitely generated $\mathcal{O}$-free $\mathcal{O}G$-module $\hat{M}$ with $\hat{M}/p^{a-r}\hat{M} \cong \tilde{M}/p^{a-r}\tilde{M}$.*

PROOF. Suppose M has dimension n over k. Since $2(a-r) \geq a+1$, the kernel of the map

$$GL_n(\mathcal{O}/p^{a+1}) \to GL_n(\mathcal{O}/p^{a-r})$$

is an abelian group, isomorphic to $\mathrm{End}_{\mathcal{O}}(\tilde{M}/p^{r+1}\tilde{M})$. Now consider the following diagram of groups and homomorphisms.

$$\begin{array}{ccccccccc}
 & & & & & & G & & \\
 & & & & & \swarrow & \downarrow & & \\
1 & \to & \mathrm{End}_{\mathcal{O}}(\tilde{M}/p^{r-1}\tilde{M}) & \to & GL_n(\mathcal{O}/p^{a+r+1}) & \to & GL_n(\mathcal{O}/p^a) & \to & 1 \\
 & & \downarrow{\scriptstyle p^r} & & \downarrow & & \downarrow & & \\
1 & \to & \mathrm{End}_{\mathcal{O}}(\tilde{M}/p^{r-1}\tilde{M}) & \to & GL_n(\mathcal{O}/p^{a+1}) & \to & GL_n(\mathcal{O}/p^{a-r}) & \to & 1
\end{array}$$

The obstruction to lifting $G \to GL_n(\mathcal{O}/p^a)$ to a map $G \to GL_n(\mathcal{O}/p^{a+r+1})$ lies in $H^2(G, \mathrm{End}_{\mathcal{O}}(\tilde{M}/p^{r+1}\tilde{M}))$. Since p^r is the p-part of $|G|$, by Proposition 3.6.17, multiplication by p^r annihilates this cohomology group. Thus the composite $G \to GL_n(\mathcal{O}/p^a) \to GL_n(\mathcal{O}/p^{a-r})$ lifts to a map $G \to GL_n(\mathcal{O}/p^{a+1})$. Now proceed by induction and use completeness to obtain an appropriate map $G \to GL_n(\mathcal{O})$ lifting the given map $G \to GL_n(\mathcal{O}/p^{a-r})$. □

3.8. Stable elements

In the last section (Corollary 3.6.18), we saw that if $|G : H|$ is invertible in R then $\mathrm{res}_{G,H} : \mathrm{Ext}^n_{RG}(M, M') \to \mathrm{Ext}^n_{RH}(M, M')$ is injective. More generally (cf. Corollary 3.6.9) we shall see in this section that if M or M' is projective relative to H then the above map $\mathrm{res}_{G,H}$ is injective, and we identify the image as the **stable elements** in $\mathrm{Ext}^n_{RH}(M, M')$.

DEFINITION 3.8.1. If H is a subgroup of G and M, M' are RG-modules, an element α of $\mathrm{Ext}^n_{RH}(M, M')$ is said to be **stable** with respect to G if for all $g \in G$

$$\mathrm{res}_{H, H \cap {}^g H}(\alpha) = \mathrm{res}_{{}^g H, H \cap {}^g H}(g\alpha).$$

PROPOSITION 3.8.2. *Suppose H is a subgroup of G, and M, M' are RG-modules with either M or M' projective relative to H. Then the map*

$$\mathrm{res}_{G,H} : \mathrm{Ext}^n_{RG}(M, M') \to \mathrm{Ext}^n_{RH}(M, M')$$

is injective, and its image is the set of stable elements with respect to G.

PROOF. We shall deal with the case where M' is projective relative to H; the case where M is projective relative to H is similar. Let $\theta \in$

$\mathrm{End}_{RH}(M')$ with $\mathrm{Tr}_{H,G}(\theta) = 1$ as in Higman's criterion (Proposition 3.6.4). By Lemma 3.6.16 (i), if $\alpha \in \mathrm{Ext}^n_{RG}(M, M')$ then

$$\mathrm{Tr}_{H,G}(\mathrm{res}_{G,H}(\alpha)\theta) = \alpha\mathrm{Tr}_{H,G}(\theta) = \alpha.$$

This proves that $\mathrm{res}_{G,H}$ is injective, since α may be recovered from $\mathrm{res}_{G,H}(\alpha)$.

Conversely, suppose that $\beta \in \mathrm{Ext}^n_{RH}(M, M')$ is stable with respect to G. Then by Lemma 3.6.16 we have

$$\begin{aligned}
\mathrm{res}_{G,H}\mathrm{Tr}_{H,G}(\theta\beta) &= \sum_{HgH} \mathrm{Tr}_{H\cap {}^gH,H}\mathrm{res}_{{}^gH,H\cap {}^gH}(g(\theta\beta)) \\
&= \sum_{HgH} \mathrm{Tr}_{H\cap {}^gH,H}(\mathrm{res}_{{}^gH,H\cap {}^gH}(g(\theta))\mathrm{res}_{{}^gH,H\cap {}^gH}(g(\beta))) \\
&= \sum_{HgH} \mathrm{Tr}_{H\cap {}^gH,H}(\mathrm{res}_{{}^gH,H\cap {}^gH}(g(\theta))\mathrm{res}_{H,H\cap {}^gH}(\beta)) \\
&= \left(\sum_{HgH} \mathrm{Tr}_{H\cap {}^gH,H}\mathrm{res}_{{}^gH,H\cap {}^gH}(g(\theta))\right)\beta = \mathrm{Tr}_{H,G}(\theta).\beta = \beta
\end{aligned}$$

and so β is in the image of $\mathrm{res}_{G,H}$. □

Of course, the same description may be applied to $H^n(G, M)$ since this is just $\mathrm{Ext}^n_{RG}(R, M)$.

DEFINITION 3.8.3. A subgroup H of G is said to **control p-fusion** in G if it contains a Sylow p-subgroup of G, and whenever K_1 and K_2 are p-subgroups of H and ${}^gK_1 = K_2$ for some $g \in G$ then there is an $h \in H$ with $gk_1g^{-1} = hk_1h^{-1}$ for all $k_1 \in K_1$.

PROPOSITION 3.8.4. *If H controls p-fusion in G, and the p' part of $|G|$ is invertible in R, then the restriction map*

$$\mathrm{res}_{G,H} : H^n(G, R) \to H^n(H, R)$$

is an isomorphism.

PROOF. Let S be a Sylow p-subgroup of G. We must show that if we calculate $H^n(G, R)$ or $H^n(H, R)$ as stable elements in $H^n(S, R)$ we get the same result in both cases.

If $g \in G$, let $K_1 = {}^{g^{-1}}S \cap S$ and $K_2 = S \cap {}^gS$. Then ${}^gK_1 = K_2$, so there is an element $h \in H$ with $gk_1g^{-1} = hk_1h^{-1}$ for all $k_1 \in K_1$. In particular $h^{-1}g$ centralises K_1, and so if $\alpha \in H^n(K_1, R)$ then $g\alpha = h\alpha \in H^n(K_2, R)$ so that the stable element calculations do give the same answer. □

REMARK. The above proof does not work for the restriction map from $\mathrm{Ext}^n_{RG}(M_1, M_2)$ to $\mathrm{Ext}^n_{RH}(M_1, M_2)$ if M_1 and M_2 are non-trivial. The problem is that $C_G(K_1)$ does not necessarily act trivially on $\mathrm{Ext}^n_{RK_1}(M_1, M_2)$.

3.9. Relative cohomology

LEMMA 3.9.1 (Relative Schanuel's lemma). *Suppose* $0 \to M_1 \to X_1 \to M \to 0$ *and* $0 \to M_2 \to X_2 \to M \to 0$ *are* H*-split short exact sequences of modules with* X_1 *and* X_2 *projective relative to* H. *Then* $M_1 \oplus X_2 \cong M_2 \oplus X_1$.

PROOF. Form a pullback diagram just as in the proof of Schanuel's lemma 1.5.3. □

DEFINITION 3.9.2. A **relatively H-projective resolution** of an RG-module M is a long exact sequence

$$\cdots \to X_2 \xrightarrow{\partial_2} X_1 \xrightarrow{\partial_1} X_0$$

of RG-modules with the following properties.
(i) $X_0/\mathrm{Im}(\partial_1) \cong M$.
(ii) Each X_i is projective relative to H.
(iii) Each sequence

$$0 \to X_{n+1}/\mathrm{Ker}(\partial_{n+1}) \to X_n \to \mathrm{Im}(\partial_n) \to 0$$

as well as

$$0 \to \mathrm{Im}(\partial_1) \to X_0 \to M \to 0$$

splits on restriction to H.

Since the surjection $M\downarrow_H\uparrow^G \to M$ always splits on restriction to H, relatively H-projective resolutions always exist.

THEOREM 3.9.3 (Comparison theorem). *Given a map of modules* $M \to M'$ *and relatively* H*-projective resolutions of* M *and* M', *we can extend to a map of chain complexes*

$$\begin{array}{ccccccccc}
\cdots \longrightarrow & X_2 & \xrightarrow{\partial_2} & X_1 & \xrightarrow{\partial_1} & X_0 & \longrightarrow & M & \longrightarrow 0 \\
& \downarrow{\scriptstyle f_2} & & \downarrow{\scriptstyle f_1} & & \downarrow{\scriptstyle f_0} & & \downarrow & \\
\cdots \longrightarrow & Y_2 & \xrightarrow{\partial'_2} & Y_1 & \xrightarrow{\partial'_1} & Y_0 & \longrightarrow & M' & \longrightarrow 0
\end{array}$$

Given any two such maps $\{f_n\}$ *and* $\{f'_n\}$, *there is a chain homotopy* $h_n : X_n \to Y_{n+1}$, *so that* $f_n - f'_n = \partial_{n+1} \circ h_n + h_{n-1} \circ \partial_n$. □

If M' is an RG-module and

$$\cdots \to X_2 \xrightarrow{\partial_2} X_1 \xrightarrow{\partial_1} X_0$$

is a relatively H-projective resolution of an RG-module M, we have a chain complex

$$\mathrm{Hom}_{RG}(X_0, M) \xrightarrow{\delta^0} \mathrm{Hom}_{RG}(X_1, M') \xrightarrow{\delta^1} \mathrm{Hom}_{RG}(X_2, M') \to \cdots$$

where δ^n is given by composition with ∂_{n+1}. By the comparison theorem this complex is independent of choice of relatively H-projective resolution, up to chain homotopy equivalence. Thus its cohomology groups are independent of this choice, and we define

$$\mathrm{Ext}^n_{G,H}(M, M') = H^n(\mathrm{Hom}_{RG}(\mathbf{X}, M'), \delta^*).$$

We can now define the **relative cohomology** of G to be

$$H^n(G, H, M) = \mathrm{Ext}^n_{G,H}(R, M).$$

Just as in Section 2.4, we may view elements of $\mathrm{Ext}^n_{G,H}(M, M')$ as equivalence classes of H-split n-fold extensions

$$0 \to M' \to M_{n-1} \to \cdots \to M_0 \to M \to 0.$$

Two such are equivalent if there is a map of H-split n-fold extensions taking one to the other. We complete this as usual to an equivalence relation. Note that two H-split n-fold extensions can be equivalent as n-fold extensions without being equivalent as H-split n-fold extensions, so that the natural map

$$\mathrm{Ext}^n_{G,H}(M, M') \to \mathrm{Ext}^n_{RG}(M, M')$$

is not necessarily injective.

Yoneda composition gives an associative bilinear product

$$\mathrm{Ext}^m_{G,H}(M', M'') \times \mathrm{Ext}^n_{G,H}(M, M') \to \mathrm{Ext}^{m+n}_{G,H}(M, M'')$$

and hence a ring structure on $\mathrm{Ext}^*_{G,H}(M, M)$. Note that in general the ring $\mathrm{Ext}^n_{G,H}(R, R)$ is not necessarily Noetherian, even when R is a field. This is in contrast to the case of $\mathrm{Ext}^n_{RG}(R, R)$, which is known to be Noetherian whenever R is (see Chapter 4 of Volume II).

LONG EXACT SEQUENCES.

LEMMA 3.9.4 (Relative horseshoe lemma). *If $0 \to M' \to M \to M'' \to 0$ is an H-split short exact sequence of left Λ-modules, then given relatively H-projective resolutions*

$$\cdots \to X'_2 \to X'_1 \to X'_0, \quad \cdots \to X''_2 \to X''_1 \to X''_0$$

of M' and M'', we may complete to a short exact sequence of chain complexes

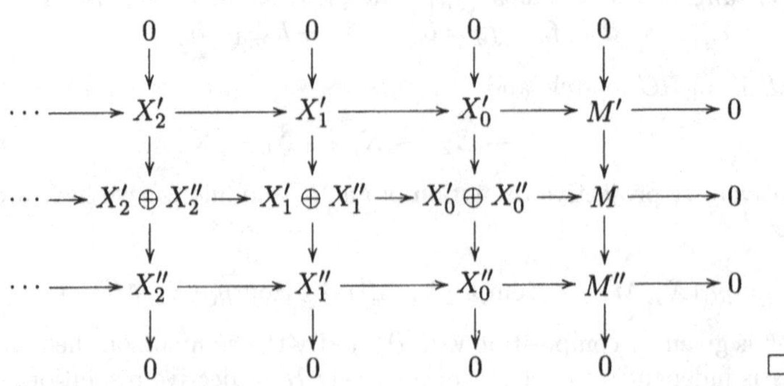

$$\begin{array}{ccccccccc} & & 0 & & 0 & & 0 & & 0 & \\ & & \downarrow & & \downarrow & & \downarrow & & \downarrow & \\ \cdots & \to & X'_2 & \to & X'_1 & \to & X'_0 & \to & M' & \to 0 \\ & & \downarrow & & \downarrow & & \downarrow & & \downarrow & \\ \cdots & \to & X'_2 \oplus X''_2 & \to & X'_1 \oplus X''_1 & \to & X'_0 \oplus X''_0 & \to & M & \to 0 \\ & & \downarrow & & \downarrow & & \downarrow & & \downarrow & \\ \cdots & \to & X''_2 & \to & X''_1 & \to & X''_0 & \to & M'' & \to 0 \\ & & \downarrow & & \downarrow & & \downarrow & & \downarrow & \\ & & 0 & & 0 & & 0 & & 0 & \end{array}$$

□

Thus as in Section 2.4, if M_0 is another RG-module we have long exact sequences

$$0 \to \mathrm{Hom}_{RG}(M'', M_0) \to \mathrm{Hom}_{RG}(M, M_0) \to \mathrm{Hom}_{RG}(M', M_0) \to \mathrm{Ext}^1_{G,H}(M'', M_0) \to \cdots \to \mathrm{Ext}^n_{G,H}(M'', M_0) \to \mathrm{Ext}^n_{G,H}(M, M_0) \to \mathrm{Ext}^n_{G,H}(M', M_0) \to \mathrm{Ext}^{n+1}_{G,H}(M'', M_0) \to \cdots$$

and

$$0 \to \mathrm{Hom}_{RG}(M_0, M') \to \mathrm{Hom}_{RG}(M_0, M) \to \mathrm{Hom}_{RG}(M_0, M'') \to \mathrm{Ext}^1_{G,H}(M_0, M') \to \cdots \to \mathrm{Ext}^n_{G,H}(M_0, M') \to \mathrm{Ext}^n_{G,H}(M_0, M) \to \mathrm{Ext}^n_{G,H}(M_0, M'') \to \mathrm{Ext}^{n+1}_{G,H}(M_0, M') \to \cdots$$

3.10. Vertices and sources

In this section, R is a ring of coefficients such that the Krull–Schmidt theorem holds for finitely generated RG-modules (e.g. R a field or a complete local domain).

DEFINITION 3.10.1. Let M be an indecomposable RG-module. Then D is a **vertex** of M if M is projective relative to D but not relative to D' for any proper subgroup D' of D. A **source** of M is an indecomposable RD-module M_0, where D is a vertex of M, such that M is a direct summand of $M_0 \uparrow^G$ (cf. Proposition 3.6.4).

PROPOSITION 3.10.2 (Green). *Let M be an indecomposable RG-module.*

(i) *The vertices of M are conjugate in G.*

(ii) *Let M_0 and M_1 be two RD-modules which are both sources of M. Then there is an element $g \in N_G(D)$ such that $M_0 \cong {}^gM_1$.*

(iii) *If the p' part of $|G|$ is invertible in R, then the vertices of M are p-subgroups.*

PROOF. (i) If M is a summand of $M\downarrow_H\uparrow^G$ and also of $M\downarrow_K\uparrow^G$ then it is a summand of

$$M\downarrow_H\uparrow^G\downarrow_K\uparrow^G \cong \bigoplus_{HgK} M\downarrow_{H\cap {}^gK}\uparrow^G$$

by the Mackey decomposition theorem, so that if H and K are both vertices of M then $H = {}^gK$ for some $g \in G$.

(ii) By Proposition 3.6.4, some summand M_2 of $M\downarrow_D$ is a source of M. Thus M_2 is also a summand of

$$M_0\uparrow^G\downarrow_D \cong \bigoplus_{DgD} {}^gM_0\downarrow_{D\cap {}^gD}\uparrow^D .$$

Thus for some $g \in N_G(D)$, $M_2 \cong {}^gM_0$.

(iii) This follows from Corollary 3.6.9. □

EXERCISE. Use Rosenberg's lemma, Higman's criterion and Lemma 3.6.3 (v) to give an alternative proof of part (i) of the above proposition, under the assumption that the endomorphism ring $\mathrm{End}_{RG}(M)$ is a local ring.

3.11. Trivial source modules

As we shall see in Chapter 5, trivial source modules control quite a lot of the structure of representation rings.

Throughout this section, R is an arbitrary commutative ring of coefficients such that the Krull–Schmidt theorem holds for finitely generated RG-modules, and $(K, \mathcal{O}, k)$ is a p-modular system.

DEFINITION 3.11.1. An RG-module M is a **trivial source module** if each indecomposable summand of M has the trivial module R as its source.

LEMMA 3.11.2. *An indecomposable RG-module M has trivial source if and only if it is a direct summand of a permutation module.*

PROOF. Suppose M is a summand of $R_H\uparrow^G$. Let D be a vertex of M, and M_0 be a source. Then by the Mackey decomposition theorem, M_0 is a summand of

$$R_H\uparrow^G\downarrow_D = \bigoplus_{HgD} R_{D\cap {}^gH}\uparrow^D .$$

Since D is a vertex, $M_0 \cong R_D$. □

One of the principal properties of trivial source modules is that they lift from characteristic p to characteristic zero.

THEOREM 3.11.3 (Scott). *Let M_1 and M_2 be the $\mathcal{O}G$-permutation modules on the cosets of H_1 and H_2 respectively. Then the natural homomorphism from $\mathrm{Hom}_{\mathcal{O}G}(M_1, M_2)$ to $\mathrm{Hom}_{kG}(\bar{M}_1, \bar{M}_2)$ given by reduction modulo $\mathfrak{p}$ is surjective.*

PROOF. By the Mackey decomposition theorem, the free $\mathcal{O}$-module $\mathrm{Hom}_{\mathcal{O}G}(M_1, M_2)$ and the k-vector space $\mathrm{Hom}_{kG}(\bar{M}_1, \bar{M}_2)$ have the same rank, namely the number of double cosets H_2gH_1. □

COROLLARY 3.11.4. (i) *Every trivial source kG-module lifts to a trivial source $\mathcal{O}G$-module, unique up to isomorphism.*

(ii) *If M_1 and M_2 are trivial source $\mathcal{O}G$-modules, then the natural map*

$$\mathrm{Hom}_{\mathcal{O}G}(M_1, M_2) \to \mathrm{Hom}_{kG}(\bar{M}_1, \bar{M}_2)$$

given by reduction modulo $\mathfrak{p}$ is surjective.

PROOF. By the theorem, reduction modulo $\mathfrak{p}$ is surjective on endomorphism rings of permutation modules. A trivial source module corresponds to an idempotent in such an endomorphism ring, and so by the Idempotent Refinement Theorem 1.9.4, trivial source modules lift.

Applying the theorem again, we see that homomorphisms between trivial source modules lift. Uniqueness of the lifts follows from the conjugacy statement in Theorem 1.9.4. □

3.12. Green correspondence

For the purpose of this section, R is a commutative ring of coefficients such that the Krull–Schmidt theorem holds for finitely generated RG-modules, and such that the p'-part of $|G|$ is invertible in R.

Let D be a fixed p-subgroup of G and let H be a subgroup of G containing $N_G(D)$. In this situation, Green correspondence is a tool for reducing questions about representations of G to questions about representations of H, modulo stuff "coming from below". Naïvely, this means induced from proper subgroups of D, but we can refine the statement to see exactly what subgroups are involved. Of course, the main tool here is Mackey decomposition.

Let

$$\mathcal{X} = \{X \leq G \mid X \leq {}^gD \cap D \text{ for some } g \in G \setminus H\}$$
$$\mathcal{Y} = \{Y \leq G \mid Y \leq {}^gD \cap H \text{ for some } g \in G \setminus H\}.$$

Note that $\mathcal{X} \subseteq \mathcal{Y}$ and $D \notin \mathcal{Y}$.

LEMMA 3.12.1. *Let M be an indecomposable RH-module which is projective relative to D.*

(i) *Let $M \uparrow^G \downarrow_H \cong M \oplus M'$. Then M' is a sum of modules projective relative to subgroups $Y \in \mathcal{Y}$.*

(ii) *Let $M \uparrow^G \cong V \oplus V'$ with V indecomposable and M a summand of $V \downarrow_H$. Then V' is a sum of modules projective relative to subgroups $X \in \mathcal{X}$.*

PROOF. (i) Let U be an indecomposable RD-module with $U \uparrow^H \cong M \oplus M_0$ for some M_0. Then

$$U \uparrow^G \downarrow_H \cong M \uparrow^G \downarrow_H \oplus M_0 \uparrow^G \downarrow_H .$$

But by the Mackey decomposition theorem,

$$U \uparrow^G \downarrow_H \cong U \uparrow^H \oplus U'$$

with U' a sum of modules projective relative to subgroups $Y \in \mathcal{Y}$. Thus

$$M \uparrow^G \downarrow_H \oplus M_0 \uparrow^G \downarrow_H \cong M \oplus M_0 \oplus U'$$

and so by the Krull–Schmidt theorem, $M \uparrow^G \downarrow_H \cong M \oplus M'$ with M' a sum of modules projective relative to subgroups $Y \in \mathcal{Y}$.

(ii) It is clear that V' is projective relative to D. Suppose it has an indecomposable summand V_1 which is not projective relative to a subgroup in $\mathcal{X}$, and suppose $D_1 \leq D$ is a vertex of V_1. Let U_1 be a source of V_1. Then U_1 is a summand of $V_1 \downarrow_{D_1}$, and so for some indecomposable summand M_1 of $V_1 \downarrow_H$, U_1 is a summand of $M_1 \downarrow_{D_1}$. Thus M_1 is not projective relative to a subgroup in $\mathcal{Y}$, and hence neither is $V' \downarrow_H$, contradicting (i). □

THEOREM 3.12.2 (Green Correspondence). *Suppose as above that H is a subgroup of G containing $N_G(D)$. Then there is a one–one correspondence between indecomposable RG-modules with vertex D and indecomposable RH-modules with vertex D given as follows.*

(i) *If V is an indecomposable RG-module with vertex D, then $V\downarrow_H$ has a unique indecomposable summand $f(V)$ with vertex D, and the remaining summands have vertices in $\mathcal{Y}$.*

(ii) *If M is an indecomposable RH-module with vertex D, then $M\uparrow^G$ has a unique indecomposable summand $g(M)$ with vertex D, and the remaining summands have vertices in $\mathcal{X}$.*

(iii) *We have $f(g(M)) \cong M$ and $g(f(V)) \cong V$.*

(iv) *The correspondences f and g take trivial source modules to trivial source modules.*

(v) *If V_1 and V_2 are RG-modules with vertex D, then the trace map $\mathrm{Tr}_{H,G}$ induces an isomorphism*

$$(f(V_1), f(V_2))^{H,\mathcal{X}} \xrightarrow{\mathrm{Tr}_{H,G}} (V_1, V_2)^{G,\mathcal{X}}$$

(see Definition 3.6.2 for the notation).

PROOF. (i) Let S be a source of V and let $S\uparrow^H \cong M \oplus M'$ with M an indecomposable module such that V is a summand of $M\uparrow^G$. By part (i) of the above lemma, M is the only summand of $M\uparrow^G\downarrow_H$ with D as vertex, and the rest have vertices in $\mathcal{Y}$. But some summand of $V\downarrow_H$ has vertex D, since V is a summand of $V\downarrow_H\uparrow^G$, and so we take $f(V) = M$.

(ii) Choose an indecomposable summand V of $M\uparrow^G$ such that M is a summand of $V\downarrow_H$. Then by part (ii) of the above lemma, $M\uparrow^G \cong V \oplus V'$ with V' a sum of modules with vertices in $\mathcal{X}$. We take $g(M) = V$.

(iii) and (iv) are clear from (i) and (ii).

(v) According to Proposition 3.3.1 and the exercise following it, the exterior trace map $\widehat{\mathrm{Tr}}_{H,G}$ induces an isomorphism

$$(V_1\downarrow_H, f(V_2))^H \xrightarrow{\widehat{\mathrm{Tr}}_{H,G}} (V_1, f(V_2)\uparrow^G)^G.$$

By Lemma 3.6.3 (vi), this takes homomorphisms in $(V_1\downarrow_H, f(V_2))^H_{\mathcal{X}}$ to homomorphisms in $(V_1, f(V_2)\uparrow^G)^G_{\mathcal{X}}$ and induces an isomorphism

$$(V_1\downarrow_H, f(V_2))^{H,\mathcal{X}} \xrightarrow{\widehat{\mathrm{Tr}}_{H,G}} (V_1, f(V_2)\uparrow^G)^{G,\mathcal{X}}.$$

Since $f(V_2)\uparrow^G$ is a direct sum of V_2 and modules projective relative to subgroups in $\mathcal{X}$ we have

$$(V_1, f(V_2)\uparrow^G)^{G,\mathcal{X}} \cong (V_1, V_2)^{G,\mathcal{X}}.$$

Since $\mathrm{Tr}_{H,G}$ is the composite of $\widehat{\mathrm{Tr}}_{H,G}$ and the map induced by the natural surjection $V_2\downarrow_H\uparrow^G \to V_2$, it follows that if we regard $f(V_2)$ as a summand of $V_2\downarrow_H$ then $\mathrm{Tr}_{H,G}$ induces an isomorphism

$$(V_1\downarrow_H, f(V_2))^{H,\mathcal{X}} \xrightarrow{\mathrm{Tr}_{H,G}} (V_1, V_2)^{G,\mathcal{X}}.$$

Finally, $V_1\downarrow_H$ is a direct sum of $f(V_1)$ and modules projective relative to subgroups in $\mathcal{Y}$. Since $f(V_2)$ is projective relative to D, applying Lemma 3.6.3 (v) shows that any homomorphism from a module projective relative to a subgroup in $\mathcal{Y}$ to $f(V_2)$ is a sum of transfers from subgroups in $\mathcal{X}$. Thus

$$(V_1\downarrow_H, f(V_2))^{H,\mathcal{X}} \cong (f(V_1), f(V_2))^{H,\mathcal{X}}. \qquad \square$$

The following theorem gives us more information in the situation where we have Green correspondence.

THEOREM 3.12.3 (D. Burry and J. F. Carlson [**47**], L. Puig [**161**]). *Suppose that H is a subgroup of G containing $N_G(D)$. Let V be an indecomposable RG-module such that $V\downarrow_H$ has a direct summand M with vertex D. Then V has vertex D, and V is the Green correspondent $g(M)$.*

PROOF. Let $e = \mathrm{Tr}_{D,H}(\alpha) \in (V,V)^H_D$ be the idempotent corresponding to the summand M of $V\downarrow_H$. By Lemma 3.6.3, we have

$$\mathrm{Tr}_{D,G}(\alpha) = \sum_{HgD} \mathrm{Tr}_{H\cap {}^gD,H}(g\alpha) = e + \sum_{\substack{HgD \\ g\notin H}} \mathrm{Tr}_{H\cap {}^gD,H}(g\alpha) \equiv e \bmod (V,V)^H_{\mathcal{Y}}.$$

Since M is not $\mathcal{Y}$-projective, $e \notin (V,V)^H_{\mathcal{Y}}$, and so $\mathrm{Tr}_{D,G}(\alpha)$ is an idempotent in $(V,V)^G/((V,V)^G \cap (V,V)^H_{\mathcal{Y}})$. Since $(V,V)^G$ is a local ring, this means that $(V,V)^G = \mathrm{Tr}_{D,G}(\alpha)(V,V)^G \subseteq (V,V)^G_D$, and so V is projective relative to D. Hence V has vertex D and M is its Green correspondent. $\square$

Once Green correspondence has been applied, we are left with a module which is projective relative to a normal p-subgroup. Analysis of such a situation is the subject of the next section.

3.13. Clifford theory

Clifford theory is concerned with the relationship between modules and normal subgroups. In this section we present a primitive version of the Clifford theory developed by Clifford, Cline, Conlon, Dade, Green, Knörr, Puig, Ward, Willems and others. Throughout this section, R is a ring of coefficients with the property that finitely generated RG-modules satisfy the Krull–Schmidt theorem for all finite groups G.

Suppose N is a normal subgroup of a finite group G, and M_0 is an indecomposable RG-module which is projective relative to N. Then there is a RN-module M such that M_0 is a summand of $M\uparrow^G$. Now by the Mackey decomposition theorem

$$M\uparrow^G\downarrow_N \cong \bigoplus_{g\in G/N} {}^gM$$

is a sum of conjugates of M. So by the Krull–Schmidt theorem, $M_0\downarrow_N$ is a direct sum of conjugates of M.

DEFINITION 3.13.1. If M is an indecomposable RN-module with $N \trianglelefteq G$, we define the **inertia group** $T = T(M)$ to be the set of all $g \in G$ such that $M \cong {}^gM$.

Thus in particular $N.C_G(N) \leq T$, and ${}^gM \cong {}^{g'}M$ if and only if g and g' are in the same left coset of T in G. If $T = G$, we say M is **inertial**.

PROPOSITION 3.13.2. *Suppose M is an indecomposable RN-module with $N \trianglelefteq G$, and with inertia group T. Let*

$$M\!\uparrow^T = M_1 \oplus \cdots \oplus M_r$$

with each M_i indecomposable. Then each $M_i \uparrow^G$ is indecomposable, and $M_i\!\uparrow^G \cong M_j\!\uparrow^G$ if and only if $M_i \cong M_j$.

PROOF. Since T is the inertia group, $M_i\!\downarrow_N$ is a sum of copies of M, say $M_i\!\downarrow_N \cong n_i.M$. Then

$$M_i\!\uparrow^G\downarrow_N \cong \bigoplus_{g\in G/T} (g \otimes M_i)\!\downarrow_N \cong \bigoplus_{g\in G/T} n_i.{}^gM.$$

Now M_i is a summand of $M_i\!\uparrow^G\downarrow_T$, and so if $M_i\!\uparrow^G$ is decomposable one of the summands, X say, has the property that M_i is a summand of $X\!\downarrow_T$, and so $n_i.M$ is a summand of $X\!\downarrow_N$. But then since G acts on X, $n_i.{}^gM$ is also a summand of $X\!\downarrow_N$. Since ${}^gM \not\cong {}^{g'}M$ if g and g' are in different left cosets of T in G, this implies that all summands of $M_i\!\uparrow^G\downarrow_N$ appear in $X\!\downarrow_N$, so that counting summands we must have $X = M_i\!\uparrow^G$.

If $M_i\!\uparrow^G \cong M_j\!\uparrow^G$ but $M_i \not\cong M_j$, then by the Krull–Schmidt theorem M_i is a summand of Y, where $M_j\!\uparrow^G\downarrow_T = M_j \oplus Y$. Then $Y\!\downarrow_N$ is a sum of copies of ${}^gM$ for $g \notin T$, while $M_i\!\downarrow_N$ is a sum of copies of M. Since ${}^gM \not\cong M$ for $g \notin T$, this is impossible. □

The effect of this proposition is that from now on we may assume that M is inertial, so that M_0 is a summand of $M\!\uparrow^G$, and $M\!\uparrow^G\downarrow_N$ is a sum of copies of M. We shall also assume that $R = k$ is an algebraically closed field, and under these hypotheses we examine idempotents in $E = \mathrm{End}_{kG}(M\!\uparrow^G)$. By Frobenius reciprocity we have

$$E \cong \mathrm{Hom}_{kN}(M, M\!\uparrow^G\downarrow_N) = \bigoplus_{g\in G/N} \mathrm{Hom}_{kN}(M, {}^gM).$$

Since $M \cong {}^gM$ for all $g \in G/N$, we may decompose E into a sum of pieces of the same dimension

$$E = \bigoplus_{g\in G/N} E_g, \qquad E_g = \mathrm{Hom}_{kN}(M, {}^gM).$$

We have $E_gE_{g'} = E_{gg'}$, $E_1 = \mathrm{End}_{kN}(M)$ is a local ring, and $J(E_1)E_g = E_gJ(E_1)$ is of codimension one in E_g, consisting of the non-isomorphisms. This is an example of what Dade calls a **strongly G/N-graded algebra**. In particular we have $J(E_1)E \subseteq J(E)$, and each $E_g/J(E_1)E_g$ is one dimensional, so that $E/J(E_1)E$ is a twisted group algebra for G/N. Thus by Proposition 3.7.5 and the remarks following it, there is a p'-cyclic central extension $1 \to Z \to \widehat{G/N} \to G/N \to 1$ and an injective map $Z \hookrightarrow k^\times$ such that $E/J(E_1)E$ is isomorphic to the summand of $k\widehat{G/N}$ on which Z acts as scalars via this map. Thus by the Idempotent Refinement Theorem 1.7.3 the summands of $M\uparrow^G$ are in one–one correspondence with the projective indecomposable $k\widehat{G/N}$-modules lying in this summand.

As an application of this set-up, we prove the following theorem.

THEOREM 3.13.3 (Green's indecomposability theorem). *Suppose that N is a normal subgroup of G such that G/N is a p-group, and M is an* **absolutely indecomposable** *kN-module (i.e., $k'\otimes_k M$ is indecomposable for all extension fields k' of k). Then $M\uparrow^G$ is absolutely indecomposable.*

PROOF. Without loss of generality k is algebraically closed. By applying induction on $|G:N|$, we may suppose $|G:N|=p$. If M is not inertial, then by Proposition 3.13.2 $M\uparrow^G$ is indecomposable, and so we may suppose M is inertial. Letting

$$E=\mathrm{End}_{kG}(M\uparrow^G)=\bigoplus_{g\in G/N} E_g$$

as above, $E/J(E_1)E$ is a twisted group algebra for G/N. But every p'-cyclic central extension of a cyclic group of order p splits, and so $E/J(E_1)E$ is isomorphic to the group algebra of G/N. Since $J(E_1)E\subseteq J(E)$, this implies that $E/J(E)\cong k$ and so $M\uparrow^G$ is indecomposable. □

Returning to the set-up discussed before the theorem, the projective indecomposable $k\widehat{G/N}$-module corresponding to a summand M_0 of $M\uparrow^G$ is called the **multiplicity module**, and another equivalent description of it is as follows. If we let $M_0\downarrow_N=M\otimes_k V$ for some k-vector space V, then

$$\mathrm{End}_{kN}(M_0\downarrow_N)/J\mathrm{End}_{kN}(M_0\downarrow_N)\cong\mathrm{End}_k(V)$$

admits an action of G/N as algebra automorphisms, so that by Corollary 3.7.6, there is a finite p'-central extension $\widehat{G/N}$ of G/N and an action of $\widehat{G/N}$ on V compatible with this. This $k\widehat{G/N}$-module is the multiplicity module. Its dimension gives the multiplicity of M as a summand of $M_0\downarrow_N$.

If $\mathrm{End}_k(M)=M\otimes_k M^*$ extends to a kG-module, then by Corollary 3.7.6, there is a finite p'-central extension $\hat{G}$ of G and a $k\hat{G}$-module $\hat{M}$ such that $\mathrm{End}_k(\hat{M})\cong\mathrm{End}_k(M)$ as algebras with G-action. This central extension splits on N, so that $\hat{G}$ is a pullback

$$\begin{array}{ccccccccc}
1 & \to & Z & \to & \hat{G} & \to & G & \to & 1\\
 & & \| & & \downarrow & & \downarrow & & \\
1 & \to & Z & \to & \widehat{G/N} & \to & G/N & \to & 1
\end{array}$$

and $\hat{M}\downarrow_{N\times Z}\cong M\otimes_k\varepsilon^{-1}$ for an appropriate one dimensional representation ε of Z. Thus if we regard $M\uparrow^G$ as a representation of $\hat{G}$ with Z in its kernel then

$$\begin{aligned} M\uparrow^G &\cong (M \otimes_k \varepsilon^{-1} \otimes_k \varepsilon)_{N\times Z}\uparrow^{\hat{G}} \cong (\hat{M}\downarrow_{N\times Z} \otimes_k \varepsilon)\uparrow^{\hat{G}} \\ &\cong \hat{M} \otimes_k \varepsilon_{N\times Z}\uparrow^{\hat{G}} \cong \hat{M} \otimes_k [kG/N]_\alpha \end{aligned}$$

where $[kG/N]_\alpha$ is the twisted group algebra for the 2-cocycle α corresponding to the one dimensional representation ε of Z (see the remarks after Proposition 3.7.5). Thus as a module for $\hat{G}$, the original module M_0 breaks up as a tensor product $M_0 \cong \hat{M} \otimes_k V$ where V is the multiplicity module as before.

The extension $\hat{M}$ does not always exist, as is easily seen by looking at the example where $G = \mathbb{Z}/p^n$, $N = \mathbb{Z}/p$, and M is any kN-module with non-trivial action.

In general, the problem of whether M can be extended to a module $\hat{M}$ for a suitable $\hat{G}$ can be rephrased as follows. Consider for each $g \in G$ the set of all possible linear maps $\phi_g : M \to M$ such that $g \otimes \phi_g : {}^gM = g \otimes M \to M$ is an isomorphism of kN-modules. Then for $g = 1$ this gives the multiplicative group $(\mathrm{End}_{kN}(M)^{\mathrm{op}})^\times$. If we endow the set of all such pairs (g, ϕ_g) with the composition $(g, \phi_g).(g', \psi_{g'}) = (gg', \phi_g \circ \psi_{g'})$, then we obtain a (usually infinite) group $\tilde{G}$, which fits into the following diagram:

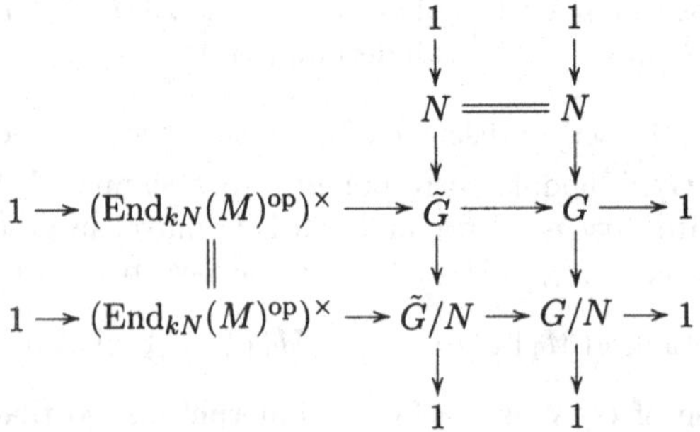

and the kN-module M extends in an obvious way to a $k\tilde{G}$-module $\tilde{M}$. The question is, can this be reduced to a central extension? In other words, if we let K be the kernel of the determinant map $(\mathrm{End}_{kN}(M)^{\mathrm{op}})^\times \to k^\times$ and set $\hat{G}/N = (\tilde{G}/N)/K = \tilde{G}/(N \times K)$, then we have a diagram of the form

$$\begin{array}{ccccccccc} & & & & 1 & & 1 & & \\ & & & & \downarrow & & \downarrow & & \\ 1 & \to & K & \to & (\mathrm{End}_{kN}(M)^{\mathrm{op}})^\times & \to & k^\times & \to & 1 \\ & & \| & & \downarrow & & \downarrow & & \\ 1 & \to & K & \to & \tilde{G}/N & \to & \hat{G}/N & \to & 1 \\ & & & & \downarrow & & \downarrow & & \\ & & & & G/N & = & G/N & & \\ & & & & \downarrow & & \downarrow & & \\ & & & & 1 & & 1 & & \end{array}$$

and the question is, does the exact sequence $1 \to K \to \tilde{G}/N \to \hat{G}/N \to 1$ split? If it does, then the $k\tilde{G}$-module $\tilde{M}$ restricts to give a $k\hat{G}$-module M as required.

There are two cases of interest where the answer is yes. The first is the case $K = 1$, in other words where $\mathrm{End}_{kN}(M) \cong k$. In this case we have $\hat{G} = \tilde{G}$ and so we can take $\hat{M} = \tilde{M}$. The second case is the case where G/N is a p'-group. In this case, K is a (usually infinite non-abelian) nilpotent group in which every element has order a power of p, and so the splitting follows by repeated application of Corollary 3.7.4.

We summarise what we have proved in the following theorem:

THEOREM 3.13.4. *Let N be a normal subgroup of a finite group G and k an algebraically closed field. Suppose M is an inertial kN-module, and either*
(i) $\mathrm{End}_{kN}(M) \cong k$, *or*
(ii) $\mathrm{char}\, k = 0$ *or* $\mathrm{char}\, k$ *does not divide* $|G/N|$.
Then there is a p'-cyclic central extension

$$\begin{array}{ccccccccc} 1 & \to & Z & \to & \hat{G} & \to & G & \to & 1 \\ & & \| & & \downarrow & & \downarrow & & \\ 1 & \to & Z & \to & \widehat{G/N} & \to & G/N & \to & 1 \end{array}$$

such that M extends to a module $\hat{M}$ for $\hat{G}$.

If M_0 is a direct summand of $M\uparrow^G$, then regarding M_0 as a module for $\hat{G}$ with Z in the kernel, we have

$$M_0 \cong \hat{M} \otimes_k V.$$

Here, V is the multiplicity module for M_0, regarded as a module for $\hat{G}$ with N in the kernel. □

3.14. Modules for p-groups

For a p-group P over a field k of characteristic p, we shall see that there is only one simple module, namely the one dimensional trivial module. But all the complications of how simple composition factors glue together to form an arbitrary module are already present in the case of a p-group.

LEMMA 3.14.1. *Suppose P is a p-group and k is a field of characteristic p. Then there is only one simple kP-module, namely the one dimensional trivial module k.*

PROOF. If M is a kP-module, we shall show that P has a non-zero fixed subspace in its action on M. Since the fixed subspace is a submodule, it then follows that if M is simple then P acts trivially on M and M is one dimensional.

Choose a non-zero element m of M and consider the (additive) abelian subgroup of M generated by the images $g(m)$, $g \in P$. This is a finite abelian p-group which admits the action of P. Since all orbits must have size a power of p, the number of fixed points is divisible by p. Since the zero element is fixed, there must be some other fixed point. □

Thus the decomposition matrix has only one column, and its entries are the dimensions of the ordinary irreducible representations. The Cartan matrix has only one entry, which is the group order.

Since there is only one projective indecomposable kP-module, namely the projective cover of the trivial module, this must be equal to the regular representation. Thus kP has a unique minimal left ideal, also isomorphic to the trivial module. It is equal to the last non-zero power of the radical, and is hence a two sided ideal. We write $\mathrm{Soc}(kP)$ for this minimal ideal. Since the projective indecomposable kP-module is also injective, whenever it is a submodule of another kP-module it is a summand. Thus every non-projective indecomposable kP-module has $\mathrm{Soc}(kP)$ in its kernel. So studying the representation theory of kP is almost the same as studying the representation theory of $kP/\mathrm{Soc}(kP)$.

More generally, every transitive permutation module for kP has a one dimensional socle, since the sum of the elements being permuted spans the unique minimal submodule. Thus a transitive permutation module for kP is automatically indecomposable.

Since every kP-module is obtained by gluing together one dimensional composition factors, it is important to understand $\mathrm{Ext}^1_{kP}(k,k)$.

PROPOSITION 3.14.2. *There is a natural isomorphism*

$$H^1(P,k) = \mathrm{Ext}^1_{kP}(k,k) \cong \mathrm{Hom}(P/\Phi(P), k^+)$$

where k^+ denotes the additive group of k. Thus if $P/\Phi(P)$ is elementary abelian of rank n then $\mathrm{Ext}^1_{kP}(k,k)$ is an n-dimensional vector space over k.

PROOF. An extension

$$0 \to k \to M \to k \to 0$$

of kP-modules is the same as a matrix representation of the form

$$g \mapsto \begin{pmatrix} 1 & \alpha(g) \\ 0 & 1 \end{pmatrix}$$

where $\alpha : P \to k^+$ is a homomorphism of groups from P to the additive group of k. The kernel of α must contain $\Phi(P)$, since k^+ is abelian of exponent p. □

In fact the same argument shows that for any finite group G, any element of $\mathrm{Ext}^1_{kG}(k,k)$ corresponds to a two dimensional representation of G with $O^p(G)$ in its kernel, and so we have the following.

PROPOSITION 3.14.3. *For any finite group G we have*

$$H^1(G,k) \cong H^1(G/O^p(G),k).$$ □

The next observation about cohomology of p-groups is that we can tell whether a module is projective just by looking at its cohomology.

PROPOSITION 3.14.4. *Suppose P is a p-group and M is a kP-module. Then the following are equivalent.*
(i) *M is projective*
(ii) *$H^n(P, M) = 0$ for some $n > 0$*
(iii) *$H^n(P, M) = 0$ for all $n > 0$.*

PROOF. By Corollary 2.5.4 we have

$$H^n(P, M) = \mathrm{Ext}^n_{kP}(k, M) \cong \mathrm{Hom}_{kP}(k, \Omega^{-n}M),$$

which is non-zero whenever $\Omega^{-n}M$ is non-zero, namely whenever M is not projective (remember kP is a self injective algebra). □

COROLLARY 3.14.5. *Suppose P is a p-group and M and N are kP-modules. Then the following are equivalent.*
(i) *$M^* \otimes_k N$ is projective*
(ii) *$\mathrm{Ext}^n_{kP}(M, N) = 0$ for some $n > 0$*
(iii) *$\mathrm{Ext}^n_{kP}(M, N) = 0$ for all $n > 0$.*

PROOF. By Proposition 3.1.8 we have

$$\mathrm{Ext}^n_{kP}(M, N) \cong \mathrm{Ext}^n_{kP}(k, M^* \otimes_k N) \cong H^n(P, M^* \otimes_k N).$$

Now apply the above proposition. □

JENNINGS' THEOREM. We now give an account of Jennings' theorem [**131**], which describes the radical layers of the group algebra of a p-group. Our presentation is closer to the one given in Quillen [**165**].

Suppose P is a p-group of order p^n, and k is a field of characteristic p. We define the **dimension subgroups** of P to be

$$F_r(P) = \{g \in P \mid g - 1 \in J^r(kP)\}$$

where $J^r(kP)$ is the rth power of the Jacobson radical of the group algebra kP. Since

$$(gh - 1) = (g - 1) + (h - 1) + (g - 1)(h - 1)$$
$$(g^{-1} - 1) = -g^{-1}(g - 1), \quad (hgh^{-1} - 1) = h(g - 1)h^{-1}$$

$F_r(P)$ is a normal subgroup of P. Also since

$$(g^p - 1) = (g - 1)^p, \quad ([g, h] - 1) = ((g - 1)(h - 1) - (h - 1)(g - 1))g^{-1}h^{-1}$$

the pth power of an element of $F_r(P)$ is in $F_{pr}(P)$, and the commutators $[F_r(P), F_s(P)]$ are contained in $F_{r+s}(P)$. Also it is easy to see that $F_1(P) = P$ and $F_2(P) = \Phi(P)$, the Frattini subgroup of P.

We denote by $\Gamma_r(P)$ the lowest central series with the above properties. Namely we begin with $\Gamma_1(P) = P$, and then $\Gamma_r(P)$ is generated by all commutators $[\Gamma_s(P), \Gamma_t(P)]$ with $s + t \geq r$, and all pth powers of elements in $\Gamma_s(P)$ with $ps \geq r$. This series is called the **Jennings series** of P. It is clear from the above discussion that $F_r(P) \supseteq \Gamma_r(P)$ for all $r > 0$. We shall prove below that $F_r(P) = \Gamma_r(P)$ for all $r > 0$.

It follows from the definition that each $\Gamma_r(P)/\Gamma_{r+1}(P)$ is an elementary abelian p-group, which we regard as a vector space over $\mathbb{F}_p$, say of dimension n_r. We form the associated graded object and tensor with k

$$\mathrm{Jen}_*(P) = \bigoplus_{r \geq 1} k \otimes_{\mathbb{F}_p} (\Gamma_r(P)/\Gamma_{r+1}(P)).$$

This is a Lie algebra, with the Lie bracket given by commutators. The Jacobi identity follows from Philip Hall's identity

$$({}^y[x,[y^{-1},z]])({}^z[y,[z^{-1},x]])({}^x[z,[x^{-1},y]]) = 1.$$

This Lie algebra also comes equipped with a pth power operation, the pth power of an element of degree r being an element in degree pr. Such an object is called a **p-restricted Lie algebra** (it is not necessary for our purposes to know the exact axioms).

We define the restricted **universal enveloping algebra** $\mathcal{U}\mathrm{Jen}_*(P)$ as follows. It is the (associative) graded algebra over k generated by the graded vector space $\mathrm{Jen}_*(P)$, subject to the relations that $xy - yx$ is equal to the commutator $[x, y]$, and the pth power x^p agrees with the pth power operation defined above. By definition, the identity element of $\mathcal{U}\mathrm{Jen}_*(P)$ spans a copy of the field in degree zero.

We now filter kP by the powers of the radical, and define the associated graded object

$$\mathrm{gr}_* kP = \bigoplus_{i \geq 0} J^i(kP)/J^{i+1}(kP)$$

where $J^0(kP) = kP$. Since the product of elements in $J^i(kP)$ and $J^j(kP)$ lies in $J^{i+j}(kP)$, $\mathrm{gr}_* kP$ inherits the multiplication to form an associative graded algebra over k.

Since $\Gamma_r(P) \subseteq F_r(P)$, we have a vector space homomorphism

$$\begin{aligned} \phi : \mathrm{Jen}_*(P) &\to \mathrm{gr}_* kP \\ \bar{g} &\mapsto \bar{g} - 1 \end{aligned}$$

induced by sending an element $g \in \Gamma_r(P)$ to $(g-1) \in J^r(kP)$. This has the property that the Lie bracket $[\bar{g}, \bar{h}]$ goes to $\phi(\bar{g})\phi(\bar{h}) - \phi(\bar{h})\phi(\bar{g})$, and $\overline{g^p}$ goes to $\phi(\bar{g})^p$. So by the definition of the restricted universal enveloping algebra, the map ϕ extends to an algebra homomorphism

$$\psi : \mathcal{U}\mathrm{Jen}_*(P) \to \mathrm{gr}_* kP.$$

Since $\mathrm{Jen}_1(P) = \Gamma_1(P)/\Gamma_2(P) = P/\Phi(P)$, the map ψ is an isomorphism in degree one. Now $\mathrm{gr}_* kP$ is generated by elements of degree one, and so ψ is surjective. In fact, it is an isomorphism, as we now show by counting dimensions. This dimension count also gives us the dimensions of the radical layers of kP.

To count dimensions, we give a basis of $\mathcal{U}\mathrm{Jen}_*(P)$, called the Poincaré–Birkhoff–Witt basis. First, we choose a basis $x_1, \dots, x_n$ of $\mathrm{Jen}_*(P)$, in such a way that the first n_1 vectors form a basis of $\mathrm{Jen}_1(P)$, the next n_2 vectors form a basis of $\mathrm{Jen}_2(P)$, and so on. It may happen that some of these are the zero space, but this does not worry us. We have $n_r = \dim_k \mathrm{Jen}_r(P)$, $n = \sum n_r$, and $|P| = p^n$. We claim that the elements $x_1^{\alpha_1} \dots x_n^{\alpha_n}$ with $0 \leq \alpha_j < p$ span $\mathcal{U}\mathrm{Jen}_*(P)$ as a vector space. This is clear, because the generators in $\mathrm{Jen}_*(P)$ are certainly in this span, and because of the commutator and pth power formulas, a product of such elements is again a linear combination of such elements.

This proves that the dimension of $\mathcal{U}\mathrm{Jen}_*(P)$ is at most p^n. But we have a surjective map $\psi : \mathcal{U}\mathrm{Jen}_*(P) \to \mathrm{gr}_* kP$, and the dimension of $\mathrm{gr}_* kP$ is p^n. This proves that the above elements are linearly independent, and that ψ is an isomorphism. Moreover, an element of $\Gamma_r(P)$ not in $\Gamma_{r+1}(P)$ gives rise to a non-zero element of $\mathrm{Jen}_r(P)$ and hence has non-zero image in $\mathrm{gr}_r(kP)$. It follows that $\Gamma_r(P) = F_r(P)$.

Finally the above basis allows us to give the dimension of $\mathrm{gr}_r kP$. This is easiest to write down in terms of Poincaré series.

$$\sum_{i\geq 0} t^i \dim_k(J^i kP/J^{i+1}kP) = \prod_r (1 + t^r + \dots + t^{(p-1)r})^{n_r} = \prod_r \left(\frac{1-t^{pr}}{1-t^r}\right)^{n_r}$$

We summarise what we have proved in the following theorem.

THEOREM 3.14.6 (Jennings, Quillen). *Suppose that P is a finite p-group. Then the dimension subgroups $F_r(P)$ are equal to the Jennings subgroups $\Gamma_r(P)$. The map*

$$\psi : \mathcal{U}\mathrm{Jen}_*(P) \to \mathrm{gr}_* kP$$

defined above is a isomorphism. In particular, the Poincaré series of the radical layers of kP is given by

$$\sum_{i\geq 0} t^i \dim_k(J^i kP/J^{i+1}kP) = \prod_r \left(\frac{1-t^{pr}}{1-t^r}\right)^{n_r}$$

where $|\Gamma_r(P)/\Gamma_{r+1}(P)| = p^{n_r}$. Thus the radical length of kP is

$$1 + \sum_r (p-1)rn_r. \qquad \square$$

COROLLARY 3.14.7. *The radical layers of kP are equal to the socle layers.*

PROOF. The Poincaré series given above is symmetric. Namely, if we set

$$l = 1 + \sum_r (p-1)rn_r,$$

the radical length of kP, then the above formula shows that

$$\dim_k(J^i kP/J^{i+1}kP) = \dim_k(J^{l-i-1}kP/J^{l-i}kP).$$

By Proposition 3.1.2, kP is self-dual as a kP-module, so that

$$\dim_k(J^ikP/J^{i+1}kP) = \dim_k(\mathrm{Soc}^{i+1}kP/\mathrm{Soc}^ikP).$$

Since $\mathrm{Soc}^ikP \supseteq J^{l-i}kP$, it follows by induction that we have equality. □

3.15. Tensor induction

We saw in Section 2.8 that if Γ is a subring of a ring Λ then we may induce modules from Γ to Λ. The particular case of induction and restriction for group algebras was briefly discussed in Section 3.3. The corresponding notion of tensor induction for modules for group algebras, however, does not easily generalise.

As in Section 3.3, if $H \leq G$ and M is an RH-module we can write

$$M\uparrow^G = RG \otimes_{RH} M = \bigoplus_{g \in G/H} g \otimes M.$$

One way of checking that this gives a well defined RG-module which is independent of the choice of coset representatives of H in G is to introduce the following construction. Let $n = |G : H|$. Recall that the wreath product $\Sigma_n \wr H$ consists of elements $(\pi; h_1, \dots, h_n)$ with $\pi \in \Sigma_n$ and $h_1, \dots, h_n \in H$, and with multiplication given by

$$(\pi'; h'_1, \dots, h'_n)(\pi; h_1, \dots, h_n) = (\pi'\pi; h'_{\pi(1)}h_1, \dots, h'_{\pi(n)}h_n).$$

Given a choice of coset representatives $g_1, \dots, g_n$ of H in G, we obtain an injective group homomorphism $i : G \to \Sigma_n \wr H$ as follows. For a given $g \in G$, we can write $g.g_j = g_{\pi(j)}h_j$ for uniquely defined elements $\pi \in \Sigma_n$ and $h_1, \dots, h_n \in H$ depending on g. We then set $i(g) = (\pi; h_1, \dots, h_n)$. A different choice of coset representatives gives rise to a conjugate embedding of G in $\Sigma_n \wr H$.

Now given an RH-module M, we can make a direct sum M^n of n copies of M into a module for $\Sigma_n \wr H$ via

$$(\pi; h_1, \dots, h_n)(m_1, \dots, m_n) = (h_{\pi^{-1}(1)}m_{\pi^{-1}(1)}, \dots, h_{\pi^{-1}(n)}m_{\pi^{-1}(n)}).$$

The induced module is then

$$M\uparrow^G = i^*(M^n) = (M^n)\downarrow_G.$$

In a similar way, since tensor product is commutative, we can make the tensor product $M^{\otimes n}$ of n copies of M into a module for $\Sigma_n \wr H$ via

$$(\pi; h_1, \dots, h_n)(m_1 \otimes \dots \otimes m_n) = h_{\pi^{-1}(1)}m_{\pi^{-1}(1)} \otimes \dots \otimes h_{\pi^{-1}(n)}m_{\pi^{-1}(n)}.$$

DEFINITION 3.15.1. The **tensor induced** module $M\Uparrow^G$ is defined to be

$$M\Uparrow^G = i^*(M^{\otimes n}) = (M^{\otimes n})\downarrow_G.$$

The basic properties of tensor induction are given in the following proposition.

Proposition 3.15.2. *Suppose $H < G$ and M_1, M_2 and M_3 are RH-modules.*

(i) $(M_1 \otimes_R M_2){\uparrow}^{G}_{\otimes} \cong M_1{\uparrow}^{G}_{\otimes} \otimes_R M_2{\uparrow}^{G}_{\otimes}$.

(ii) *If $H' \leq H \leq G$ and M is an RH'-module then $M{\uparrow}^{H}_{\otimes}{\uparrow}^{G}_{\otimes} \cong M{\uparrow}^{G}_{\otimes}$.*

(iii) $(M_1 \oplus M_2){\uparrow}^{G}_{\otimes} \cong M_1{\uparrow}^{G}_{\otimes} \oplus M_2{\uparrow}^{G}_{\otimes} \oplus M'$, *where M' is a direct sum of modules induced from proper subgroups K containing the intersection of the conjugates of H.*

(iv) *If $H' \leq H$ and M is an RH'-module then $M \uparrow^H {\uparrow}^{G}_{\otimes}$ is a direct sum of modules induced from subgroups K containing the intersection of the conjugates of H', and with $K \cap H \leq H'$.*

(v) *(Mackey formula) If K, $H \leq G$ and M is an RK-module then*

$$M{\uparrow}^{G}_{\otimes}\downarrow_H \cong \bigotimes_{H\backslash G/K} {}^gM \downarrow_{H\cap {}^gK} {\uparrow}^{H}_{\otimes}.$$

(vi) *(Dress* [**89**]*) Suppose X is a permutation representation of G and $0 \to M_1 \to M_2 \to M_3 \to 0$ is an X-split short exact sequence of RH-modules (see Definition 3.6.13). Then*

$$RX \otimes_R (M_2{\uparrow}^{G}_{\otimes}) \cong RX \otimes_R ((M_1 \oplus M_3){\uparrow}^{G}_{\otimes}).$$

Proof. (i) and (ii) are clear from the definitions.

(iii) As modules for $\Sigma_n \wr H$ we have

$$(M_1 \oplus M_2)^{\otimes n} \cong M_1^{\otimes n} \oplus M_2^{\otimes n} \oplus M'$$

where

$$M' = \bigoplus_{1\leq r\leq n-1} (M_1^{\otimes r} \otimes_R M_2^{\otimes(n-r)})_{(\Sigma_r\times\Sigma_{n-r})\wr H} \uparrow^{\Sigma_n\wr H},$$

and the result follows by restricting to G.

(iv) The module

$$(M\uparrow^H)^{\otimes n} = \left(\bigoplus_{g\in H/H'} g\otimes M\right)^{\otimes n}$$

is a direct sum of submodules corresponding to the $\Sigma_n \wr H$-orbits of ways of choosing an n-tuple of coset representatives of H' in H. The stabiliser in H of such an n-tuple is the intersection of the corresponding conjugates of H', and is therefore contained in some conjugate of H'. The result follows by restricting to G.

(v) This follows by partitioning the set of left coset representatives of K in G as a disjoint union of orbits of H corresponding to the double cosets.

(vi) Regard M_1 as a submodule of M_2 by abuse of notation. As an RG-module, $M_2{\uparrow}^{G}_{\otimes}$ has a natural filtration

$$M_1{\uparrow}^{G}_{\otimes} = U_0 \leq U_1 \leq \cdots \leq U_n = M_2{\uparrow}^{G}_{\otimes}$$

($n = |G : H|$), where U_j is the linear span of the tensors $m_1 \otimes \cdots \otimes m_n$ for which at most j of the m_i do not lie in M_1. It is easy to see that

$$(M_1 \oplus M_3)\mathop{\uparrow}\limits^{\otimes G} \cong U_0 \oplus U_1/U_0 \oplus \cdots \oplus U_n/U_{n-1}.$$

Thus we must show that tensoring with RX splits the above filtration of $M_2\mathop{\uparrow}\limits^{\otimes G}$.

Suppose $f : RX \otimes_R M_3 \to RX \otimes_R M_2$ is an X-splitting of the sequence $0 \to M_1 \to M_2 \to M_3 \to 0$ of the form

$$f(\sum_{x \in X} x \otimes m_x) = \sum_{x \in X} x \otimes f_x(m_x)$$

with f_x linear maps from M_3 to M_2 (if there is any splitting of the form $f(\sum_{x \in X} x \otimes m_x) = \sum_{x,y \in X} x \otimes f_{xy}(m_y)$, then by dropping the terms with $x \neq y$, we obtain a splitting of the desired form). Since f is an RH-module homomorphism, we have $hf_x(m_x) = f_{hx}(hm_x)$ for $h \in H$. Now a splitting ϕ for the map

$$RX \otimes_R U_j \twoheadrightarrow (RX \otimes_R U_j)/(RX \otimes_R U_{j-1}) \cong RX \otimes_R (U_j/U_{j-1})$$

is given as follows. The typical generator for the right hand side is given as $x \otimes (m_1 \otimes \cdots \otimes m_n)$, where j of the m_i are in M_3 and $n-j$ are in M_1. To apply ϕ, we leave the x and those m_i which are in M_1 alone, and we replace those m_i which are in M_3 by $f_{g_i^{-1}x}(m_i)$, where g_i is the coset representative of H in G labelling the ith copy of M_2 in $M_2\mathop{\uparrow}\limits^{\otimes G} = M_2^{\otimes n}$. It is easily checked that ϕ is an RG-module homomorphism which splits the above surjection. □

The reader is referred to Chapter 4 of Volume II for an extensive discussion of the tensor induction construction at the level of chain complexes. In particular this gives rise to the Evens norm map and Steenrod operations in cohomology. The algebraic proof by Evens that the cohomology ring of a finite group is finitely generated depends heavily on this construction.

CHAPTER 4

Methods from the representations of algebras

4.1. Representations of quivers

It follows from Morita theory (Section 2.2) that to study the representations of a finite dimensional algebra over an algebraically closed field, it suffices to consider the case where every irreducible module is one dimensional. We shall see that such an algebra is expressible as a quotient of the path algebra of a quiver (directed graph) by an ideal contained in the ideal of paths of length at least two. More generally, a finite dimensional basic algebra over any field can be expressed essentially uniquely as a quotient of a "modulated quiver" by such an ideal, provided certain sequences of bimodules over division rings split (this condition is always satisfied over a perfect field). This makes the representations of quivers important to the study of representations of finite dimensional algebras.

DEFINITION 4.1.1. A **quiver** is a directed graph, possibly with multiple arrows and loops.

If Q is a quiver and k is a field, we define the **path algebra** kQ as follows. It is an algebra over k, which as a vector space has a basis consisting of the **paths** $\bullet \to \bullet \to \cdots \to \bullet$ in Q. Multiplication is given on basis elements by composition of paths *in reverse order* (because we are dealing with left rather than right modules) if the paths are composable in this way, and zero otherwise. Thus for example corresponding to each vertex x there is a path of length zero giving rise to an idempotent basis element denoted e_x. A free algebra is an example of a path algebra, for a graph with only one vertex. Clearly kQ is finitely generated if and only if Q has only finitely many vertices and arrows, and finite dimensional if and only if in addition it has no oriented cycles.

A **representation** of a quiver Q associates to each vertex x of Q a vector space V_x, and to each arrow $x \to y$ a linear transformation $V_x \to V_y$ between the corresponding vector spaces.

There is a natural one–one correspondence between representations of Q and kQ-modules given as follows. Given a representation of Q, we form a kQ-module whose underlying vector space is $\bigoplus_x V_x$, and where the action of a basis element $x_1 \to \cdots \to x_2$ is as the composite of the corresponding maps:

$$\bigoplus_x V_x \twoheadrightarrow V_{x_1} \to \cdots \to V_{x_2} \hookrightarrow \bigoplus_x V_x.$$

So for example the action of the idempotent e_x is to project onto V_x.

Conversely, given a kQ-module V, we form a representation of Q by setting $V_x = e_xV$. If u is the basis element of kQ corresponding to an arrow $x \to y$, then $e_yu = u = ue_x$, and so u maps V_x to V_y. These are the maps we use to define the representation of Q. It is clear that the above procedures are mutually inverse.

There is a simple kQ-module S_x of dimension one corresponding to each vertex x of Q. It consists of a one dimensional vector space at the vertex x, and a zero dimensional vector space at each other vertex. In case kQ is finite dimensional, these are the only simple kQ-modules, but otherwise there are others. For example, for a free algebra with at least two generators, there are simple modules of every dimension.

EXAMPLE. Let Q be the following quiver.

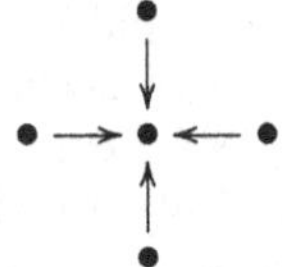

If any of the maps corresponding to the four arrows is not injective, then the kernel splits off as a direct summand. Thus apart from four simple modules, the indecomposable kQ-modules are in one–one correspondence with indecomposable **four subspace systems**; namely a vector space V together with four given subspaces V_1, V_2, V_3, V_4. There is an obvious notion of direct sum for such systems, namely $V \oplus W$; $V_1 \oplus W_1, \dots, V_4 \oplus W_4$. The **four subspace problem** is the problem of classifying the *indecomposable* four subspace systems. This problem was solved in 1970 by Gel'fand and Ponomarev [**114**]. It turns out that for at most three subspaces, there are only finitely many isomorphism types of indecomposable system. For four subspaces, there are infinitely many, but they are classifiable (this situation is called **tame**), while for at least five they are in some sense unclassifiable (**wild**). We shall discuss all this in more detail later.

The projective module $P_x = kQ.e_x$ corresponding to an idempotent e_x may be described as follows. The basis elements correspond to the paths in Q which begin with the vertex x, and the action of kQ is given by (reverse) composition of paths as before. We now show that these are essentially the only projective modules, and that every submodule of a projective module is projective.

DEFINITION 4.1.2. A ring Λ is (left) **hereditary** if every submodule of a projective (left) Λ-module is projective.

Thus for example the ring of integers in an algebraic number field is hereditary, and the next theorem shows that the path algebra of a quiver with finitely many vertices is hereditary.

DEFINITION 4.1.3. We write $kQ_{(n)}$ for the linear span in kQ of the paths of length at least n. It is a two sided ideal in kQ.

As a (left) kQ-module, $kQ_{(n)}/kQ_{(n+1)}$ is semisimple. It is a direct sum of simple modules of the form S_x, and the copies of a particular S_x correspond to the paths of length n ending at x. That is, $e_x.kQ_{(n)}/kQ_{(n+1)}$ has a basis consisting of the paths of length n ending at x. For example, if kQ is finite dimensional then $kQ_{(n)} = J^n(kQ)$.

Now a path of length m, $x \to \cdots \to y$, induces an injective map

$$e_x.kQ_{(n)}/kQ_{(n+1)} \to e_y.kQ_{(n+m)}/kQ_{(n+m+1)}$$

and the images of these maps for distinct paths are linearly independent.

If F is any free kQ-module (not necessarily finitely generated) with a chosen basis, then it inherits a filtration $F_{(n)}$ from the above filtration on kQ. Again if $x \to \cdots \to y$ is a path of length m then the induced map

$$e_x.F_{(n)}/F_{(n+1)} \to e_y.F_{(n+m)}/F_{(n+m+1)}$$

is injective, and the images of these maps for distinct paths are linearly independent.

THEOREM 4.1.4. *Suppose Q has only finitely many vertices. Then every submodule of a free kQ-module is isomorphic to a direct sum of modules of the form $kQ.e_x$, so that kQ is a hereditary algebra. The Krull–Schmidt theorem holds for finitely generated projective kQ-modules.*

PROOF. We shall only prove the first statement in case kQ is finite dimensional. The general case may be found in Bergman [**24**, Cor. 2.6], and is quite hard.

Suppose P is a submodule of a free kQ-module F. Define $F_{(n)}$ as above, and $P_{(n)} = P \cap F_{(n)}$. Since $\bigcap_n F_{(n)} = \{0\}$ we have $\bigcap_n P_{(n)} = \{0\}$. As above, if $x \to \cdots \to y$ is a path of length m then the induced map $e_x.P_{(n)}/P_{(n+1)} \to e_y.P_{(n+m)}/P_{(n+m+1)}$ is injective.

Since kQ is finite dimensional, we have $kQ_{(1)} = J(kQ)$. Now $P/kQ_{(1)}.P$ is a module for $kQ/kQ_{(1)}$, which is a finite direct sum of copies of k. Thus we may write $P/kQ_{(1)}.P$ as a direct sum of modules isomorphic to some S_x, say

$$P/kQ_{(1)}.P = \bigoplus_{x,\alpha} S_{x,\alpha}$$

with $S_{x,\alpha} \cong S_x$. Letting $P_{x,\alpha} \cong kQ.e_x$ be the corresponding projective kQ-module, we have a surjective kQ-module homomorphism

$$\phi : \bigoplus_{x,\alpha} P_{x,\alpha} \to P/kQ_{(1)}.P$$

which may be lifted to a kQ-module homomorphism

$$\psi : \bigoplus_{x,\alpha} P_{x,\alpha} \to P.$$

We claim that ψ is an isomorphism. The injectivity of ψ follows from the injectivity of $e_x.P_{(n)}/P_{(n+1)} \to e_y.P_{(n+m)}/P_{(n+m+1)}$ above. Write X for $\mathrm{Coker}(\psi)$. Tensoring the short exact sequence

$$0 \to \bigoplus_{x,\alpha} P_{x,\alpha} \xrightarrow{\psi} P \to X \to 0$$

over kQ with $kQ/kQ_{(1)}$, we obtain

$$\bigoplus_{x,\alpha} S_{x,\alpha} \to P/kQ_{(1)}.P \to X/kQ_{(1)}.X \to 0$$

so that $X/J(kQ).X = X/kQ_{(1)}.X = 0$, and hence by Nakayama's lemma $X = 0$, so that ψ is an isomorphism.

To prove the Krull–Schmidt theorem for finitely generated projective kQ-modules, we note that

$$kQ/kQ_{(1)} \otimes_{kQ} \bigoplus_{x,\alpha} P_{x,\alpha} = \bigoplus_{x,\alpha} S_{x,\alpha}$$

with $S_{x,\alpha} \cong S_x$. Thus if the module is finitely generated, the number of copies of of each $P_{x,\alpha}$ is finite, and hence well defined by applying the Krull–Schmidt theorem for finitely generated $kQ/kQ_{(1)}$-modules. □

COROLLARY 4.1.5. *Every projective module over a free algebra (over a field) is free.* □

EXAMPLE. The following is an example where Q has infinitely many vertices, so that kQ is a ring without identity. Let Q be the quiver

$$\cdots \to \bullet \to \bullet \to \bullet$$

Then a representation V of Q is the same thing as an inverse system of vector spaces $\cdots \to V_2 \to V_1 \to V_0$. A kQ-module is projective if and only if every map involved is injective, and the inverse limit is zero. Thus every submodule of a projective kQ-module is projective, and so kQ is hereditary.

Let $\mathbf{k}$ denote the representation $\cdots \to k \to k \to k$ of Q in which each arrow is the identity map. Then giving a homomorphism $\mathbf{k} \to V$ is the same thing as choosing an element v_i in each V_i in such a way that v_i goes to v_{i-1} in V_{i-1}. Thus $\mathrm{Hom}_{kQ}(\mathbf{k}, V) = \varprojlim V_i$. Since every submodule of a projective module is projective, every module has a projective resolution of length one, and so $\mathrm{Ext}^n_{kQ}(\mathbf{k}, V) = 0$ for $n > 1$. The functor $\mathrm{Ext}^1_{kQ}(\mathbf{k}, V)$ is usually written $\varprojlim^1 V_i$ or $R^1 \varprojlim V_i$, since it is the first (and only) right derived functor of $\varprojlim$. By Proposition 2.5.2, given a short exact sequence $0 \to V' \to V \to V'' \to 0$ of kQ-modules, there is a six term exact sequence

$$0 \to \varprojlim V_i' \to \varprojlim V_i \to \varprojlim V_i'' \to \varprojlim{}^1 V_i' \to \varprojlim{}^1 V_i \to \varprojlim{}^1 V_i'' \to 0.$$

Of course, the above discussion applies equally well when the field k is replaced by $\mathbb{Z}$, because a subgroup of a free abelian group is again free abelian. Thus we have a functor $\varprojlim^1$ for inverse systems of abelian groups, and a six term sequence as before. For further discussion of $\varprojlim^1$, see Milnor [**151**] and Bousfield–Kan [**35**], Sections IX.2 and XI.6.

DEFINITION 4.1.6. Suppose Λ is a finite dimensional algebra over an algebraically closed field k. Let $S_1, \dots, S_r$ be the isomorphism classes of simple Λ-modules with projective covers $P_i = \Lambda e_i$. The **Ext-quiver** $Q(\Lambda)$ has vertices $x_1, \dots, x_r$ corresponding to these simple modules, and the number of arrows from x_i to x_j is $\dim_k \mathrm{Ext}^1_\Lambda(S_i, S_j)$. Note that according to the remarks in Section 2.4, this is the same as the dimension of

$$\mathrm{Hom}_\Lambda(P_j, \mathrm{Rad}(P_i))/\mathrm{Hom}_\Lambda(P_j, \mathrm{Rad}^2(P_i)) = e_j J(\Lambda) e_i / e_j J^2(\Lambda) e_i.$$

PROPOSITION 4.1.7 (Gabriel [**112**]). *Suppose Λ is a finite dimensional basic algebra over an algebraically closed field k, and let $Q = Q(\Lambda)$ be its Ext-quiver. Then there is a surjective map of algebras $\phi : kQ \twoheadrightarrow \Lambda$ such that the kernel of ϕ is contained in the ideal of paths of length at least two. In particular, in case kQ is finite dimensional, this latter ideal is equal to the square of the radical, and ϕ induces a bijection between the simple Λ-modules and the simple kQ-modules, and between the blocks of Λ and the blocks of kQ.*

PROOF. Since Λ is a basic algebra, we can choose the e_i with $e_1 + \dots + e_r = 1$ and $e_i e_j = e_j e_i = 0$, and send the idempotent corresponding to x_i in kQ to the idempotent e_i in Λ. Choose a complement to $e_j J^2(\Lambda) e_i$ in $e_j J(\Lambda) e_i$ as a vector space, and choose a basis for it. Send the basis elements of kQ corresponding to the arrows from x_i to x_j to these basis elements. Every relation in kQ says that products of non-composable paths are zero. These relations are satisfied by the corresponding products in Λ by the relations $e_i e_j = e_j e_i = 0$, and so this map extends to a well defined map $kQ \to \Lambda$. It is surjective modulo $J^2(\Lambda)$, and so by Proposition 1.2.8 it is surjective. In case kQ is finite dimensional, the bijection of blocks comes from Proposition 1.8.4. □

DEFINITION 4.1.8. A **system of linear relations** on a quiver Q is a two sided ideal I contained in the ideal of paths of length at least two. This is the same as assigning to each pair of vertices x and y in Q a subspace of the space of paths of length at least two from x to y, called the **relations**, in such a way that composing a relation on either side with any path gives another relation.

We call the pair (Q, I) a **quiver with relations**. The **path algebra** of a quiver with relations is the algebra kQ/I.

The importance of the above definition is that by the proposition, every basic algebra over an algebraically closed field is isomorphic to the path algebra of a quiver with relations.

EXAMPLE. Let Q be the quiver

$$\cdots \to \bullet \to \bullet \to \bullet \to \cdots$$

and let I be the ideal generated by the paths of length at least two. Then a kQ-module is the same thing as a chain complex of vector spaces. The projective kQ-modules are the exact complexes.

More generally, for any ring Λ, one may form the path algebra ΛQ in the obvious way, and a ΛQ-module is the same as a chain complex of Λ-modules. The projective ΛQ-modules are the split exact sequences of projective Λ-modules.

We now indicate what modifications have to be made to the above discussion in case k is not algebraically closed. We confidently leave the details to the reader.

DEFINITION 4.1.9. A **labelled graph** is an undirected graph together with a pair of positive integers $({}_xd_y^\gamma, {}_yd_x^\gamma)$ for each edge $x \overset{\gamma}{—} y$. We usually omit to write in the label in case ${}_xd_y^\gamma = {}_yd_x^\gamma = 1$. An **orientation** of a labelled graph assigns a direction $x \to y$ or $x \leftarrow y$ to each edge $x — y$.

A **valued graph** is a labelled graph with the property that there exist positive integers f_x, one for each vertex, with ${}_xd_y^\gamma f_y = {}_yd_x^\gamma f_x$ for each edge $x \overset{\gamma}{—} y$. Thus for example a labelled graph with no cycles is always a valued graph.

A **modulation** of a valued graph consists of an assignment of a division ring Δ_x to each vertex x, and a Δ_x–Δ_y-bimodule ${}_xM_y^\gamma$ to each edge $x \overset{\gamma}{—} y$ satisfying

(i) ${}_yM_x^\gamma \cong \mathrm{Hom}_{\Delta_x}({}_xM_y^\gamma, \Delta_x) \cong \mathrm{Hom}_{\Delta_y}({}_xM_y^\gamma, \Delta_y)$

(ii) $\dim_{\Delta_y}({}_xM_y^\gamma) = {}_xd_y^\gamma$.

Finally, a **modulated quiver** consists of a valued graph together with an orientation and a modulation.

If Λ is a finite dimensional algebra over a field k, which is not necessarily algebraically closed, its Ext-quiver is defined as a modulated quiver as follows. Again the vertices x_i correspond to the isomorphism classes of simple modules S_i, with $\Delta_i = \mathrm{End}_\Lambda(S_i)^{\mathrm{op}}$. There is an arrow $x_i \xrightarrow{\gamma} x_j$ if and only if $\mathrm{Ext}^1_\Lambda(S_i, S_j) \neq 0$, and

$$\begin{aligned} {}_jM_i^\gamma &= M_{ij} = \mathrm{Ext}^1_\Lambda(S_i, S_j) \quad \text{as a } \Delta_j\text{–}\Delta_i\text{-bimodule,} \\ {}_iM_j^\gamma &= M'_{ij} = \mathrm{Hom}_k({}_jM_i^\gamma, k) \cong \mathrm{Hom}_{\Delta_i}({}_jM_i^\gamma, \Delta_i) \cong \mathrm{Hom}_{\Delta_j}({}_jM_i^\gamma, \Delta_j) \\ {}_jd_i^\gamma &= d_{ij} = \dim_{\Delta_i}({}_jM_i^\gamma) \\ {}_id_j^\gamma &= d'_{ij} = \dim_{\Delta_j}({}_iM_j^\gamma) \quad \text{and} \quad f_i = \dim_k(\Delta_i). \end{aligned}$$

Note that if $\mathrm{Ext}^1_\Lambda(S_i, S_j) \neq 0$ and $\mathrm{Ext}^1_\Lambda(S_j, S_i) \neq 0$ then there are two distinct arrows $x_i \xrightarrow{\gamma} x_j$ and $x_j \xrightarrow{\gamma'} x_i$ with separate bimodules.

A **representation** V of a modulated quiver assigns to each vertex x a Δ_x-module V_x and to each arrow $x \xrightarrow{\gamma} y$ a homomorphism

$${}_yM_x^\gamma \otimes_{\Delta_x} V_x \to V_y.$$

Just as before, representations of a modulated quiver Q are in natural one–one correspondence with modules for the **path algebra** kQ. We give a more abstract definition of this path algebra than in the unmodulated case, but it amounts to the same thing. Namely we set $\Delta = \bigoplus_x \Delta_x$, $M = \bigoplus_{x \xrightarrow{\gamma} y} {}_xM_y^\gamma$ as a Δ-Δ-bimodule, and then we define the space $kQ_{(n)}/kQ_{(n+1)}$ of paths of length n to be the n-fold tensor product $M \otimes_\Delta M \otimes_\Delta \cdots \otimes_\Delta M$ (and $kQ_{(0)}/kQ_{(1)} = \Delta$). Then the path algebra kQ is the direct sum of the $kQ_{(n)}/kQ_{(n+1)}$ as a tensor algebra.

Imitating the proof of Proposition 4.1.7, we find that there are two possible obstructions, both of which disappear over a perfect field.

PROPOSITION 4.1.10. *Suppose that Λ is a finite dimensional basic algebra over a field k. Suppose that*

(i) *the map of algebras $\Lambda \to \Lambda/J(\Lambda)$ splits, so that we may choose a copy of Δ_i in $\mathrm{End}_\Lambda(P_i)$ complementary to $J\mathrm{End}_\Lambda(P_i)$, and*

(ii) *the short exact sequence*

$$0 \to e_iJ^2(\Lambda)e_j \to e_iJ(\Lambda)e_j \to e_iJ(\Lambda)/J^2(\Lambda)e_j \to 0$$

splits as a sequence of Δ_j-Δ_i-bimodules.

Then Λ is a quotient of the path algebra of its Ext-quiver by an ideal contained in the ideal of paths of length at least two. □

COROLLARY 4.1.11. *Suppose that Λ is a finite dimensional basic algebra over a perfect field k. Then Λ is a quotient of the path algebra of its Ext-quiver by an ideal contained in the ideal of paths of length at least two.*

PROOF. Over a perfect field, there are no inseparable extensions, and so $\Delta_i \otimes \Delta_j^{\mathrm{op}}$ is semisimple. Thus every exact sequence of Δ_i–Δ_j-bimodules splits. In particular the sequence

$$0 \to J(\Lambda)/J^2(\Lambda) \to \Lambda/J^2(\Lambda) \to \Lambda/J(\Lambda) \to 0$$

of $\Lambda/J(\Lambda)$-$\Lambda/J(\Lambda)$-bimodules splits and so the Δ_i's lift to $\Lambda/J^2(\Lambda)$. Continuing inductively, the map $\Lambda \to \Lambda/J(\Lambda)$ splits, and so we may apply the proposition. □

REMARKS. (i) There should be a way of modifying the definition of the Ext-quiver of Λ to contain sufficient cocycle information so that a suitable "path algebra" will always map onto Λ. To the best of my knowledge no-one has attempted to do this.

(ii) Let k_0 be a field of characteristic p. Over the field $k = k_0(x)$, which is not perfect, let Λ be the commutative algebra $k[y,\varepsilon]/(\varepsilon^p, y^p - x - \varepsilon)$ of dimension p^2. Then $\Lambda/J(\Lambda) = k_0(y)$ is an inseparable extension of k of degree p, and the map $\Lambda \to \Lambda/J(\Lambda)$ does not split as a map of algebras over k. This example shows that the above corollary does not extend to non-perfect fields.

(iii) Since the group algebra of a finite group over the field of p elements has a finite splitting field, the above corollary is true for a block of such a group algebra over any field.

4.2. Finite dimensional hereditary algebras

In the last section, we saw that the path algebra of a quiver with finitely many vertices is hereditary. Conversely, we shall see in this section that over an algebraically closed field, every finite dimensional hereditary algebra is Morita equivalent to the path algebra of a quiver with finitely many vertices and no oriented cycles.

LEMMA 4.2.1 (Eilenberg and Nakayama [**94**]). *Let Λ be a ring and I a two sided ideal contained in $J^2(\Lambda)$ which is finitely generated both as a left ideal and as a right ideal. If Λ/I is hereditary then $I = 0$.*

PROOF. Since I is finitely generated as a right ideal, by Nakayama's lemma, if $I = I.J(\Lambda)$ then $I = 0$. Thus it suffices to work modulo $I.J(\Lambda)$, and so we assume $I.J(\Lambda) = 0$. Let $\Gamma = \Lambda/I$, so that Γ is hereditary. Since I annihilates the Λ-module $J(\Lambda)$, we may regard $J(\Lambda)$ as a Γ-module. The map of Γ-modules $J(\Lambda) \to J(\Lambda)/I = J(\Gamma)$ splits, as $J(\Gamma)$ is a projective Γ-module, so $J(\Lambda) \cong J(\Lambda)/I \oplus I$. But $I \subseteq J^2(\Lambda) = (J(\Lambda)/I).J(\Lambda) = \mathrm{Rad}(J(\Lambda))$ as a Γ-module, so by another application of Nakayama's lemma (this time on the left), $I = 0$. □

LEMMA 4.2.2. *Suppose Λ is a finite dimensional hereditary algebra and S_i, S_j are simple modules with $\mathrm{Ext}^1_\Lambda(S_i, S_j) \neq 0$. Then the projective cover P_i of S_i contains a copy of the projective cover P_j of S_j as a proper submodule.*

PROOF. Since $\mathrm{Ext}^1_\Lambda(S_i, S_j) \neq 0$, there is a non-zero homomorphism $P_j \to P_i$ whose image is in $J(P_i)$ but not in $J^2(P_i)$. The image of this map is a submodule of a projective module, and hence projective, and so the map splits. Since P_j is indecomposable, this implies that this homomorphism is injective. □

LEMMA 4.2.3. *If Λ is a finite dimensional hereditary algebra then the Ext-quiver of Λ contains no oriented cycles.*

PROOF. Suppose $S_1 \to S_2 \to \cdots \to S_n \to S_1$ is an oriented cycle. Then by the previous lemma we have $P_1 \subset P_2 \subset \cdots \subset P_n \subset P_1$. These are strict inclusions of finite dimensional modules, and so this situation is impossible. □

PROPOSITION 4.2.4. *Suppose Λ is a finite dimensional hereditary basic algebra over an algebraically closed field. Then $\Lambda \cong kQ_\Lambda$, the path algebra of its Ext-quiver.*

PROOF. By Lemma 4.2.3, kQ_Λ is finite dimensional, and so $J^n kQ_\Lambda$ is the ideal of paths of length at least n. By Proposition 4.1.7, there is a surjective map $\phi : kQ_\Lambda \to \Lambda$ whose kernel is contained in $J^2 kQ_\Lambda$. Thus by Lemma 4.2.1 ϕ is an isomorphism. □

The same arguments apply in the case of a modulated quiver, using Corollary 4.1.11.

PROPOSITION 4.2.5. *Suppose Λ is a finite dimensional hereditary basic algebra over a perfect field. Then $\Lambda \cong kQ_\Lambda$, the path algebra of its Ext-quiver.* □

REMARK. The above proposition is no longer true over a field which is not perfect, since the splitting condition in Proposition 4.1.10 may fail. An example of a finite dimensional hereditary algebra which is not isomorphic to the path algebra of its Ext-quiver may be found in Dlab and Ringel [**79**].

4.3. Representations of the Klein four group

In this section we work through an example in detail to illustrate the concepts introduced in the last two sections.

Denote by V_4 the Klein four group (Kleinsche Vierergruppe), namely a direct product of two cyclic groups of order two

$$V_4 = \langle x, y \mid x^2 = y^2 = [x, y] = 1 \rangle.$$

Let k be a field of characteristic two. We shall investigate the representations of kV_4. According to Section 3.14, there is only one simple kV_4-module, namely the trivial one dimensional module k, and we have

$$\dim_k \operatorname{Ext}^1_{kV_4}(k, k) = 2.$$

Thus the Ext-quiver for kV_4 is the following graph:

$$a \circlearrowleft \bullet \circlearrowright b$$

The arrows a, b correspond to elements $x - 1$ and $y - 1$ of $J(kV_4)$ complementing $J^2(kV_4)$. The relations are $a^2 = 0$, $b^2 = 0$ and $ab = ba$.

Unfortunately the kernel of the natural map from the path algebra of this graph to kV_4 is rather large, and so we try to be a bit cleverer. Now according to Section 3.14, if we want to understand indecomposable kV_4-modules, it suffices to understand Λ-modules, where $\Lambda = kV_4/\operatorname{Soc}(kV_4)$ is the three dimensional ring with basis elements 1, $x - 1$ and $y - 1$, and

$$(x-1)^2 = (y-1)^2 = (x-1)(y-1) = 0.$$

DEFINITION 4.3.1. The **Kronecker quiver** is the following graph:

$$Q = \; x_1 \bullet \underset{b}{\overset{a}{\rightrightarrows}} \bullet\, x_2$$

The path algebra of this quiver has two simple modules, S_1 and S_2, corresponding to x_1 and x_2. We have

$$\dim_k \operatorname{Ext}^1_{kQ}(S_i, S_j) = \begin{cases} 2 & \text{if } i = 1,\ j = 2 \\ 0 & \text{otherwise.} \end{cases}$$

If M is a $k\Lambda$-module, we obtain a representation of the quiver Q by letting the vector spaces V_1 and V_2 corresponding to the vertices x_1 and x_2 be $M/(\operatorname{Im}(x-1)+\operatorname{Im}(y-1))$ and $\operatorname{Im}(x-1)+\operatorname{Im}(y-1)$. The relations above for

$(x-1)$ and $(y-1)$ imply that they both act trivially on $\mathrm{Im}(x-1)+\mathrm{Im}(y-1)$, so that they induce maps a and b from V_1 to V_2.

Conversely if $a, b : V_1 \to V_2$ is a representation of Q, we obtain a $k\Lambda$-module whose underlying vector space is $V_1 \oplus V_2$, and where x acts as $1+a$, and y acts as $1+b$.

These recipes are not quite inverse, but they do set up a one–one correspondence between Λ-modules and representations of Q for which $V_2 = \mathrm{Im}(a) + \mathrm{Im}(b)$, and preserving direct sums. There is only one indecomposable representation of Q for which $V_2 \neq \mathrm{Im}(a) + \mathrm{Im}(b)$, namely the simple module S_2, and so the above is a one–one correspondence between indecomposable Λ-modules and indecomposable kQ-modules other than S_2.

Beware that the isomorphism between a Λ-module M and its image under the composite of these functors depends on a choice of complement for $\mathrm{Im}(x-1)+\mathrm{Im}(y-1)$ in M, and so it is not a natural isomorphism of functors. In particular, endomorphism rings are not preserved by the above correspondences.

THEOREM 4.3.2 (Kronecker). *Suppose $a, b : V_1 \to V_2$ is a pair of linear maps constituting a finite dimensional indecomposable representation of the Kronecker quiver over a field k. Then one of the following holds:*

(i) *The vector spaces V_1 and V_2 have the same dimension, and the determinant of $a + \lambda b$ is a non-zero element of $k[\lambda]$.*

In this case, if $\det(a) \neq 0$ then the representations can be written in the form

$$a \mapsto I, \qquad b \mapsto J$$

where I is an identity matrix and J is an indecomposable rational canonical form. A rational canonical form is indecomposable if and only if it has only one block, and is associated to a polynomial which is a power of an irreducible polynomial over k.

If $\det(a) = 0$ then the representation can be written in the form

$$a \mapsto J_0, \qquad b \mapsto I$$

where J_0 is a rational canonical form associated with a polynomial of the form λ^n. In some sense this corresponds to the "rational canonical form at infinity."

(ii) *The dimension of V_2 is one larger than the dimension of V_1, and bases may be chosen so that a and b are represented by the matrices*

$$a = \begin{pmatrix} 1 & & 0 & 0 \\ & \ddots & & \vdots \\ 0 & & 1 & 0 \end{pmatrix}, \quad b = \begin{pmatrix} 0 & 1 & & 0 \\ \vdots & & \ddots & \\ 0 & 0 & & 1 \end{pmatrix}.$$

(iii) *The dimension of V_1 is one larger than the dimension of V_2, and bases may be chosen so that a and b are represented by the transposes of the above matrices.*

PROOF. We first deal with case (i). If V_1 and V_2 have the same dimension and $\det(a) \neq 0$, then we can choose bases so that a is represented by the identity matrix. Then b is determined up to conjugation, and the result follows from the theory of rational canonical forms.

In case $\det(a + \lambda b) \neq 0$ in $k[\lambda]$ but $\det(a) = 0$ we argue as follows. We homogenise by introducing a new variable μ, so that $\det(\mu a + \lambda b)$ is a non-zero homogeneous polynomial in $k[\lambda, \mu]$. Suppose first that k is infinite. Then there is some point $(\lambda_1 : \mu_1)$ in the projective line $\mathbb{P}^1(k)$ over k so that $\det(\mu_1 a + \lambda_1 b) \neq 0$. We set $a' = \mu_1 a + \lambda_1 b$ and $b' = a$ and argue as before. Since a rational canonical form can have at most one eigenvalue in k, we deduce that $\det(b) \neq 0$ and the result follows by reversing the rôles of a and b.

In case k is finite we use Galois descent as follows. For some finite extension $\tilde{k}$ of k we can find a point $\mu_1 a + \lambda_1 b$ with $(\lambda_1 : \mu_1) \in \mathbb{P}^1(\tilde{k})$ as above. Now the representation does not have to stay indecomposable over $\tilde{k}$, but it will be a sum of Galois conjugates of the same indecomposable, and the restriction back to k will just be a direct sum of $[\tilde{k} : k]$ copies of the original representation. Thus we again deduce that $\det(b) \neq 0$ and proceed as before.

If case (i) does not occur then either V_1 and V_2 have different dimensions or $a + \lambda b$ is singular as a matrix over $k[\lambda]$. In either case, after dualising (and switching the rôles of V_1 and V_2) if necessary, we have a vector $v(\lambda) = \sum_{i=0}^{n}(-1)^i v_i \lambda^i$ satisfying $(a + \lambda b)v(\lambda) = 0$, i.e., $av_0 = 0$, $av_i = bv_{i-1}$ and $bv_n = 0$. Suppose such a $v(\lambda)$ has been chosen with n minimal. Our goal is to show that $\dim V_1 = n$, $\dim V_2 = n + 1$, and for a suitable choice of bases

$$a + \lambda b = \begin{pmatrix} 1 & \lambda & & 0 \\ & \ddots & \ddots & \\ 0 & & 1 & \lambda \end{pmatrix}.$$

Let us write $E_\lambda(n)$ for this matrix.

First we claim that the vectors $av_1, \dots, av_n$ are linearly independent. For otherwise if $\sum_{i=1}^{n} \alpha_i av_i = 0$, we set $v'_j = \sum_{i=0}^{n} \alpha_i v_{i-n+j}$ (with the convention that $v_i = 0$ if $i < 0$), and notice that

$$\begin{aligned}(a + \lambda b)\sum_{j=0}^{n}(-1)^j v'_j \lambda^j &= \sum_{j=0}^{n}(-1)^j(av'_j - bv'_{j-1})\lambda^j \\ &= \sum_{i,j=0}^{n}(-1)^j \alpha_i(av_{i-n+j} - bv_{i-n+j-1})\lambda^j = 0,\end{aligned}$$

contradicting the minimality of n. In particular it follows that $v_0, \dots, v_n$ are linearly independent.

Now if we choose our basis of V_1 to begin with $v_0, \dots, v_n$ and of V_2 to begin with $av_1, \dots, av_n$, then $a + \lambda b$ is represented by a matrix of the form

$$\begin{pmatrix} E_\lambda(n) & C + \lambda D \\ 0 & A' + \lambda B' \end{pmatrix}.$$

Thus the theorem will follow as soon as we show that we can find matrices X and Y with

$$\begin{pmatrix} I & X \\ 0 & I \end{pmatrix}\begin{pmatrix} E_\lambda(n) & C+\lambda D \\ 0 & A'+\lambda B' \end{pmatrix}\begin{pmatrix} I & Y \\ 0 & I \end{pmatrix} = \begin{pmatrix} E_\lambda(n) & 0 \\ 0 & A'+\lambda B' \end{pmatrix}$$

since one of the hypotheses was indecomposability. This equation is the same as

$$X(A'+\lambda B') + (C+\lambda D) + E_\lambda(n)Y = 0.$$

Separating out the constant term and the coefficient of λ, we obtain

$$XA' + C + \overline{Y} = 0$$
$$XB' + D + \underline{Y} = 0$$

where $\overline{Y}$ and $\underline{Y}$ denote the matrices obtained by removing the first, resp. the last row of Y. Eliminating the terms in Y, we end up with an equation of the form

$$(x_{11},\dots,x_{1m},x_{21},\dots,x_{nm})\underbrace{\begin{pmatrix} A' & & & 0 \\ -B' & A' & & \\ & -B' & \ddots & \\ & & \ddots & A' \\ 0 & & & -B' \end{pmatrix}}_{n-1}$$
$$= (d_{21}-c_{11},\dots,d_{2,m-1}-c_{1,m-1},d_{31}-c_{21},\dots,d_{n+1,m-1}-c_{n,m-1})$$

and so it remains to show that this $(n-1)(m-1)\times nm$ matrix has full column rank. But a row vector

$$(u_{11},\dots,u_{1m},u_{21},\dots,u_{nm})$$

annihilated by this matrix gives rise to a vector

$$v'(\lambda) = (u_{11}-u_{21}\lambda+\cdots\pm u_{n1}\lambda^{n-1},\dots,u_{1m}-u_{2m}\lambda+\cdots\pm u_{nm}\lambda^{n-1})$$

of degree $(n-1)$ in λ satisfying $v'(\lambda)(A'+\lambda B')=0$. Thus

$$(0,v'(\lambda))\begin{pmatrix} E_\lambda(n) & C+\lambda D \\ 0 & A'+\lambda B' \end{pmatrix} = 0$$

and so there is also a column vector of degree $(n-1)$ in λ annihilated by $a+\lambda b$, contradicting the minimality of n. □

We may now read off the classification of indecomposable modules for kV_4.

THEOREM 4.3.3 (Bašev[**12**], Heller and Reiner[**124**]).
A complete set of representatives of isomorphism classes of indecomposable kV_4-modules is given as follows.

(i) *The projective indecomposable module of dimension 4.*

(ii) *For each even dimension* $2n$ *and each indecomposable rational canonical form* J *of dimension* n *there is an indecomposable representation*

$$x \mapsto \begin{pmatrix} I & I \\ 0 & I \end{pmatrix} \qquad y \mapsto \begin{pmatrix} I & J \\ 0 & I \end{pmatrix}.$$

(iii) *For each even dimension* $2n$ *there is an indecomposable representation of the form*

$$x \mapsto \begin{pmatrix} I & J_0 \\ 0 & I \end{pmatrix} \qquad y \mapsto \begin{pmatrix} I & I \\ 0 & I \end{pmatrix}.$$

where J_0 *denotes the rational canonical form associated to the irreducible polynomial* λ^n*.*

(iv) *For each odd dimension* $2n+1$ *there are two indecomposable representations*

$$x \mapsto \left(\begin{array}{c|cccc} & 1 & & 0 & 0 \\ I & & \ddots & & \vdots \\ & 0 & & 1 & 0 \\ \hline 0 & & I & & \end{array}\right) \qquad y \mapsto \left(\begin{array}{c|cccc} & 0 & 1 & & 0 \\ I & \vdots & & \ddots & \\ & 0 & 0 & & 1 \\ \hline 0 & & I & & \end{array}\right)$$

and (v)

$$x \mapsto \left(\begin{array}{c|cccc} & 1 & & 0 \\ I & & \ddots & \\ & 0 & & 1 \\ & 0 & \cdots & 0 \\ \hline 0 & & I & \end{array}\right) \qquad y \mapsto \left(\begin{array}{c|cccc} & 0 & \cdots & 0 \\ & 1 & & 0 \\ I & & \ddots & \\ & 0 & & 1 \\ \hline 0 & & I & \end{array}\right)$$

which are isomorphic if and only if $n = 0$.

PROOF. Apart from the projective indecomposable module, the indecomposable kV_4-modules are in one–one correspondence with those for Λ. Apart from the one dimensional simple module S_2, these are in one–one correspondence with those for kQ. The modules appearing in (ii) and (iii) of the previous theorem, while cases (iv) and (v) correspond to cases (ii) and (iii). The disappearance of the module S_2 corresponds to the isomorphism between (iv) and (v) in the case $n = 0$. □

REMARK. If k is algebraically closed, the indecomposable rational canonical forms are powers of a linear factor $(\lambda - \alpha)^n$. We write $V_{n,\alpha}$ for the corresponding indecomposable representation of type (ii) in this case. The representations of type (iii) are written $V_{n,\infty}$. Thus for each $n \geq 1$ there is a family of modules parametrised by $\mathbb{P}^1(k)$. The modules of type (iv) are the modules $\Omega^{-n}(k)$, and those of type (v) are the modules $\Omega^n(k)$.

We can also read off the classification of indecomposable modules for the alternating group A_4 in characteristic two using Clifford theory. Namely, every such module is a summand of a module induced from V_4. Let us suppose for convenience that k is algebraically closed, and let $\{1,\omega,\bar{\omega}\}$ be the cube roots of unity in k. The action of an element t of order three on the indecomposable kV_4-modules is easy to see. Namely, it fixes the modules $\Omega^{\pm n}(k)$, so that these extend in three ways to form the modules $\Omega^{\pm n}$ of the three one dimensional simples k, ω, $\bar{\omega}$. The action on $\mathbb{P}^1(k)$ is via $\alpha \mapsto 1/(1-\alpha)$. Thus the fixed points correspond to the cube roots of unity ω and $\bar{\omega}$. Fo a non-fixed α, $W_{n,\alpha} = V_{n,\alpha}\uparrow^{A_4}$ is indecomposable. For $\alpha \in \{\omega,\bar{\omega}\}$ $V_{n,\alpha}\uparrow^{A_4}$ has three summands, $W_{n,\alpha}(1)$, $W_{n,\alpha}(\omega) = W_{n,\alpha}(1)\otimes\omega$ and $W_{n,\alpha}(\bar{\omega}) = W_{n,\alpha}(1)\otimes\bar{\omega}$. For example, $W_{n,\omega}(1)$ is given in terms of matrices as follows.

$$x \mapsto \begin{pmatrix} I & X \\ 0 & I \end{pmatrix} \qquad y \mapsto \begin{pmatrix} I & Y \\ 0 & I \end{pmatrix}$$

$$t \mapsto \operatorname{diag}(\omega^{n-2}, \omega^{n-4}, \dots, \omega^{-n}, \omega^{n}, \omega^{n-2}, \dots, \omega^{-n+2})$$

where X (resp. Y) is an $n\times n$ matrix with $\bar{\omega}$ (resp. ω) on every diagonal entry, ω (resp. $\bar{\omega}$) on every entry just below the diagonal, and zeros elsewhere.

LINEAR RELATIONS. Following Ringel [**175**], we use Kronecker's classification to classify the linear relations on a vector space. We shall use this later in the classification of representations of dihedral groups.

DEFINITION 4.3.4. A **linear relation** on a vector space V is a subspace of $V\times V$. For example, if $\lambda : V \to V$ is an endomorphism then we also write λ for the **graph** of λ, namely the relation $\{(v,\lambda(v))\} \le V\times V$. If C is a linear relation and U is a subspace of V, we write CU for $\{v \in V \mid \text{for some } u \in U, (u,v) \in C\}$. We write C^{-1} for the linear relation $\{(v,w) \in V\times V \mid (w,v)\in C\}$. If C and D are relations, we write CD for the linear relation $\{(v,w) \in V\times V \mid \text{ for some } x \in V,\ (v,x) \in C \text{ and } (x,w) \in D\}$.

If $a, b : W \to V$ is a representation of the Kronecker quiver, then we obtain a linear relation $\{(a(v),b(v)), v \in W\} \le V \times V$. Conversely, if $C \le V\times V$ is a linear relation then the projections $\pi_1,\pi_2 : C \to V$ give us a representation of the Kronecker quiver. These recipes set up a one–one correspondence between the representations of the Kronecker quiver for which $\operatorname{Ker}(a)\cap\operatorname{Ker}(b) = 0$, and linear relations. The only indecomposable representation of the Kronecker quiver for which $\operatorname{Ker}(a)\cap\operatorname{Ker}(b) \ne 0$ is the simple module S_1. Thus we may read off the list of indecomposable linear relations from Theorem 4.3.2.

One particular consequence of this classification which we shall need later is the following. If C is a linear relation on V, let $C' = \bigcup_n C^n 0_V$, and $C'' = \bigcap_n C^n V$. Then $C' \le C''$, and C''/C' is called the **regular part** of the relation. The following proposition asserts that the regular part of a relation splits off:

PROPOSITION 4.3.5. *If C is a linear relation on V, then there are subspaces U and W of V with $V = U \oplus W$, and $C = (C \cap (U \times U)) \oplus (C \cap (W \times W))$, with $C \cap (U \times U)$ the graph of an automorphism of U, and $C' \oplus U = C''$. The regular part of $C \cap (W \times W)$ is zero.*

PROOF. This follows from the classification. □

Note in particular that C induces an automorphism ϕ of C''/C', defined by

$$\phi(x + C') = (Cx \cap C'') + C'.$$

COROLLARY 4.3.6. *If C is a linear relation on V, and U is chosen as above, then $(C^{-1})' \oplus U = (C^{-1})''$.* □

INTEGRAL REPRESENTATIONS. The integral representations of the Klein four group were first classified by L. A. Nazarova [**154**]. We shall follow the approach of M. C. R. Butler [**49**]. In both these approaches, the problem is reduced to the four subspace problem, which had already been solved by Gel'fand and Ponomarev [**114**]. We shall give an outline of this reduction here. For a list of the four subspace configurations, see Brenner [**38**].

Suppose R is a principal ideal domain in which $2 \neq 0$ is prime (e.g. $R = \mathbb{Z}$), let K be the field of fractions of R, a field of characteristic zero, and $k = R/(2)$, a field of characteristic two. Then we can make R into an RV_4-module in four different ways, corresponding to the four different choices of signs. We call these rank one modules L_1, L_2, L_3 and L_4, and we write $e_1, \dots, e_4$ for the corresponding idempotents in KG (a typical one is $(1 + x - y - xy)/4$), so that $e_ie_j = \delta_{ij}e_j$ and $e_1 + \dots + e_4 = 1$.

DEFINITION 4.3.7. If M is an R-torsion free RG-module, we say M is **reduced** if it has no summands isomorphic to $L_1, \dots, L_4$ or RG.

If M is a reduced RG-module, we define $e_*M = e_1M + \dots + e_4M \subseteq KM$. Since $e_1 + \dots + e_4 = 1$, we have $e_*M = M + e_iM + e_jM + e_kM$ for any $\{i, j, k\} \subseteq \{1, 2, 3, 4\}$. Thus letting

$$V = e_*M/M, \quad V_i = (e_iM + M)/M \subseteq V$$

we have a four subspace system $V; V_1, \dots, V_4$ in which V is the sum of any three of the V_i.

DEFINITION 4.3.8. A four subspace system $V; V_1, \dots, V_4$ is **reduced** if V is the sum of any three of the V_i.

It turns out that all but five of the indecomposable four subspace systems are reduced.

Conversely, given a reduced four subspace system $V; V_1, \dots, V_4$, we form a reduced RV_4-module as follows. Let M_i be a direct sum of $\dim(V_i)$ copies of L_i, so that $M_i/2M_i \cong V_i$. Then we have a map $M_1 \oplus \dots \oplus M_4 \to V$ given by reduction modulo two followed by inclusion, and we define M to be the kernel of this map. It is easy to see that this is a reduced RV_4-module.

THEOREM 4.3.9. *The above processes set up a one–one correspondence between the reduced R-torsion free RV_4-modules and the reduced four subspace systems. This correspondence preserves direct sums.*

PROOF. See Butler [**49**]. □

COROLLARY 4.3.10. *The Krull–Schmidt theorem holds for R-torsion free RV_4-lattices.* □

4.4. Representation type of algebras

In this section, we describe without proof the trichotomy theorem for the representation type of finite dimensional algebras. We then investigate the representation type of finite dimensional hereditary algebras. There are many variations of the following definitions in the literature. The idea is always the same. An algebra is of finite representation type if there are only finitely many indecomposables; otherwise it is of infinite representation type. It is of tame representation type if the indecomposables in each dimension come in finitely many one parameter families with finitely many exceptions. In some sense this is supposed to represent classifiability of the representations, although in particular cases this can be a very hard problem. It is of wild representation type if the representation theory "includes" that of a free algebra in two variables. The latter in some sense includes the representation theory of an arbitrary finite dimensional algebra, and the consensus feeling is that the representations of a wild algebra are in some sense unclassifiable. This is the same as the problem of finding a canonical form for pairs of not necessarily commuting matrices. The definition of tame does not make sense over a finite field, for obvious reasons.

DEFINITION 4.4.1. Suppose k is an infinite field.

A finite dimensional algebra Λ is of **finite representation type** if there are only finitely many isomorphism classes of indecomposable Λ-modules.

A finite dimensional algebra Λ is of **tame representation type** if it is not of finite representation type, and for any dimension n, there is a finite set of Λ–$k[T]$-bimodules M_i which are free as right $k[T]$-modules, with the property that all but a finite number of indecomposable Λ-modules of dimension n are of the form $M_i \otimes_{k[T]} M$ for some i, and for some indecomposable $k[T]$-module M. If the M_i may be chosen independently of n, we say Λ has **domestic representation type**. Note that the indecomposable $k[T]$-modules are classified by their rational canonical forms, which are the powers of irreducible polynomials over k.

A finite dimension algebra Λ has **wild representation type** if there is a finitely generated Λ–$k\langle X,Y\rangle$-bimodule M which is free as a right $k\langle X,Y\rangle$-module, such that the functor $M \otimes_{k\langle X,Y\rangle} -$ from finite dimensional $k\langle X,Y\rangle$-modules to finite dimensional Λ-modules preserves indecomposability and isomorphism class. Here, $k\langle X,Y\rangle$ is the free algebra on two variables; namely the path algebra of the graph with one vertex and two arrows.

We shall omit the proof of the following trichotomy theorem, which is rather technical.

THEOREM 4.4.2 (Drozd [**91, 92**]; Crawley-Boevey [**63**]).
Over an algebraically closed field, every finite dimensional algebra is of finite, tame or wild representation type, and these types are mutually exclusive. □

A precise statement about the unclassifiability of finite dimensional modules over a free algebra on two variables is the following. The theory of finite dimensional Λ-modules is said to be **decidable** if there is a Turing machine algorithm which will decide the truth or falsehood of any sentence in the language of finite dimensional Λ-modules.

THEOREM 4.4.3 (Baur [**13**]; Kokorin and Mart'yanov [**142**]).
Let k be any field. Then the theory of finite dimensional $k\langle X, Y\rangle$-modules is undecidable. □

The proof consists of encoding the word problem for finitely presented groups into the module theory of $k\langle X, Y\rangle$. It is conjectured that for finite dimensional algebras Λ, the theory of finite dimensional Λ-modules is undecidable if and only if Λ has wild representation type. For an extensive discussion of this question, see Prest [**160**], where a proof is also given for the above theorem. The conjecture has been verified for path algebras of quivers (without relations).

For group algebras the trichotomy theorem is much easier to prove. The following is a more precise statement:

THEOREM 4.4.4 (Bondarenko and Drozd [**34**]; see also Ringel [**174**]).
Let G be a finite group and k an infinite field of characteristic p.

(i) *kG has finite representation type if and only if G has cyclic Sylow p-subgroups.*

(ii) *kG has domestic representation type if and only if $p = 2$ and the Sylow 2-subgroups of G are isomorphic to the Klein four group.*

(iii) *kG has tame representation type if and only if $p = 2$ and the Sylow 2-subgroups are dihedral, semidihedral or generalised quaternion.*

(iv) *In all other cases kG has wild representation type.* □

We shall be investigating the indecomposable representations of some of the tame group algebras later in this chapter. We shall also see that for path algebras, there is a connection between the representation type and the positivity of a certain quadratic form defined in terms of the quiver. In fact, the representation type is finite if and only if the associated form is positive definite, and tame if and only if the associated form is positive semidefinite (and not definite). We shall only prove some of these results, and give references for the rest. The classification of quivers according to the associated quadratic form is the subject of the next two sections, and leads to the Dynkin and Euclidean diagrams.

EXERCISE. Use the theory of vertices and sources to prove part (i) of the above theorem. For a proof of a stronger version of this, see Corollary 6.3.5.

4.5. Dynkin and Euclidean Diagrams

In our discussion of representation type, and also in our discussion of tree classes of Auslander–Reiten quivers later in this chapter, we shall have need to refer to the list of diagrams given in the following definition. The infinite diagrams are only needed in the latter topic. For the moment you should ignore the numbers attached to the vertices of the Euclidean and infinite Dynkin diagrams; these will appear in the proofs of Lemma 4.5.5 and Proposition 4.5.7.

DEFINITION 4.5.1. (i) The following labelled graphs are called the (finite) **Dynkin diagrams.**

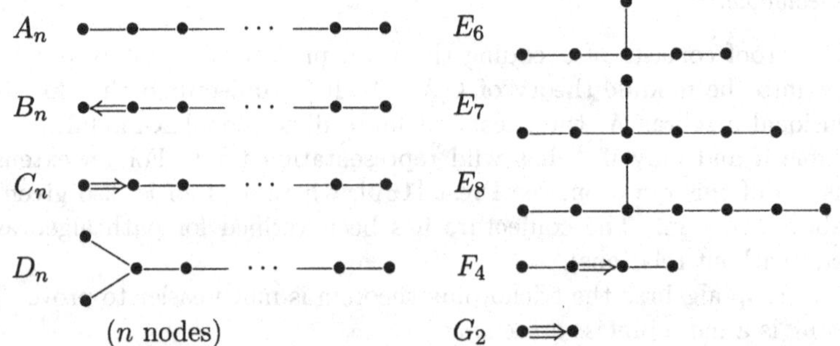

In these pictures, we have used $\bullet\!\Rightarrow\!\bullet$ and $\bullet\!\Rrightarrow\!\bullet$ to denote the labelled edges $\bullet\overset{(2,1)}{\text{———}}\bullet$ and $\bullet\overset{(3,1)}{\text{———}}\bullet$ respectively.

(ii) The following are the **infinite Dynkin diagrams.**

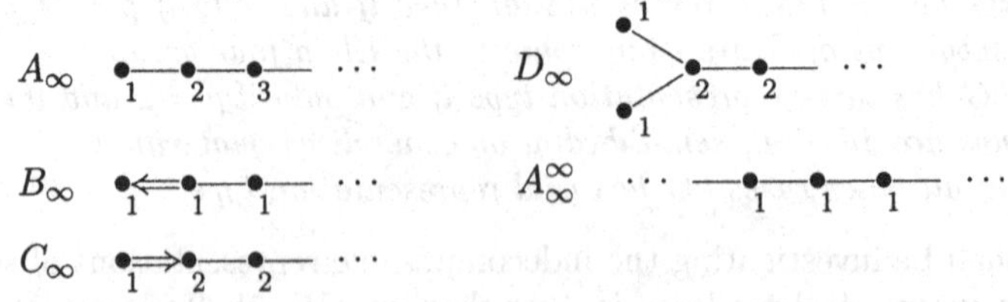

(iii) The following are the **Euclidean diagrams.**

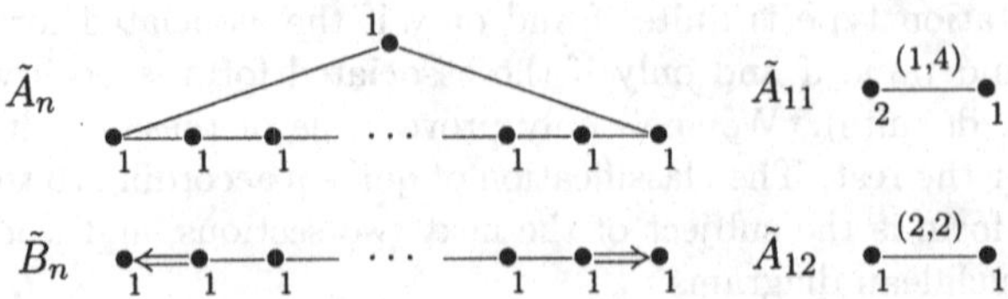

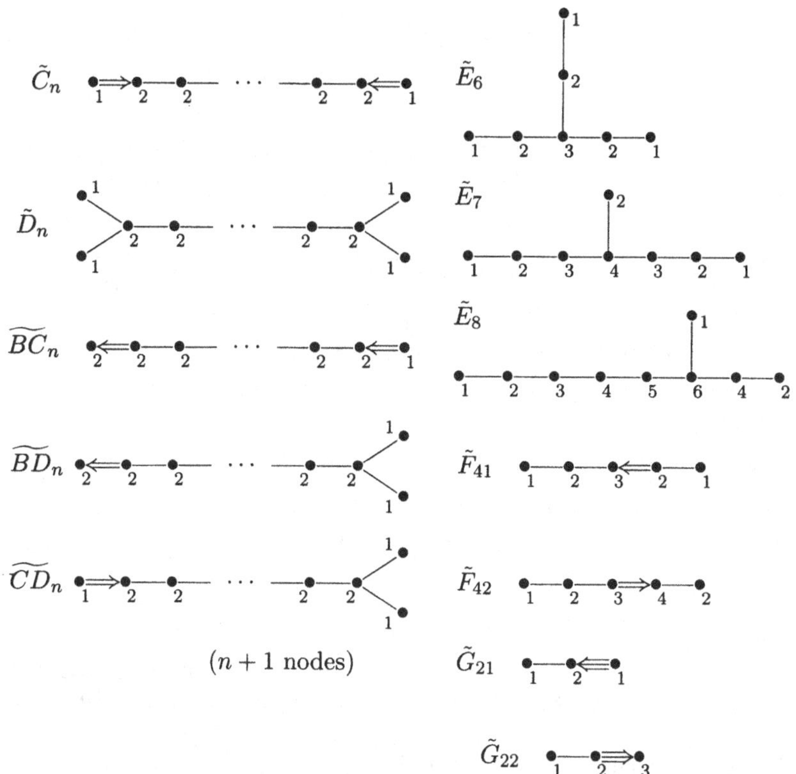

Note that $\tilde{A}_0$ consists of a single vertex with a loop, and $\tilde{A}_1$ consists of two vertices with two distinct edges going between them. Also note that $B_2 = C_2$ and $A_3 = D_3$, but that there are no further duplications in the above list.

Given two labelled graphs T_1 and T_2, we say that T_1 is **smaller** than T_2 if there is an injective morphism of graphs $\rho : T_1 \to T_2$ such that for each edge $x \overset{\gamma}{—} y$ in T_1, ${}_x d_y^\gamma \leq {}_{\rho(x)} d^\gamma_{\rho(y)}$, and **strictly smaller** if ρ can be chosen not to be an isomorphism. Note that an infinite labelled graph may be strictly smaller than itself. A_∞ is an example of this.

LEMMA 4.5.2. *Given any connected labelled graph T, either T is a (finite or infinite) Dynkin diagram or there is a Euclidean diagram which is smaller than T. Both possibilities may not occur simultaneously.*

PROOF. Suppose there is no Euclidean diagram which is smaller than T. Looking at $\tilde{A}_n$, we see that T has no cycles, and is hence a labelled tree. Looking at $\tilde{A}_{11}$ and $\tilde{A}_{12}$, all edges are of the form $(1,1)$, $(2,1)$ or $(3,1)$. Looking at $\tilde{G}_{21}$ and $\tilde{G}_{22}$, if an edge of type $(3,1)$ occurs then $T \cong G_2$. Looking at $\tilde{B}_n$, $\tilde{C}_n$ and $\widetilde{BC}_n$, there is at most one edge of type $(2,1)$. Looking at $\widetilde{BD}_n$

and $\widetilde{CD}_n$, if there is an edge of the form $(2,1)$ then

$$T = \cdots \;—\bullet—\bullet\!\Longrightarrow\!\bullet—\bullet—\; \cdots$$

and looking at $\tilde{F}_{41}$ and $\tilde{F}_{42}$ this forces $T \cong F_4$, B_n, C_n, B_∞ or C_∞. Otherwise T is a tree with all edges of type $(1,1)$. Then looking at $\tilde{D}_n$, it has at most one branch point. Finally, looking at $\tilde{E}_6$, $\tilde{E}_7$ and $\tilde{E}_8$ completes the proof. □

DEFINITION 4.5.3. The **Cartan matrix** of a labelled graph T (not to be confused with the Cartan matrix of an algebra) is the matrix whose rows and columns are indexed by the vertices of the graph, and with entries $c_{xy} = 2\delta_{xy} - \sum_\gamma {}_xd_y^\gamma$, where the sum runs over edges $x \overset{\gamma}{—} y$ in T.

The **symmetrised Cartan matrix** of a valued graph has entries $\tilde{c}_{xy} = c_{xy}f_y$, where f_y is as given in the definition of a valued graph. Note that this matrix is symmetric.

Thus for example the Cartan matrix of F_4 is the matrix

$$\begin{pmatrix} 2 & -1 & 0 & 0 \\ -1 & 2 & -2 & 0 \\ 0 & -1 & 2 & -1 \\ 0 & 0 & -1 & 2 \end{pmatrix}.$$

DEFINITION 4.5.4. A **subadditive function** on a labelled graph T is a function $x \mapsto n_x$ from the vertices of T to the positive rationals satisfying $\sum_x n_x c_{xy} \geq 0$ for all y. A subadditive function is called **additive** if $\sum_x n_x c_{xy} = 0$ for all y.

If T is a labelled graph, the **opposite** labelled graph T^{op} has the same vertices and edges, but with the label ${}_xd_y^\gamma$ replaced by ${}_yd_x^\gamma$. If T is a valued graph, then so is T^{op}, by replacing the f_x by n/f_x for some positive integer n.

REMARK. This broadens the usual definition, where the n_x are taken to be positive integers. We shall make use of this broader definition in our analysis of periodic modules.

LEMMA 4.5.5. (i) *Each Euclidean diagram admits an additive function.*

(ii) *If T^{op} admits an additive function then every subadditive function on T is additive.*

(iii) *Every subadditive function on a Euclidean diagram is additive.*

PROOF. (i) The numbers attached to the vertices of the Euclidean diagrams in the illustration form an additive function in each case.

(ii) Suppose $x \mapsto n_x$ is a subadditive function on T. By hypothesis there is a function $x \mapsto n'_x$ such that $\sum_y c_{xy}n'_y = 0$ for all x. Thus $\sum_{x,y} n_x c_{xy} n'_y = 0$, while $\sum_x n_x c_{xy} \geq 0$ and $n'_y > 0$ for each y. Hence we have equality and $x \mapsto n_x$ is additive.

(iii) follows from (i) and (ii), since the opposite of any Euclidean diagram is Euclidean. □

LEMMA 4.5.6. *Suppose T and T' are connected labelled graphs and T is strictly smaller than T'. Suppose also that $x \mapsto n_x$ is a subadditive function on T'. Then identifying T with a subgraph of T', the restriction of $x \mapsto n_x$ to T is a subadditive function on T which is not additive.*

PROOF. For x a vertex of T, we have

$$2n_y \geq \sum_{x,\gamma \in T'} n_x \, {}_x d_y^\gamma \geq \sum_{x,\gamma \in T} n_x \, {}_x d_y^\gamma,$$

where the sums are over edges $x \overset{\gamma}{—} y$, and the ${}_x d_y^\gamma$ are the values in T' in the first sum and in T in the second. Since T is strictly smaller than T', for some $y \in T$ the second inequality is strict, and so the restriction of $x \mapsto n_x$ is not additive. □

PROPOSITION 4.5.7. *Each of the infinite Dynkin diagrams admits an additive function.*

(i) *For A_∞ there are also subadditive functions which are not additive.*

(ii) *For the other infinite Dynkin diagrams every subadditive function is a multiple of a given bounded additive function.*

PROOF. The numbers attached to the vertices in the illustration form an additive function in each case.

(i) A_∞ is strictly smaller than itself, and so by Lemma 4.5.6 there is a subadditive function which is not additive.

(ii) First we show that every subadditive function on A_∞^∞ is constant and additive. We label the vertices of A_∞^∞ with the integers. Suppose $j \mapsto n_j$ is a subadditive function on A_∞^∞, and suppose $n_{j-1} < n_j$ for some j. The inequality $2n_{j-1} \geq n_j + n_{j-2}$ may be written in the form $n_{j-1} - n_{j-2} \geq n_j - n_{j-1}$, so that by induction on r we have $n_{j-r} \leq n_j - r(n_j - n_{j-1})$ for all $r \geq 0$. Choosing r large enough so that $r(n_j - n_{j-1}) > n_j$ we see that $n_{j-r} < 0$, contradicting the definition of subadditive function. Similarly if $n_{j-1} > n_j$ we find that some $n_{j+r} < 0$.

For B_∞, C_∞ and D_∞, given a subadditive function we form a subadditive function on A_∞^∞ according to the following scheme.

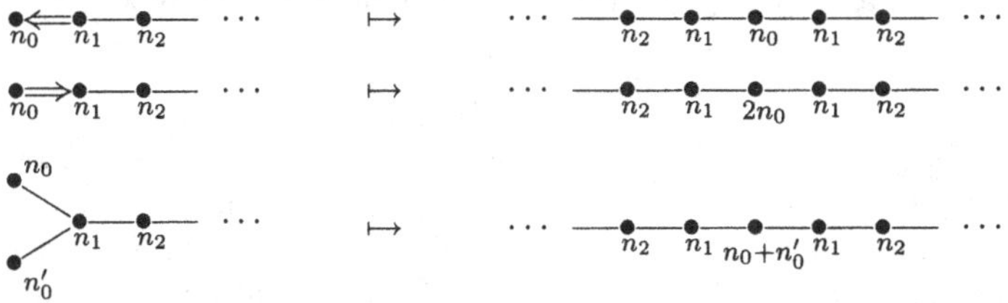

The result follows immediately for B_∞ and C_∞. For D_∞ we obtain $n_0 + n_0' = n_1$. Subadditivity forces $2n_0 \geq n_1$ and $2n_0' \geq n_1$ whence $n_0 = n_0'$, and the result follows. □

The following is a generalisation by Happel, Preiser and Ringel [**122**] of the characterisation by Vinberg [**200**] and by Berman, Moody and Wonenburger [**25**] of the finite Dynkin diagrams.

THEOREM 4.5.8. *Let T be a connected labelled graph, and $x \mapsto n_x$ a subadditive function on T. Then the following hold:*

(i) *T is either a Dynkin diagram (finite or infinite) or a Euclidean diagram.*

(ii) *If $x \mapsto n_x$ is not additive then T is a finite Dynkin diagram or A_∞.*

(iii) *If $x \mapsto n_x$ is additive then T is an infinite Dynkin diagram or a Euclidean diagram.*

(iv) *If $x \mapsto n_x$ is unbounded then $T \cong A_\infty$.*

PROOF. (i) Suppose this is false. Then by Lemma 4.5.2 there is a Euclidean diagram which is strictly smaller than T. Thus by Lemma 4.5.6 there is a subadditive function on this Euclidean diagram which is not additive. This contradicts Lemma 4.5.5 (iii).

(ii) This follows from Lemma 4.5.5 (iii) and Proposition 4.5.7 (ii).

(iii) Suppose this is false. Then T is a finite Dynkin diagram by (i), and hence so is T^{op}. Thus T^{op} is strictly smaller than some Euclidean diagram, and so by Lemma 4.5.6 and Lemma 4.5.5 (i), T^{op} admits a subadditive function which is not additive, contradicting Lemma 4.5.5 (ii).

(iv) If $x \mapsto n_x$ is unbounded then T is infinite, and so by (i) it is an infinite Dynkin diagram. Hence by Proposition 4.5.7 (ii), $T \cong A_\infty$. □

4.6. Weyl groups and Coxeter transformations

In this section, we examine the geometry of a certain real vector space associated with the graphs discussed in the last section.

DEFINITION 4.6.1. Given a valued graph T with no loops (edges from a vertex to itself), we form the real vector space $\mathbb{R}^T$ with $\mathbb{R}$-basis elements v_x corresponding to the vertices x in T, and with symmetric bilinear form given by the symmetrised Cartan matrix

$$(v_x, v_y) = \tilde{c}_{xy}.$$

The **Weyl group** $W(T)$ is the group generated by the reflections

$$w_x(v) = v - \frac{2(v, v_x)}{(v_x, v_x)} v_x.$$

It is easy to check that the w_x are transformations of order two preserving the bilinear form, and fixing the hyperplane perpendicular to x. Note that if Q is a (modulated) quiver with underlying valued graph T, and V is a representation of Q, then the dimension of V is in a natural way a vector in $\mathbb{R}^T$. Namely the coefficient of v_x is the dimension over Δ_x of V_x. This vector is called the **dimension vector** of V.

LEMMA 4.6.2. *For $x \neq y$, the product $w_x w_y$ has order 2, 3, 4, 6 or ∞ according as $c_{xy}c_{yx}$ is 0, 1, 2, 3 or at least 4 respectively.*

PROOF. This is an easy exercise in two dimensional geometry. □

PROPOSITION 4.6.3. *Let T be a finite connected valued graph without loops.*

(i) *T is a finite Dynkin diagram if and only if the bilinear form $(\ ,\ )$ is positive definite on $\mathbb{R}^T$.*

(ii) *T is a Euclidean diagram if and only if $(\ ,\ )$ is positive semidefinite on $\mathbb{R}^T$. In this case every null vector is a multiple of the vector given by the additive function shown in Definition 4.5.1.*

(iii) *If T is neither Dynkin nor Euclidean, then there is a vector $v \in \mathbb{R}^T$ with non-negative integral coordinates with respect to the basis vectors v_x, such that $(v, v) < 0$.*

PROOF. Suppose T is a Euclidean diagram. By Lemma 4.5.5 (i), there is an additive function $x \mapsto n_x$ on T. For a particular value of x, we have $\sum_{y \neq x} n_y c_{yx}/n_x = -2$, and so the norm of a typical vector $\sum_x a_x v_x$ is

$$
\begin{aligned}
(\sum_x a_x v_x, \sum_x a_x v_x) &= 2\sum_x a_x^2 f_x + \sum_{x \neq y} a_x a_y c_{xy} f_y \\
&= -\frac{1}{2}\sum_{x\neq y}\left(\frac{a_x^2 n_y c_{yx} f_x}{n_x} + \frac{a_y^2 n_x c_{xy} f_y}{n_y}\right) + \sum_{x\neq y} a_x a_y c_{xy} f_y \\
&= -\frac{1}{2}\sum_{x\neq y} n_x n_y c_{xy} f_y (a_x/n_x - a_y/n_y)^2.
\end{aligned}
$$

which is positive semidefinite since the c_{xy} are negative for $x \neq y$, and the n_x and f_x are positive. Moreover for a null vector, all the a_x/n_x must have the same value, so that the null space is one dimensional.

Since every Dynkin diagram is strictly smaller than a Euclidean diagram, it follows that $(\ ,\ )$ is positive definite on the Dynkin diagrams.

If T is neither Euclidean nor Dynkin then by Lemma 4.5.2 there is a Euclidean diagram T' which is strictly smaller than T. If T' contains all the points of T, then a null vector for T' has negative norm for T. Otherwise choose a point of T adjacent to a point of T', and add a small enough positive rational multiple of the corresponding basis element to the null vector for T', to obtain a vector of negative norm. A suitable multiple of this vector has non-negative integral coordinates with respect to the basis vectors v_x. □

PROPOSITION 4.6.4. (i) *Suppose T is a finite Dynkin diagram. Then $W(T)$ is a finite group of automorphisms of $\mathbb{R}^T$. There is no non-zero vector in $\mathbb{R}^T$ fixed by the whole of $W(T)$.*

(ii) *Suppose T is a Euclidean diagram. Let $\mathbf{n}$ be the null vector given by the additive function $x \mapsto n_x$ shown in Definition 4.5.1. Then $W(T)$ preserves $\langle \mathbf{n} \rangle$ and acts as a finite group of automorphisms of $\mathbb{R}^T/\langle \mathbf{n} \rangle$.*

PROOF. (i) Since the matrices in $W(T)$ have integer entries with respect to our basis v_x, $W(T)$ is a discrete subgroup of the compact orthogonal group. Thus it is finite. A vector in $\mathbb{R}^T$ fixed by the whole of $W(T)$ would have to have zero inner product with each basis vector v_x, and hence it would have to be zero.

(ii) Since $\langle \mathbf{n} \rangle$ is the radical of $(\, , \,)$, $\langle \mathbf{n} \rangle$ is preserved by $W(T)$. Now $W(T)$ acts as a discrete subgroup of the compact orthogonal group on $\mathbb{R}^T/\langle \mathbf{n} \rangle$, and this action is therefore finite. □

DEFINITION 4.6.5. A **Coxeter transformation** on $\mathbb{R}^T$ is a linear transformation obtained by applying all the w_x once each, in some order.

If T is a finite Dynkin or Euclidean diagram and c is a Coxeter transformation, then by the above Proposition, c has finite order h on $\mathbb{R}^T$ in the first case, and on $\mathbb{R}^T/\langle \mathbf{n} \rangle$ in the second. This number h is called the **Coxeter number**.

If T is Euclidean and $v \in \mathbb{R}^T$ then the **defect** $\partial_c(v)$ of v with respect to a Coxeter transformation c is defined by

$$c^h(v) = v + \partial_c(v)\mathbf{n}.$$

Thus ∂_c is a linear form $\mathbb{R}^T \to \mathbb{R}$, and the map ∂_c gives us a splitting $\mathbb{R}^T = \operatorname{Ker}(\partial_c) \oplus \langle \mathbf{n} \rangle$.

EXERCISE. Show that if T is a tree then all the Coxeter transformations are conjugate in $W(T)$.

LEMMA 4.6.6. *Suppose T is a Euclidean diagram. The following two conditions on a vector $v \in \mathbb{R}^T$ are equivalent:*

(i) *v has infinitely many images under c.*

(ii) $\partial_c(v) \neq 0$.

If (i) *and* (ii) *are satisfied then some image of v under some power of c has negative coordinates.*

PROOF. This is clear from the previous discussion. □

DEFINITION 4.6.7. If T is a finite Dynkin diagram, then the **root system** associated to T is the finite subset Φ of $\mathbb{R}^T$ consisting of the images under $W(T)$ of the basis vectors v_x. These are exactly the root systems in the sense of for example Humphreys [**127**, Chapter III]. The elements of Φ are called the **roots**. Since the matrices in $W(T)$ have integer entries, all roots are integral combinations of the basis vectors. Reflection in the hyperplane perpendicular to a root permutes the set of roots. If v is a root we write w_v for this reflection.

A non-zero vector which is a non-negative linear combination of the v_x is called **positive**, and a vector v is **negative** if $-v$ is positive.

LEMMA 4.6.8. *Suppose T is a Dynkin diagram.*

(i) *If v is a root then so is $-v$.*

(ii) *Suppose v and v' are roots, neither a multiple of the other, and suppose $(v, v') > 0$. Then either $(v', v') = 2(v, v')$ or $(v, v) = 2(v, v')$. The vector $v - v'$ is also a root.*

(iii) *If v is a positive root and w_x is a reflection then either $w_x(v)$ is positive or $v = v_x$ and $w_x(v) = -v_x$.*

(iv) *Every root is either positive or negative.*

PROOF. (i) $w_v(v) = -v$.

(ii) The reflection perpendicular to v of v' is a root, as is the reflection of v' perpendicular to v, and $2(v,v')/(v,v)$ and $2(v,v')/(v',v')$ are integers. If θ is the angle between v and v', then

$$\left(\frac{2(v,v')}{(v,v)}\right)\left(\frac{2(v,v')}{(v',v')}\right) = \frac{4(v,v')^2}{(v,v)(v',v')} = 4\cos^2\theta < 4.$$

Thus if $(v,v') > 0$ then either $(v',v') = 2(v,v')$ or $(v,v) = 2(v,v')$. In the former case $w_{v'}(v) = v - v'$, while in the latter case $w_v(v') = v' - v$ and we use (i).

(iii) Suppose v is an image $w(v_y)$ under $W(T)$ of the basis vector v_y. If $v \neq v_x$ and $w_x(v)$ is not positive, then $(v, v_x) > 0$. As in (ii), we have two cases according as $2(v,v_x) = (v_x,v_x)$ or $2(v,v_x) = (v,v) = (v_y,v_y)$. In the former case, $w_x(v) = v - v_x$. Since v must involve a strictly positive multiple of v_x in order that $(v,v_x) > 0$, we conclude that $w_x(v)$ is again positive.

In the latter case, we argue as follows. There is an isomorphism $\rho : \mathbb{R}^T \to \mathbb{R}^{T^{\text{op}}}$ given by setting $\rho(\sum_x a_x v_x) = \sum_x a_x f_x v_x$, and it is easy to check that this isomorphism commutes with the actions of the reflections w_x, so that the two Weyl groups are isomorphic. In fact the rôle of ρ is to make long roots short and vice versa. We now look at the vector $v' = w(v_y) \in \mathbb{R}^{T^{\text{op}}}$. If $2(v,v_x) = (v,v)$ in $\mathbb{R}^T$ then $2(v',v_x) = (v_x,v_x)$ in $\mathbb{R}^{T^{\text{op}}}$ and so we may apply the argument of the previous paragraph to deduce that $w_x(v')$ is positive, and hence $w_x(v)$ is positive.

(iv) This easily follows from (iii). □

LEMMA 4.6.9. *Suppose T is a Dynkin diagram, with vertices $x_1, \dots, x_n$ and corresponding basis elements $v_1, \dots, v_n$ of $\mathbb{R}^T$, reflections $w_1, \dots, w_n$ on $\mathbb{R}^T$ and Coxeter transformation $c = w_1 \dots w_n$.*

(i) *The transformation c has no non-zero fixed points on $\mathbb{R}^T$.*

(ii) *Given any vector $v \in \mathbb{R}^T$, for some value of $m \geq 0$ the vector $c^m(v)$ is not positive.*

PROOF. (i) Suppose $v = \sum_i r_i v_i$ and $c(v) = v$. Then

$$w_1(v) = w_2 \dots w_n(v)$$

and so the multiple of v_1 in $w_1(v)$ is still r_1, and hence $w_1(v) = v$. Repeating this argument, we see that

$$w_2(v) = w_3 \dots w_n(v)$$

and hence $w_2(v) = v$. Continuing this way, we see that v is fixed by all the w_i, and is hence zero.

(ii) If $c^m(v)$ is positive for all m, then $\sum_{i=0}^{h-1} c^i(v)$ is a positive vector fixed by c, contradicting (i). □

4.7. Path algebras of finite type

The main theorem of this section is Gabriel's theorem, which says that the path algebra of a (modulated) quiver Q is of finite representation type if and only if Q is a Dynkin diagram (with some orientation). In this case, the indecomposable kQ-modules are in one–one correspondence with the positive roots in the associated root system. The proofs we shall give for these statements are due to Tits and Bernstein–Gel'fand–Ponomarev.

PROPOSITION 4.7.1. (i) *If Q is a modulated quiver of finite representation type then the underlying valued graph T of Q is a Dynkin diagram.*

(ii) *If Q is of tame representation type then T is either Dynkin or Euclidean. (In fact T is Euclidean, but this requires the proof, given later in this section, that the Dynkin diagrams are indeed of finite representation type.)*

PROOF. (i) We shall first give the proof in the case of a quiver (i.e., the case where all the ${}_xd_y^{\gamma}$ are equal to one) over an infinite field, and then indicate what alterations are needed for the general case.

Suppose T is not a Dynkin diagram. Then by Proposition 4.6.3, there is a vector $v = \sum_x n_x v_x \in \mathbb{R}^T$ with the n_x non-negative integers, such that $(v, v) \leq 0$. In other words,

$$2\sum_x n_x^2 \leq \sum_{x-y} n_x n_y.$$

Each edge gets counted both ways round in this sum, so this can be written as

$$\sum_x n_x^2 \leq \sum_{x\to y} n_x n_y.$$

Let V_x be a vector space of dimension n_x. We shall show that there are infinitely many isomorphism classes of representations of T using these particular V_x, and hence with dimension vector (n_x). Such a representation is determined by assigning a linear map $V_x \to V_y$ to each arrow $x \to y$, and two such representations are isomorphic if and only if there are automorphisms in $\prod_x GL(V_x)$ taking one to the other. In other words we are interested in the orbits of $\prod_x GL(V_x)$ on $\prod_{x\to y} \mathrm{Hom}(V_x, V_y)$. The former is an algebraic group of dimension $\sum_x n_x^2$ while the latter is an algebraic variety of dimension $\sum_{x\to y} n_x n_y$. Moreover, the diagonally embedded scalars act trivially, so that we are really looking at the action of $\prod_x GL(V_x)/\text{scalars}$, of dimension $(\sum_x n_x^2) - 1$. According to the above inequality, this algebraic group has strictly smaller dimension than the variety, and so there must be infinitely many orbits and T is not of finite representation type.

If k is not infinite, then we apply the above argument over the algebraic closure $\bar{k}$ of k. Each representation over the algebraic closure is defined over some finite extension of k, and so gives rise to a representation of larger dimension over k by restriction of scalars. Only finitely many representations over $\bar{k}$ give rise to each representation over k this way, and so we have infinitely many representations over k.

If Q is a modulated quiver then we must replace $GL(V_x)$ by $GL_{\Delta_x}(V_x)$ where $n_x = \dim_{\Delta_x}(V_x)$ so that this has dimension $n_x^2 f_x$. We replace the space $\mathrm{Hom}(V_x, V_y)$ by $\mathrm{Hom}_{\Delta_y}({}_yM_x^\gamma \otimes_{\Delta_x} V_x, V_y)$ of k-dimension $n_x n_y\, {}_yd_x^\gamma f_x$. The corresponding inequality is

$$\sum_x n_x^2 f_x \le \sum_{x \xrightarrow{\gamma} y} n_x n_y\, {}_yd_x^\gamma f_x$$

and so the proof proceeds as before.

(ii) If T is neither Dynkin nor Euclidean, then there is a Euclidean diagram strictly smaller than T. Thus by Proposition 4.6.3, there is a non-zero vector v with non-negative integral coördinates with respect to the v_x, and with $(v, v) < 0$. The same argument as in (i) then shows that the space of orbits with this particular dimension vector is at least two dimensional. This is not quite enough to complete the proof, because a direct sum of two one parameter families can be used to give a two parameter family. However, the point is that $(nv, nv) = n^2(v, v)$ is strictly negative, and grows quadratically in size with n. It follows that the spaces of orbits with these dimension vectors are also growing quadratically in dimension with n, whereas finite sums of one parameter families can only account for a number of parameters which grows linearly with n. □

This means that we must now concentrate on the representations of Dynkin diagrams. For this purpose, we introduce the concept of a reflection functor, which is a means of reversing the orientations on some of the edges of a quiver, while almost giving an equivalence of categories.

DEFINITION 4.7.2. A vertex x of a (modulated) quiver is a **sink** (resp. **source**) if all the arrows between x and another vertex point towards (resp. away from) x.

If x is any vertex of Q we define a new (modulated) quiver s_xQ with the same underlying (labelled) graph as Q but with the orientations of the edges meeting x reversed.

An ordering $x_1, \dots, x_n$ of the vertices of Q is **admissible** if for each i, x_i is a sink for $s_{i+1} \dots s_nQ$. There exists an admissible ordering for the vertices of Q if and only if there are no oriented cycles in Q.

Now suppose y is a sink in Q. We define functors

$$S_y^+ : {}_{kQ}\mathbf{mod} \to {}_{k(s_yQ)}\mathbf{mod}, \quad S_y^- : {}_{k(s_yQ)}\mathbf{mod} \to {}_{kQ}\mathbf{mod}$$

as follows. Given a representation $V = \{V_x\}$ of Q, we define a representation $S_y^+(V) = W = \{W_x\}$ of s_yQ by letting $W_x = V_x$ for $x \neq y$, and letting W_y

be the kernel of the sum of the maps going towards V_y:

$$0 \to W_y \to \bigoplus_{x \xrightarrow{\gamma} y \text{ in } Q} {}_yM_x^\gamma \otimes_{\Delta_x} V_x \xrightarrow{\phi} V_y.$$

There are obvious maps

$$W_y \to {}_yM_x^\gamma \otimes_{\Delta_x} V_x$$

and hence

$${}_xM_y^\gamma \otimes_{\Delta_y} W_y \to W_x$$

since ${}_xM_y^\gamma \cong \mathrm{Hom}_{\Delta_y}({}_yM_x^\gamma, \Delta_y)$. Thus the W_y form a representation of s_yQ. A map of representations $V \to V'$ of Q gives rise in an obvious way to a map of representations $S_y^+V \to S_y^+V'$ of s_yQ.

The functor S_y^- is constructed dually. Given a representation $\{W_x\}$ of s_yQ, we let $V_x = W_x$ for $x \neq y$, and we let V_y be the cokernel of the sum of the maps going away from W_y. Namely, each map ${}_xM_y^\gamma \otimes_{\Delta_y} W_y \to W_x$ gives rise as above to a map $W_y \to {}_yM_x^\gamma \otimes_{\Delta_x} W_x$, and we have an exact sequence

$$W_y \xrightarrow{\psi} \bigoplus_{y \xrightarrow{\gamma} x \text{ in } s_yQ} {}_yM_x^\gamma \otimes_{\Delta_x} W_x \to V_y \to 0.$$

If we start off with a representation V of Q for which the map ϕ above is surjective, it is clear that $S_y^-S_y^+(V) \cong V$. Thus S_y^- and S_y^+ give an equivalence between the subcategory of ${}_{kQ}\mathbf{mod}$ for which the map ϕ is surjective, and the subcategory of ${}_{k(s_yQ)}\mathbf{mod}$ for which the map ψ is injective.

Now every representation of Q breaks up as a direct sum of the cokernel of ϕ concentrated at x, and a representation for which ϕ is surjective. Similarly, every representation of s_yQ breaks up as a direct sum of the kernel of ψ concentrated at x, and a representation for which ψ is injective. Thus we have established the following proposition.

PROPOSITION 4.7.3. *The functors S_y^+ and S_y^- establish a bijection between the indecomposable representations of Q and the indecomposable representations of s_yQ, with the exception of the simple module S_x corresponding to x in each case, which is killed by these functors.* □

Now let us examine the effect of these functors on dimension vectors. If V is a representation of Q for which the map ϕ is surjective, then the dimension over Δ_y of W_y is equal to

$$\dim_{\Delta_y} W_y = \sum_{x \xrightarrow{\gamma} y \text{ in } Q} {}_yd_x^\gamma f_x \dim_{\Delta_x} V_x - \dim_{\Delta_y} V_y.$$

The dimension of the remaining W_x is the same as the dimension of V_x. Therefore the effect of S_y^+ on the dimension vector is the same as the effect of applying the reflection w_x.

Conversely if W is a representation of s_yQ for which the map ψ is injective, then the effect of S_y^- on the dimension vector of W is again the same as the effect of applying the reflection w_x.

DEFINITION 4.7.4. Suppose $x_1, \ldots, x_n$ is an admissible ordering for the vertices of a (modulated) quiver Q. Then the **Coxeter functor** with respect to this ordering is the functor

$$C^+ = S_{x_1}^+ \cdots S_{x_n}^+ : {}_{kQ}\mathbf{mod} \to {}_{kQ}\mathbf{mod}.$$

Note that since each arrow gets reversed twice, $s_1 \ldots s_n Q = Q$. We also set

$$C^- = S_{x_n}^- \cdots S_{x_1}^- : {}_{kQ}\mathbf{mod} \to {}_{kQ}\mathbf{mod}.$$

LEMMA 4.7.5. *Given any indecomposable kQ-module V, either*

(i) $C^-C^+(V) \cong V$, *and the effect of C^+ on the dimension vector of V is the same as the effect of the Coxeter transformation $c = w_1 \ldots w_n$, or*

(ii) $C^+(V) = 0$.

PROOF. This is clear from the above discussion. □

We are now ready to classify the indecomposable representations of a Dynkin diagram.

THEOREM 4.7.6 (Gabriel). *Suppose Q is a modulated quiver whose underlying labelled graph T is a Dynkin diagram. Then there is a natural one–one correspondence between the indecomposable representations of Q and the positive roots in $\mathbb{R}^T$, in such a way that each indecomposable is associated to its dimension vector. In particular, Q has finite representation type.*

PROOF. (Bernstein–Gel'fand–Ponomarev [**26**]) Choose an admissible ordering for the vertices of Q (this may be done since every Dynkin diagram is a tree). Let C^+ be the corresponding Coxeter functor on ${}_{kQ}\mathbf{mod}$ and c the corresponding Coxeter transformation on $\mathbb{R}^T$. Suppose V is an indecomposable representation of Q, with dimension vector $v \in \mathbb{R}^T$. By Lemma 4.6.9 (ii), for some $m \geq 1$ the vector $c^m(v)$ is not positive. Thus by Lemma 4.7.5 we have $(C^+)^m(V) = 0$. Choose m as small as possible with $(C^+)^m(V) = 0$. Thus for some i, $S_{i+1}^+ \cdots S_n^+(C^+)^{m-1}(V) \neq 0$ but $S_i^+ S_{i+1}^+ \cdots S_n^+(C^+)^{m-1}(V) = 0$. So by Proposition 4.7.3 we have $S_{i+1}^+ \cdots S_n^+(C^+)^{m-1}(V) \cong S_i$, and

$$V \cong (C^-)^{m-1} S_n^- \cdots S_{i+1}^-(S_i).$$

Thus the dimension vector of V is $c^{-m+1} w_n \ldots w_{i+1}(v_i)$, a positive root. This argument also shows that any indecomposable representation with the same dimension vector as V is isomorphic to V.

Conversely, if v is a positive root then for some $m \geq 1$ the vector $c^m(v)$ is not positive. Choose the shortest expression of the form

$$w_i \ldots w_n (w_1 \ldots w_n)^{m-1}(v)$$

which is not a positive root. Then by Lemma 4.6.8 (iii), $w_{i+1} \ldots w_n c^{m-1}(v) = v_i$ and so the representation $(C^-)^{m-1} S_n^- \cdots S_{i+1}^- S_i$ has dimension vector v. □

COROLLARY 4.7.7. *If Q is a modulated quiver of tame representation type then T is a Euclidean diagram.*

PROOF. This follows from Proposition 4.7.1 and Theorem 4.7.6. □

We shall not prove that if T is a Euclidean diagram then Q is in fact tame. A proof of this can be found in Dlab and Ringel [**78**]. A complete classification of the indecomposables in the case of the Euclidean diagrams of types $\tilde{A}_n$, $\tilde{D}_n$ and $\tilde{E}_n$ may be found in Ringel [**177**].

4.8. Functor categories

In the last section, we classified the indecomposable representations of the Dynkin diagrams by using the geometry of root systems. In the next few sections, we present another method of classification, first formulated explicitly by Gabriel. This is the method of functorial filtrations, and is based on Auslander's work on the structure of functor categories. The idea is as follows. We consider functors from the module category in question to vector spaces. There are obvious notions of subfunctor and quotient functor, and it turns out that the simple functors are in one–one correspondence with the indecomposable modules. It is this observation of Auslander that initiated this circle of ideas. It follows that to find all the indecomposable modules, it suffices to find all the simple functors. We look at any functor which reflects isomorphisms (see Definition 4.10.1), for example the underlying vector space functor, and find its simple composition factors. Of course, in practice this is easier said than done, but we shall give some examples where this method has proved effective. We shall study in detail the case of the group algebras of the dihedral groups (Ringel [**175**]), and mention without proof the corresponding answer for the semidihedral groups (Crawley-Boevey [**62**]). Our exposition is broadly based on Gabriel [**111**] and Ringel [**175**].

DEFINITION 4.8.1. If Λ is a finite dimensional algebra over a field k, we denote by $\mathbf{Fun}(\Lambda)$ (resp. $\mathbf{Fun}^{\circ}(\Lambda)$) the category whose objects are the covariant(resp. contravariant) additive functors ${}_{\Lambda}\mathbf{mod} \to {}_{k}\mathbf{Vec}$ from finitely generated Λ-modules to k-vector spaces, and whose morphisms are the natural transformations of functors.

EXAMPLES. (i) There is the forgetful functor $\rho \in \mathbf{Fun}(\Lambda)$ which assigns to each Λ-module its underlying vector space.

(ii) Duality is a contravariant functor $D : {}_{k}\mathbf{Vec} \to {}_{k}\mathbf{Vec}$. Composing with this duality functor gives contravariant functors also denoted $D : \mathbf{Fun}(\Lambda) \to \mathbf{Fun}^{\circ}(\Lambda)$ and $D : \mathbf{Fun}^{\circ}(\Lambda) \to \mathbf{Fun}(\Lambda)$.

(iii) For any Λ-module M there is a covariant representable functor

$$(M, -) = \mathrm{Hom}_{\Lambda}(M, -) \in \mathbf{Fun}(\Lambda)$$

and a contravariant representable functor

$$(-, M) = \mathrm{Hom}_{\Lambda}(-, M) \in \mathbf{Fun}^{\circ}(\Lambda).$$

There are also the dual functors

$$D(M,-) \in \mathbf{Fun}^{\circ}(\Lambda) \quad \text{and} \quad D(-,M) \in \mathbf{Fun}(\Lambda).$$

It is easy to check directly that $\mathbf{Fun}(\Lambda)$ satisfies the axioms for an abelian category, and we shall see later that it has a natural interpretation as a module category for the **Auslander algebra**.

The kernel of a natural transformation $F_1 \rightsquigarrow F_2$ of functors is the functor assigning to each module M the vector space kernel of $F_1(M) \to F_2(M)$ and to each homomorphism $M \to M'$ the map of kernels making the following diagram commute.

$$\begin{array}{ccccccc} 0 & \longrightarrow & V & \longrightarrow & F_1(M) & \longrightarrow & F_2(M) \\ & & \downarrow & & \downarrow & & \downarrow \\ 0 & \longrightarrow & V' & \longrightarrow & F_1(M') & \longrightarrow & F_2(M') \end{array}$$

Cokernels are constructed dually. A natural transformation $F_1 \rightsquigarrow F_2$ is a monomorphism if and only if $F_1(M) \to F_2(M)$ is injective for each M. In this case we may identify F_1 as a **subfunctor** of F_2. The **quotient functor** F_2/F_1 is defined by $(F_2/F_1)(M) = F_2(M)/F_1(M)$. A **simple functor** is defined to be a non-zero functor with no proper subfunctors. A **finitely generated** functor is one which is isomorphic to a quotient of $(M,-)$ for some M.

Similarly $\mathbf{Fun}^{\circ}(\Lambda)$ is an abelian category and the notions of subfunctor, quotient functor, simple functor and finitely generated functor and defined analogously.

THEOREM 4.8.2 (Auslander). *The finitely generated projective objects in the category* $\mathbf{Fun}(\Lambda)$ *(resp.* $\mathbf{Fun}^{\circ}(\Lambda)$*) are the representable functors* $(M,-)$ *(resp.* $(-,M)$*).*

If M is indecomposable then the functor $(M,-)$ (resp. $(-,M)$) has a unique maximal subfunctor, written $\mathrm{Rad}(M,-)$ *(resp.* $\mathrm{Rad}(-,M)$*), consisting of those homomorphisms which are not split monomorphisms (resp. split epimorphisms). Every simple functor in* $\mathbf{Fun}(\Lambda)$ *(resp.* $\mathbf{Fun}^{\circ}(\Lambda)$*) is of the form* $S_M = (M,-)/\mathrm{Rad}(M,-)$ *(resp.* $S_M = (-,M)/\mathrm{Rad}(-,M)$*) for some indecomposable Λ-module M.*

PROOF. We shall prove these statements in $\mathbf{Fun}(\Lambda)$; the statements in $\mathbf{Fun}^{\circ}(\Lambda)$ are proved dually.

First we shall prove that representable functors are projective. To show that $(M,-)$ is projective, we must show that given natural transformations λ and μ as in the following diagram

$$\begin{array}{ccccc} & & (M,-) & & \\ & \overset{\nu}{\swarrow} & \wr\downarrow \lambda & & \\ F_1 & \overset{\mu}{\rightsquigarrow} & F_2 & \rightsquigarrow & 0 \end{array}$$

we may find ν such that $\mu \circ \nu = \lambda$. But by Yoneda's Lemma 2.1.4,

$$\mathrm{Nat}((M,-),F_i) \cong F_i(M)$$

so that the map

$$\mathrm{Nat}((M,-),F_1) \xrightarrow{\mu_*} \mathrm{Nat}((M,-),F_2)$$

is epi and so we may find ν such that $\mu \circ \nu = \mu_*(\nu) = \lambda$.

Conversely, if F is a finitely generated projective object in $\mathbf{Fun}(\Lambda)$ then F is a quotient of $(M,-)$ for some M. Thus there is a short exact sequence of functors

$$0 \to F' \to (M,-) \to F \to 0.$$

Since F is projective, this sequence splits and $(M,-) \cong F \oplus F'$. So there is an idempotent natural transformation from $(M,-)$ to itself whose image is F. But by Yoneda's Lemma, all natural transformations from $(M,-)$ to itself come from endomorphisms of M, and so $F \cong (M',-)$ for some direct summand M' of M. This completes the determination of the finitely generated projectives in $\mathbf{Fun}(\Lambda)$.

Now if M is indecomposable, every proper subfunctor F of $(M,-)$ is contained in $\mathrm{Rad}(M,-)$, since if a split monomorphism $f : M \to M'$ with splitting $f' : M' \to M$ (so that $f' \circ f = \mathrm{id}_M$) is in $F(M')$, then given any map $f'' : M \to N$ we have

$$f'' = f'' \circ f' \circ f = F(f'' \circ f')(f) \in F(N).$$

Now if S is a simple functor, then choose an indecomposable module M with $S(M) \neq 0$. By Yoneda's Lemma, there is a non-zero natural transformation from $(M,-)$ to S. Since S is simple, this natural transformation is an epimorphism. The kernel is a proper subfunctor of $(M,-)$, and is hence contained in the radical. Since S is simple, the kernel is equal to the radical. □

4.9. The Auslander algebra

DEFINITION 4.9.1. The **Auslander algebra** of a finite dimensional algebra Λ is defined to be

$$\mathrm{Aus}(\Lambda) = \mathrm{End}_\Lambda(\Xi)$$

where $\Xi = \bigoplus_\alpha M_\alpha$ is a direct sum of one Λ-module from each isomorphism class of finitely generated indecomposable Λ-modules.

The following lemma is a direct consequence of the definitions.

LEMMA 4.9.2. *The following are equivalent:*
(i) Λ *has finite representation type.*
(ii) Ξ *is a finitely generated* Λ*-module.*
(iii) $\mathrm{Aus}(\Lambda)$ *is a finite dimensional algebra.* □

Now given a covariant additive functor $F : {}_\Lambda\mathbf{mod} \to {}_k\mathbf{Vec}$ we produce a module for $\mathrm{Aus}(\Lambda)$ by taking the direct sum $\bigoplus_\alpha F(M_\alpha)$ as the underlying vector space, and letting $\mathrm{Aus}(\Lambda)$ act in the obvious way. Namely, an endomorphism of $\bigoplus_\alpha M_\alpha$ is specified by giving a homomorphism $M_\alpha \to M_\beta$ for each pair of indices α, β in such a way that for each α all but finitely many are zero. This gives a homomorphism $F(M_\alpha) \to F(M_\beta)$ for each pair, with the same restriction, and hence an endomorphism of $\bigoplus_\alpha F(M_\alpha)$. Of course, if Λ has finite representation type then this module is just $F(\Xi)$. We shall write $F(\Xi)$ for this $\mathrm{Aus}(\Lambda)$-module even when Λ does not have finite representation type, despite the fact that in this case Ξ is not in ${}_\Lambda\mathbf{mod}$. All we are really doing is extending F in a natural way to the category of Λ-modules which are (possibly infinite) direct sums of finite dimensional ones.

Conversely, if X is an $\mathrm{Aus}(\Lambda)$-module then we define a covariant additive functor $\Phi_X : {}_\Lambda\mathbf{mod} \to {}_k\mathbf{Vec}$ as follows. If M is a finitely generated Λ-module, then $\mathrm{Hom}_\Lambda(M, \Xi)$ is an $\mathrm{Aus}(\Lambda)$-module and we set

$$\Phi_X(M) = \mathrm{Hom}_{\mathrm{Aus}(\Lambda)}(\mathrm{Hom}_\Lambda(M, \Xi), X).$$

LEMMA 4.9.3. *If $F : {}_\Lambda\mathbf{mod} \to {}_k\mathbf{Vec}$ is a covariant additive functor then there is a natural isomorphism*

$$\mathrm{Hom}_{\mathrm{Aus}(\Lambda)}(\mathrm{Hom}_\Lambda(M, \Xi), F(\Xi)) \cong F(M).$$

PROOF. Since both sides are additive in M, we may assume without loss of generality that M is indecomposable. Choose a split surjection $\pi : \Xi \twoheadrightarrow M$, with splitting $i : M \hookrightarrow \Xi$, so that $\pi \circ i = \mathrm{id}_M$. If ϕ is an element of the left hand side of the above equation, then $F(\pi)(\phi(i)) \in F(M)$. Conversely if $x \in F(M)$ then the map taking $\rho : M \to \Xi$ to $F(\rho)(x) \in F(\Xi)$ is an $\mathrm{Aus}(\Lambda)$-module homomorphism. It is easy to check that these processes are mutually inverse. In particular, any other split surjection gives rise to a map with the same inverse, which is therefore the same map. □

PROPOSITION 4.9.4. *There is an equivalence of categories between* $\mathbf{Fun}(\Lambda)$ *and the category of* $\mathrm{Aus}(\Lambda)$*-modules, given in one direction by $F \mapsto F(\Xi)$ and in the other direction by $X \mapsto \Phi_X$.*

The finitely generated functors correspond to $\mathrm{Aus}(\Lambda)$*-modules which are quotients of* $\mathrm{Hom}_\Lambda(M, \Xi)$ *for some finitely generated Λ-module M. Note that this is not the same as the category of finitely generated* $\mathrm{Aus}(\Lambda)$*-modules unless Λ has finite representation type.*

PROOF. If X is an $\mathrm{Aus}(\Lambda)$-module then

$$\Phi_X(\Xi) = \mathrm{Hom}_{\mathrm{Aus}(\Lambda)}(\mathrm{Hom}_\Lambda(\Xi, \Xi), X) = \mathrm{Hom}_{\mathrm{Aus}(\Lambda)}(\mathrm{Aus}(\Lambda), X) \cong X.$$

Conversely if $F \in \mathbf{Fun}(\Lambda)$ then

$$\Phi_{F(\Xi)}(M) = \mathrm{Hom}_{\mathrm{Aus}(\Lambda)}(\mathrm{Hom}_\Lambda(M, \Xi), F(\Xi)) \cong F(M)$$

by the lemma.

The statement about finitely generated functors follows immediately from the definitions. □

COROLLARY 4.9.5. *There is a natural one-one correspondence between simple* $\mathrm{Aus}(\Lambda)$*-modules and finitely generated indecomposable* Λ*-modules.*

PROOF. This follows from Theorem 4.8.2 and Proposition 4.9.4. □

REMARKS. (i) In the case of finite representation type, a similar equivalence exists between $\mathbf{Fun}^{\circ}(\Lambda)$ and $\mathrm{Aus}(\Lambda)^{\mathrm{op}}$-modules, but this breaks down for infinite representation type.

(ii) The simple $\mathrm{Aus}(\Lambda)$-module corresponding to an indecomposable Λ-module M is $\mathrm{Hom}_\Lambda(M, \Xi)$ modulo those homomorphisms which are not split monomorphisms. We shall give this the same name S_M as the corresponding functor. As a vector space, this is just $\mathrm{End}_\Lambda(M)/J\mathrm{End}_\Lambda(M)$. Endomorphisms of M give rise to endomorphisms of S_M, and so we have

$$\mathrm{End}_{\mathrm{Aus}(\Lambda)}(S_M) \cong \mathrm{End}_\Lambda(M)/J\mathrm{End}_\Lambda(M).$$

(iii) We shall see later that in fact the algebra $\mathrm{Aus}(\Lambda)$ has global dimension two. In other words, every $\mathrm{Aus}(\Lambda)$-module has a projective resolution of the form

$$0 \to P_2 \to P_1 \to P_0 \to X \to 0.$$

This is related to the theory of almost split sequences. The Ext-quiver of $\mathrm{Aus}(\Lambda)$ is called the **Auslander–Reiten quiver** of Λ, and this is also intimately connected with the theory of almost split sequences. But more of this later.

4.10. Functorial filtrations

DEFINITION 4.10.1. A functor $F \in \mathbf{Fun}(\Lambda)$ is said to **reflect isomorphisms** if a homomorphism $f : M \to M'$ in ${}_\Lambda\mathbf{mod}$ is an isomorphism if and only if $F(f) : F(M) \to F(M')$ is an isomorphism. For example, the underlying vector space functor reflects isomorphisms. Note in particular that if F reflects isomorphisms and $M \neq 0$ then $F(M) \neq 0$.

LEMMA 4.10.2. *If F is a functor which reflects isomorphisms then F has every simple functor S_M as a subquotient.*

PROOF. If M is a finitely generated indecomposable module, then by Yoneda's lemma

$$\mathrm{Nat}((M, -), F) \cong F(M) \neq 0$$

and so there is a non-zero natural transformation from $(M, -)$ to F. The image of this modulo the image of $\mathrm{Rad}(M, -)$ is the desired subquotient. □

DEFINITION 4.10.3. A collection of subquotients, or **intervals** F'_α/F''_α of a functor $F \in \mathbf{Fun}(\Lambda)$ is said to **cover** F if given any $M \in {}_\Lambda\mathbf{mod}$ and $x \in F(M)$ there is an index α with $x \in F'_\alpha(M)$ but $x \notin F''_\alpha(M)$. Two intervals F'_α/F''_α and F'_β/F''_β are said to **avoid** each other if either $F'_\beta \leq F''_\alpha$ or $F''_\beta \geq F'_\alpha$. A **filtration** of F is a collection of intervals covering F and avoiding each other.

LEMMA 4.10.4. *Suppose F is a functor which reflects isomorphisms and F'_α/F''_α is a collection of intervals covering F. Then $f : M \to M'$ is an isomorphism if and only if $(F'_\alpha/F''_\alpha)(f)$ is an isomorphism for each index α.*

PROOF. Suppose $(F'_\alpha/F''_\alpha)(f)$ is an isomorphism for each α. Since the F'_α/F''_α cover F, it follows that $F(f)$ is an isomorphism. Since F reflects isomorphisms, f is an isomorphism. □

LEMMA 4.10.5. *Suppose $\{M_\alpha\}$ is a complete set of representatives of the isomorphism classes of finitely generated Λ-modules, and $\{S_\alpha\}$ are the corresponding simple functors. Then we have the following:*

(i) $S_\alpha(M_\beta) \cong \begin{cases} \mathrm{End}_\Lambda(M_\beta)/J\mathrm{End}_\Lambda(M_\beta) & \text{if } \alpha = \beta \\ 0 & \text{otherwise.} \end{cases}$

(ii) *A homomorphism $f : M \to M'$ in ${}_\Lambda\mathbf{mod}$ is an isomorphism if and only if $S_\alpha(f)$ is an isomorphism for each α.*

PROOF. (i) This is clear from the discussion earlier in this section.

(ii) If $f : M \to M'$ is not an isomorphism then it is either not injective or not surjective. If it is not injective then some indecomposable summand M_α of M intersects the kernel non-trivially, and then $S_\alpha(f)$ is not an isomorphism. Similarly if f is not surjective then some indecomposable summand M_β of M' is not contained in the image, and then $S_\beta(f)$ is not an isomorphism. □

REMARK. The number of times a simple functor S_M occurs in a filtration of the underlying vector space functor is equal to the dimension of M (over the division ring $\mathrm{End}_\Lambda(M)/J\mathrm{End}_\Lambda(M)$) by part (i) of the above lemma. So when we filter the underlying vector space functor we should expect to have this many repetitions of the simple functors, and it is important to discard all except one copy at some stage so that the list of functors obtained can satisfy (i).

We now present a naïve form of the functorial filtration method, and then we state the method in the generality we need. For convenience we state the following proposition over an algebraically closed field.

PROPOSITION 4.10.6. *Let Λ be a finite dimensional algebra over an algebraically closed field k.*

Suppose $\{M_\alpha\}$ is a collection of finitely generated Λ-modules and $\{S_\alpha\}$ is a collection of functors in $\mathbf{Fun}(\Lambda)$ *such that*

(i) $S_\alpha(M_\beta) \cong \begin{cases} k & \text{if } \alpha = \beta \\ 0 & \text{otherwise.} \end{cases}$

(ii) *$S_\alpha(M)$ is finite dimensional for all finitely generated Λ-modules M, so that $S_\alpha(M) \otimes_k M_\alpha$ makes sense as a finitely generated Λ-module and has the property that $S_\alpha(S_\alpha(M) \otimes M_\alpha) \cong S_\alpha(M)$. Moreover, for every finitely generated Λ-module M there is a map $\gamma_{\alpha,M} : S_\alpha \otimes M_\alpha \to M$ such that $S_\alpha(\gamma_{\alpha,M})$ is an isomorphism.*

(iii) *A map* $f : M \to M'$ *in* ${}_\Lambda\mathbf{mod}$ *is an isomorphism if and only if* $S_\alpha(f)$ *is an isomorphism for all indices* α.

(iv) *For any finitely generated* Λ*-module* M, *only finitely many of the* $S_\alpha(M)$ *are non-zero.*

Then the M_α *are indecomposable and form a complete set of representatives of the isomorphism classes of finitely generated indecomposable* Λ*-modules without repetitions.*

PROOF. Suppose M is a finitely generated Λ-module. By (iv), the sum $\bigoplus_\alpha S_\alpha(M) \otimes M_\alpha$ is a finitely generated Λ-module. By (ii) there is a natural map

$$\sum_\alpha \gamma_{\alpha,M} : \bigoplus_\alpha S_\alpha(M) \otimes M_\alpha \to M$$

and by (i) $S_\beta(\sum_\alpha \gamma_{\alpha,M})$ is an isomorphism for each β. So by (iii) $\sum_\alpha \gamma_{\alpha,M}$ is an isomorphism. Thus every finitely generated Λ-module is uniquely expressible as a direct sum of modules M_α from the given list. In particular, these must be indecomposable and form a complete list of representatives of the isomorphism classes of indecomposable finitely generated Λ-modules without repetition. □

REMARKS. (i) One of the remarkable things about this proposition is that we do not have to demonstrate explicitly that the M_α are indecomposable or non-isomorphic in order to satisfy the hypotheses of the proposition.

(ii) Condition (ii) of the proposition is implied by the existence of a surjective natural transformation $(M_\alpha, -) \rightsquigarrow S_\alpha$ for each α, but is sometimes easier to check. The reason for this implication is as follows. Let $x_\alpha \in S_\alpha(M_\alpha)$ correspond to the given natural transformation via Yoneda's lemma. Since the natural transformation is surjective, given a finitely generated Λ-module M and a basis $\{v_{\alpha,j}\}$ for $S_\alpha(M)$ for each α, we can find homomorphisms $\gamma_{\alpha,j,M} : M_\alpha \to M$ such that $S_\alpha(\gamma_{\alpha,j,M})(x_\alpha) = v_{\alpha,j}$. Then we take $\gamma_{\alpha,M} = \sum_j \gamma_{\alpha,j,M} : S_\alpha(M) \otimes M_\alpha \to M$.

(iii) Condition (iii) is guaranteed if there is a functor F which reflects isomorphisms and a set of intervals covering F and each isomorphic to some S_α.

(iv) If we filter the underlying vector space functor then only finitely many of the intervals can be non-zero on a given finitely generated Λ-module, since the underlying vector space is finite dimensional.

EXAMPLE. The group algebra of a cyclic p-group over an algebraically closed field of characteristic p is of the form $\Lambda = k[T]/(T^n)$, where n is the order of the group and T is of the form $g - 1$ for a generator g. It is easy to classify the modules for this algebra using Jordan canonical forms, but we shall use it as an example to illustrate the functorial filtration method.

As our modules we take $M_i = k[T]/(T^i)$, a uniserial Λ-module of dimension i for $1 \le i \le n$. We have functors

$$F_{ij} = \mathrm{Ker}(T^i) \cap (\mathrm{Im}(T^j) + \mathrm{Ker}(T^{i-1})) \in \mathbf{Fun}(\Lambda).$$

Since $\mathrm{Ker}(T^i) \supseteq \mathrm{Im}(T^j)$ if $i \geq n-j$, we have $F_{i+1,n-i}(M) = \mathrm{Ker}(T^i) = F_{i,0}(M)$, and so we only consider the F_{ij} with $i \leq n-j$. If M is a Λ-module with underlying vector space V, we have the inclusions

$$\begin{aligned} V = F_{n,0}(M) &\supseteq (F_{n,1}(M) =)F_{n-1,0}(M) \supseteq F_{n-1,1}(M) \\ &\supseteq (F_{n-1,2}(M) =)F_{n-2,0}(M) \supseteq \cdots \supseteq F_{1,n-1}(M) \supseteq F_{1,n}(M) = 0. \end{aligned}$$

Moreover, if $n-j \geq i > 0$, the action of T induces an isomorphism from $F_{i,j-1}/F_{i,j}$ to $F_{i-1,j}/F_{i-1,j+1}$. Thus if we take as our functors $S_i = F_{i,0}/F_{i,1} \cong \cdots \cong F_{1,i-1}/F_{1,i}$, then S_i appears as an interval in our filtration exactly i times. Since $(M_i, -) = \mathrm{Ker}(T^i) = F_{i,0}$, each S_i is a quotient of $(M_i, -)$ and so condition (ii) is satisfied according to remark (ii). The other conditions are easily checked using the other remarks. So we may conclude that the M_i form a complete list of representatives of the isomorphism classes of indecomposable Λ-modules, without repetitions.

The following is the form of the functorial filtration theorem used by Ringel [**175**] for his classification of the indecomposable modules for the dihedral 2-groups. The point of this version is that it allows for the classification of entire one parameter families of modules at a time. We shall give an outline of Ringel's classification in the next section.

THEOREM 4.10.7. *Let* $\mathbf{M}$ *and* $\mathbf{A}_\alpha$ *be abelian categories, and let* $S_\alpha : \mathbf{M} \to \mathbf{A}_\alpha$ *and* $T_\alpha : \mathbf{A}_\alpha \to \mathbf{M}$ *be additive functors such that the following conditions are satisfied.*

(i) $S_\alpha T_\beta = \begin{cases} \mathrm{id}_{\mathbf{A}_\alpha} & \text{if } \alpha = \beta \\ 0 & \text{otherwise.} \end{cases}$

(ii) *For every* M *in* $\mathbf{M}$ *there is a map* $\gamma_{\alpha,M} : T_\alpha S_\alpha(M) \to M$ *such that* $S_\alpha(\gamma_{\alpha,M})$ *is an isomorphism.*

(iii) *A map* $f : M \to M'$ *in* $\mathbf{M}$ *is an isomorphism if and only if* $S_\alpha(f)$ *is an isomorphism for all indices* α.

(iv) *For any* M *in* $\mathbf{M}$, *only finitely many of the* $S_\alpha(M)$ *are non-zero.*

Then the objects $T_\alpha(A)$ *with* A *indecomposable in* $\mathbf{A}_\alpha$, *are indecomposable and form a complete set of representatives of the isomorphism classes of indecomposable objects in* $\mathbf{M}$ *without repetitions.*

PROOF. The proof is exactly the same as the proof of the above proposition. □

Note that we are interested in the case where $\mathbf{M}$ is the category of finite dimensional Λ-modules for some algebra Λ. The $\mathbf{A}_\alpha$ are things like the category of finite dimensional vector spaces or the category of finite dimensional $k[T]$-modules. The latter will be used to capture entire one parameter families of modules at once.

4.11. Representations of dihedral groups

As an example of the method of functorial filtrations, in this section, we present Ringel's classification [**175**] of the finite dimensional modules for the algebra

$$\Lambda = k\langle X, Y\rangle/(X^2, Y^2),$$

namely the quotient of the free algebra on the two non-commuting variables X and Y by the ideal generated by X^2 and Y^2, over any field k.

The relationship with the finite dihedral 2-groups is as follows. If

$$G = \langle x, y : x^2 = y^2 = 1\rangle$$

is the infinite dihedral group and k is a field of characteristic 2 then $\Lambda \cong kG$ via $X \leftrightarrow x - 1$, $Y \leftrightarrow y - 1$. The finite dihedral 2-groups are the quotients

$$D_{4q} = \langle x, y : x^2 = y^2 = 1, (xy)^q = (yx)^q\rangle$$

(q a power of 2).

LEMMA 4.11.1. *For q a power of 2 we have*

$$kD_{4q} \cong \Lambda/((XY)^q - (YX)^q).$$

PROOF. In $\Lambda \cong kG$ we have

$$\begin{aligned}(xy)^q - (yx)^q &= (xy - yx)^q = ((x-1)(y-1) - (y-1)(x-1))^q \\ &= ((x-1)(y-1))^q - ((y-1)(x-1))^q = (XY)^q - (YX)^q.\end{aligned}$$

□

THE MODULES. Let $\mathcal{W}$ be the set of words in the **direct letters** a and b, and the **inverse letters** a^{-1} and b^{-1}, such that a and a^{-1} are always followed by b or b^{-1} and vice versa, together with the "zero length words" 1_a and 1_b (which are regarded as "beginning" with a and b respectively). If C is a word, we define C^{-1} as follows. $(1_a)^{-1} = 1_b$, $(1_b)^{-1} = 1_a$; and otherwise we reverse the order of the letters in the word and invert each letter according to the rule $(a^{-1})^{-1} = a$, $(b^{-1})^{-1} = b$. Let $\mathcal{W}_1$ be the set obtained from $\mathcal{W}$ by identifying each word with its inverse.

The nth power of a word of even length is obtained by juxtaposing n copies of the word. Let $\mathcal{W}'$ be the subset of $\mathcal{W}$ consisting of all words of even non-zero length which are not powers of smaller words. Let $\mathcal{V}$ denote the set of isomorphism classes of pairs (V, ϕ) where V is a finite dimensional vector space over k and ϕ is an indecomposable automorphism of V (ϕ is indecomposable if its rational canonical form has only one block, and that block is associated with a power of an irreducible polynomial over k). Let $\tilde{\mathcal{W}}_2$ be the set obtained from $\mathcal{W}' \times \mathcal{V}$ by identifying each word in $\mathcal{W}'$ with its images under cyclic permutations

$$\ell_1 \dots \ell_n \to \ell_n \ell_1 \dots \ell_{n-1},$$

and by identifying $(C, (V, \phi))$ with $(C^{-1}, (V, \phi^{-1}))$. Let $\mathcal{W}_2$ be the set obtained from $\mathcal{W}'$ by identifying a word with its cyclic permutations and its inverse.

The following is a list of all the isomorphism types of finite dimensional indecomposable Λ-modules. The rest of the section will then be devoted to proving this statement.

MODULES OF THE FIRST KIND. These are also called **string modules**. They are in one–one correspondence with the elements of $\mathcal{W}_1$. Let $C = \ell_1 \dots \ell_n \in \mathcal{W}$. Let $M(C)$ be a vector space over k with basis $z_0, \dots, z_n$ on which Λ acts according to the schema

$$kz_0 \overset{\ell_1}{\longleftarrow} kz_1 \overset{\ell_2}{\longleftarrow} kz_2 \cdots kz_{n-1} \overset{\ell_n}{\longleftarrow} kz_n$$

where X acts via a and Y acts via b. For example, if $C = ab^{-1}aba^{-1}$ then the schema is

$$kz_0 \overset{a}{\longleftarrow} kz_1 \overset{b}{\longrightarrow} kz_2 \overset{a}{\longleftarrow} kz_3 \overset{b}{\longleftarrow} kz_4 \overset{a}{\longrightarrow} kz_5$$

and the module is given by

$$X \mapsto \begin{pmatrix} 0&0&0&0&0&0\\ 1&0&0&0&0&0\\ 0&0&0&0&0&0\\ 0&0&1&0&0&0\\ 0&0&0&0&0&1\\ 0&0&0&0&0&0 \end{pmatrix} \qquad Y \mapsto \begin{pmatrix} 0&0&0&0&0&0\\ 0&0&1&0&0&0\\ 0&0&0&0&0&0\\ 0&0&0&0&0&0\\ 0&0&0&1&0&0\\ 0&0&0&0&0&0 \end{pmatrix}.$$

It is clear that $M(C) \cong M(C^{-1})$.

MODULES OF THE SECOND KIND. These are also called **band modules**. They are in one–one correspondence with the elements of $\tilde{\mathcal{W}}_2$. Let ϕ be an indecomposable automorphism of a finite dimensional k-vector space V, and let $C = \ell_1 \dots \ell_n$ be a word in $\mathcal{W}'$. Let $M(C, \phi)$ be given as a vector space by $M(C, \phi) = \bigoplus_{i=0}^{n-1} V_i$ with $V_i \cong V$, on which Λ acts according to the schema

$$V_0 \xleftarrow{\ell_1=\phi} V_1 \xleftarrow{\ell_2=\mathrm{id}} V_2 \longleftarrow \cdots \longleftarrow V_{n-2} \xleftarrow{\ell_{n-1}=\mathrm{id}} V_{n-1}$$

with an arrow $\ell_n=\mathrm{id}$ from V_0 to V_{n-1},

where again X acts as a and Y acts as b in the same sense as above. It is clear that if $(C, (V, \phi))$ and $(C', (V', \phi'))$ represent the same element of $\tilde{\mathcal{W}}_2$ then $M(C, \phi) \cong M(C', \phi')$.

One of the above modules has $(XY)^q - (YX)^q$ in its kernel if and only if one of the following holds:

(i) The module is of the first kind and the corresponding word does not contain $(ab)^q$, $(ba)^q$ or their inverses.

(ii) The module is of the second kind and no power of the corresponding word contains $(ab)^q$, $(ba)^q$ or their inverses.

(iii) The module is $M((ab)^q(ba)^{-q}, \mathrm{id}_k)$ (of the second kind). This is the projective indecomposable module for the quotient algebra $\Lambda/((XY)^q - (YX)^q)$.

THE FUNCTORS. As usual, we provide a filtration of the underlying vector space functor. Each isomorphism type will appear more than once, and so we then give lemmas which remove the repetitions. The final index set for the categories $\mathbf{A}_\alpha$ is $\mathcal{W}_1 \cup \mathcal{W}_2$.

Denote by $\mathbf{M}$ the category of finite dimensional modules for

$$\Lambda = \langle X, Y \rangle / (X^2, Y^2).$$

If C, D and $C^{-1}D$ are words in $\mathcal{W}$, we set $\mathbf{A}_{C,D} = {}_k\mathbf{mod}$, the category of finite dimensional vector spaces. The functor $T_{C,D} : \mathbf{A}_{C,D} \to \mathbf{M}$ is defined via $T_{C,D}(k) = M(C^{-1}D)$. This depends up to isomorphism only on the equivalence class of $C^{-1}D$ in $\mathcal{W}_1$.

If C is a word in $\mathcal{W}'$, we let $\mathbf{A}_C$ be the subcategory of ${}_{k[T,T^{-1}]}\mathbf{mod}$ consisting of finite dimensional $k[T,T^{-1}]$-modules. A finite dimensional $k[T,T^{-1}]$-module is the same as a vector space V together with an automorphism ϕ. The functor $T_C : \mathbf{A}_C \to \mathbf{M}$ is defined via $T_C(V,\phi) = M(C,\phi)$. This depends up to isomorphism only on the equivalence class of C in $\mathcal{W}_2$.

We now describe the functors $S_{C,D} : \mathbf{M} \to \mathbf{A}_{C,D}$ and $S_C : \mathbf{M} \to \mathbf{A}_C$ as intervals (i.e., subquotients) of the underlying vector space functor. If C is a word in $\mathcal{W}$, then there is a unique direct letter d such that Cd is again a word. If $M \in \mathbf{M}$, we define $C^-(M) = CdM$ and $C^+(M) = Cd^{-1}0_M$, where the letters a and b stand for the actions of X and Y on M, so that words are interpreted as the corresponding linear relations on M. Thus for example if $C = a^{-1}b$ then $C^-(M)$ is the subspace of elements of M whose images under X are expressible as $YX(m)$ for some $m \in M$, and $C^+(M)$ is the subspace of elements of M whose image under X are expressible as $Y(m)$ for some $m \in M$ with $X(m) = 0$. We define

$$\begin{aligned} S_{C,D} &= ((D^+ + C^-) \cap C^+)/((D^- + C^-) \cap C^+) \\ &\cong (C^+ \cap D^+)/((C^+ \cap D^-) + (C^- \cap D^+)). \end{aligned}$$

Note that if C, D and CdD are words, with d a direct letter, then

$$CdD^- \leq CdD^+ \leq C^- \leq C^+ \leq Cd^{-1}D^- \leq Cd^{-1}D^+.$$

LEMMA 4.11.2. *The isomorphism type of the functor $S_{C,D}$ only depends on the equivalence class of the word $C^{-1}D$ in $\mathcal{W}_1$.*

PROOF. The above isomorphism shows that $S_{C,D} \cong S_{D,C}$, and so it suffices to show that if d is a direct letter and C and D are words with $C^{-1}dD$ a word in $\mathcal{W}$ then $S_{C,dD} \cong S_{d^{-1}C,D}$. But this follows from the fact that multiplication by the corresponding element X or Y in Λ defines a natural isomorphism

$$\frac{((d^{-1}D^+ + C^-) \cap C^+)(M)}{((d^{-1}D^- + C^-) \cap C^+)(M)} \to \frac{((D^+ + dC^-) \cap dC^+)(M)}{((D^- + dC^-) \cap dC^+)(M)}. \qquad \square$$

If C is a word in $\mathcal{W}'$ and $M \in \mathbf{M}$, we set $C'(M) = \bigcup_n C^n 0_M$ and $C''(M) = \bigcap_n C^n M$. Then $C'(M) \leq C''(M)$, and C determines a vector space automorphism $\phi_{C,M}$ of $C''(M)/C'(M)$. Set

$$S_C(M) = (C''(M)/C'(M), \phi_{C,M}) \in \mathbf{A}_C.$$

LEMMA 4.11.3. *Suppose $C = \ell_1 \dots \ell_n$ is a word in $\mathcal{W}'$.*
(i) *Let $C_{(i)} = \ell_{i+1} \dots \ell_n \ell_1 \dots \ell_i$. Then $S_C \cong S_{C_{(i)}}$.*
(ii) *Suppose $S_C(M) = (V, \phi)$. Then $S_{C^{-1}}(M) = (V, \phi^{-1})$.*
Thus the isomorphism type of the functor S_C only depends on the equivalence class of C in $\mathcal{W}_2$, as long as we allow ourselves to apply the automorphism $T \leftrightarrow T^{-1}$ of $\mathbf{A}_C$ if necessary.

PROOF. (i) Let $V_i = C''_{(i)}(M)/C'_{(i)}(M)$. If ℓ_i is a direct letter then multiplication by the corresponding element X or Y of Λ induces an epimorphism $V_i \to V_{i-1}$. If ℓ_i is an inverse letter then multiplication by the corresponding element of Λ induces a monomorphism $V_{i-1} \to V_i$. Thus we have

$$\dim V_0 \leq \dim V_1 \leq \cdots \leq \dim V_{n-1} \leq \dim V_0$$

and so all these maps are isomorphisms. The map $\phi_{C,M}$ is the composite of these maps, and up to conjugacy only depends on the cyclic ordering.

(ii) $S_C(M)$ is the largest interval of the underlying vector space of M on which the action of the word C is invertible, in the sense that any such interval is an interval of $S_C(M)$. Thus $S_{C^{-1}}(M) \cong S_C(M)$ as vector spaces. The action of C^{-1} on this interval is inverse to the action of C. □

We now show that our intervals form a filtration of the underlying vector space functor. We start with a lemma.

LEMMA 4.11.4. *Suppose $d : V \to V$ is an endomorphism with $d^2 = 0$. If $U_1 \leq U_2$ are subspaces of V then*

$$\dim U_2/U_1 \geq \dim dU_2/dU_1 + \dim d^{-1}U_2/d^{-1}U_1.$$

PROOF. The action of d defines isomorphisms $d^{-1}U_2/d^{-1}U_1 \cong ((U_2 \cap \mathrm{Im}(d)) + U_1)/U_1$ and $U_2/(U_2 \cap (\mathrm{Ker}(d) + U_1)) \cong dU_2/dU_1$, so that the inequality follows from the inclusions

$$U_1 \leq (U_2 \cap \mathrm{Im}(d)) + U_1 \leq U_2 \cap (\mathrm{Ker}(d) + U_1) \leq U_2. \qquad \square$$

PROPOSITION 4.11.5. *The intervals $S_{C,D}$ for which C starts with $a^{\pm 1}$ and D starts with $b^{\pm 1}$, together with the intervals S_C for which C starts with a, form a filtration of the underlying vector space functor $\mathbf{M} \to {}_k\mathbf{mod}$.*

PROOF. For the purpose of proving this, we introduce the set of infinite words $\ell_1 \ell_2 \dots$ where the letters are $a^{\pm 1}$ and $b^{\pm 1}$, and the restrictions are the same as for the words in $\mathcal{W}$. In other words, if $A = \ell_1 \ell_2 \dots$ is an infinite word then every initial segment $A_{[n]} = \ell_1 \ell_2 \dots \ell_n$ is in $\mathcal{W}$. If C is a word of length m in $\mathcal{W}'$, we write C^∞ for the infinite word with $(C^\infty)_{[mn]} = C^n$. We call such infinite words **periodic**.

First, we claim that if x is an element of $M \in \mathbf{M}$ then either there is a finite word C starting with $a^{\pm 1}$ such that $x \in C^+$ but $x \notin C^-$, or there is an infinite word A starting with $a^{\pm 1}$ such that $x \in A''(M) = \bigcap_n A_{[n]}(M)$ but $x \notin A'(M) = \bigcup_n A_{[n]} 0_M$. To prove this, we totally order the (finite and infinite) words starting with $a^{\pm 1}$ as follows. We write $C \leq D$ if there is a direct letter d, a finite word C_1, and words E_1 and E_2 such that $C = C_1$ or C_1dE_1, and $D = C_1$ or $C_1d^{-1}E_2$. Note that if $C \leq D$ then $C(M) \leq D(M)$ for all M. Let $A_{[n]}$ be the smallest word with respect to this ordering of length n such that $x \in A_{[n]}(M)$. Thus if $x \in A_{[n]}d(M)$ then $A_{[n+1]} = A_{[n]}d$, and otherwise $A_{[n+1]} = A_{[n]}d^{-1}$. This way we construct an infinite word A with $x \in A''(M)$. If $x \notin A'(M)$ we are done. If $x \in A'(M)$ then $x \in A_{[n+1]}(0_M)$ for some n. Choose n minimal with this property. Then the last letter of $A_{[n+1]}$ has to be inverse, and we have $x \in A^+_{[n]}(M)$. By construction of A, $x \notin A^-_{[n]}(M)$, and we are done. Note that the total order also implies that these intervals C^+/C^- and A''/A' all avoid each other.

Next, we claim that if A is an infinite word with $A' \neq A''$, then A is periodic. For suppose M is a module with $A'(M) \neq A''(M)$. If we chop any initial segment off from A, we obtain another infinite word B with $B'(M) \neq B''(M)$. If $B_1 \neq B_2$ are different infinite words obtained this way, then the intervals $B_1''(M)/B_1'(M)$ and $B_2''(M)/B_2'(M)$ avoid each other in the sense that either $B_2'(M) \geq B_1''(M)$ or $B_2''(M) \leq B_1'(M)$. Since M is finite dimensional, this implies that there are only finitely many different words obtainable from A in this way. This shows that $A = A_{[n]}C^\infty$ for some value of n and some C in $\mathcal{W}'$. Suppose n is minimal with this property. We wish to show that $n = 0$. If not, then the last letter of $A_{[n]}$ is inverse (say d^{-1}) and the last letter of C is d. So we have $(d^{-1}C^\infty)'(M) \neq (d^{-1}C^\infty)''(M)$ and $(dC^\infty)'(M) \neq (dC^\infty)''(M)$. But by the Lemma, for any k we have

$$\dim C^kM/C^k0_M \geq \dim dC^kM/dC^k0_M + \dim d^{-1}C^kM/d^{-1}C^k0_M,$$

while for k large we have

$$\dim C^kM/C^k0_M = \dim dC^kM/dC^k0_M.$$

This proves the assertion.

Finally, we claim that the quotient C^+/C^- is covered by the intervals $((D^+ + C^-) \cap C^+)/((D^- + C^-) \cap C^+)$ for words D in $\mathcal{W}$ starting with b. Applying what we know so far (replacing words starting with a by those starting with b), it is certainly covered by these together with the intervals $((D''+C^-)\cap C^+)/((D'+C^-)\cap C^+)$ for words D in $\mathcal{W}'$ starting with b. So we must show that $(D'+C^-)\cap C^+ = (D''+C^-)\cap C^+$. Now by Proposition 4.3.5 and its corollary, the regular part of the relation D splits off, and if U is this regular part then $D' \oplus U = D''$ and $(D^{-1})' \oplus U = (D^{-1})''$. Since D^{-1} starts with a, either $C^+ \leq (D^{-1})'$ or $C^- \geq (D^{-1})''$. In either case we have the desired equality.

It is clear from the total order that the given intervals avoid each other. □

Our next task is to evaluate the functors on the modules.

PROPOSITION 4.11.6. (i) *If C, D and $C^{-1}D$ are words in $\mathcal{W}$ then the following hold.*

(a) *If $M(E)$ is a module of the first kind, then* $S_{C,D}(M(E)) = kz_i$ *if* $E = C^{-1}D$, *where* i *is the length of* C, *and* $S_{C,D}(M(E)) = 0$ *if* $C^{-1}D \neq E \neq D^{-1}C$.

(b) *If $M(E, \phi)$ is a module of the second kind then* $S_{C,D}(M(E, \phi)) = 0$.

(ii) *If C is a word in $\mathcal{W}'$ then the following hold.*

(a) *If $M(E)$ is a module of the first kind then* $S_C(M(E)) = 0$.

(b) *If $M(E, \phi)$ is a module of the second kind, where ϕ is an automorphism of a vector space V, then* $S_C(M(E, \phi)) = (V, \phi)$ *if E is of the form* $C_{(i)}$ *or* $C_{(i)}^{-1}$ *for some value of* i.

PROOF. (i) If $C = 1_a$ and M is a module of the first or second kind, it is easy to see that $C^+(M) = a^{-1}0_M$ is the sum of $C^-(M) = aM$ and those spaces occurring at the end of a schema for which the end letter is $b^{\pm 1}$. This is because $\mathrm{Ker}(a) = \mathrm{Im}(a)$ on the rest of the schema. By induction, it is now easy to see that for any C in $\mathcal{W}$, C^+ is equal to C^- on a module of the second kind. If $M = M(E)$ is a module of the first kind with E a word of length n, then $C^+(M)$ is the sum of $C^-(M)$ and those spaces kz_i for which either C^{-1} is an initial segment of E of length i or C is a final segment of E of length $n - i$. Thus if $(D^+ + C^-) \cap C^+ \neq (D^- + C^-) \cap C^+$ then either $M = M(C^{-1}D)$ or $M = M(D^{-1}C)$, and these are isomorphic modules. If $M = M(C^{-1}D)$ then the space kz_i forms a complement to $(D^- + C^-) \cap C^+$ in $(D^+ + C^-) \cap C^+$.

(ii) Without loss of generality C is a word of length n beginning with $a^{\pm 1}$. Since C is not a power of a smaller word, and is of even length, the words

$$C_{(0)}, \dots, C_{(n-1)}, C_{(0)}^{-1}, \dots, C_{(n-1)}^{-1}$$

are all distinct, while by Lemma 4.11.3 they give rise to isomorphic functors. Recall from the proof of the last proposition that we have a total order on the words of length n starting with $a^{\pm 1}$. Thus we may replace C by a rotation of C or C^{-1} in such a way that C is strictly smaller than any rotation of C or C^{-1} other than itself. We now assume this has been done.

Let $M = M(C, \phi) = \bigoplus_{i=0}^{n-1} V_i$, $V_i \cong V$. It is easy to see by induction on length that if E is any word of length at most n starting with $a^{\pm 1}$, then $E0_M$ is the sum of those V_i for which $C_{(i)}$ or $(C_{(i)})^{-1}$, whichever starts with $a^{\pm 1}$, is $> E$ in the total order. In particular, taking $E = C$, we see by the choice made above that $C0_M = 0_M$. Thus $C'(M) = 0$. On the other hand, $V_0 \leq C''(M)$, and so V_0 embeds into $(C''/C')(M)$.

Thus the n different words among the rotations of C and C^{-1} starting with $a^{\pm 1}$ define subquotients of M each of dimension at least $\dim V$. These subquotients avoid each other by the previous proposition, and so the dimension must equal $\dim V$ and all other S_E must take zero value. Thus $C''(M) = V_0$, and the automorphism induced by C on V_0 is of course just ϕ.

In a similar fashion, the intervals $S_{C,D}$ already exhaust the dimension of $M(C^{-1}D)$ so that the S_C must take zero value on these too. □

THE MAPS $\gamma_{\alpha,M}$. In order to invoke Theorem 4.10.7, it remains to construct maps $\gamma_{\alpha,M}$ for each $\alpha \in \mathcal{W}_1 \cup \mathcal{W}_2$ as in part (ii) of the Theorem.

PROPOSITION 4.11.7. *For every module $M \in \mathbf{M}$ and every pair of words C, $D \in \mathcal{W}$ such that $C^{-1}D$ is also a word, there is a map*

$$\gamma_{C,D,M} : T_{C,D}S_{C,D}(M) \to M$$

such that $S_{C,D}(\gamma_{C,D,M})$ is an isomorphism.

PROOF. For any element $x \in (C^+ \cap D^+)(M)$, there is a map

$$\gamma : M(C^{-1}D) \to M$$

such that $\gamma(z_i) = x$, where i is the length of C. Choosing a basis for a complement of $((C^+ \cap D^-) + (C^- \cap D^+))(M)$ in $(C^+ \cap D^+)(M)$, we obtain a map from a direct sum of $\dim S_{C,D}(M)$ copies of $M(C^{-1}D)$ to M such that the sum of the basis vectors corresponding to z_i map to the chosen basis for the complement. This is the required map $\gamma_{C,D,M}$. □

PROPOSITION 4.11.8. *For every module $M \in \mathbf{M}$ and every $C \in \mathcal{W}'$ there is a map $\gamma_{C,M} : T_C S_C(M) \to M$ such that $S_C(\gamma_{C,M})$ is an isomorphism.*

PROOF. We consider C as a linear relation on M. By Proposition 4.3.5, the regular part of C splits off, so that there is a subspace U of M such that $C'(M) \oplus U = C''(M)$, and C induces an automorphism ϕ on U via $Cx \cap U = \{\phi(x)\}$ for $x \in U$. Thus if we choose a basis x_i of U, and $C = \ell_1 \dots \ell_n$, we can choose elements $x_i^{(k)} \in M$ with $x_i^{(n)} = x_i$, $x_i^{(k-1)} \in \ell_k(x_i^{(k)})$ and $x_i^{(0)} = \phi(x_i)$. This then defines a map from $T_C S_C(M) = T_C(U, \phi) \to M$ taking the basis element x_i of $U_k = U$ to $x_i^{(k)}$. This is the required map $\gamma_{C,M}$. □

REMARK. Using similar but more complicated techniques, Crawley-Boevey [**62**] has classified the indecomposable representations of the semidihedral groups $SD_{2^{m+1}}$ of order 2^{m+1} $(m \geq 2)$

$$G = \langle x, y \mid x^{2^m} = y^2 = 1,\ y^{-1}xy = x^{2^{m-1}-1} \rangle.$$

in characteristic two. In this case, the appropriate algebra is

$$kG/\mathrm{Soc}(kG) \cong k\langle X, Y \rangle / (X^3, Y^2, X^2 - (YX)^n Y)$$

with $n = 2^{m-1} - 1$. The classification is similar to the dihedral case, but with added complications coming from symmetry conditions on the words involved.

It is interesting to note that the subalgebra of the Steenrod algebra generated by Sq^1 and Sq^2 (see Chapter 4 of Volume II) is the case $n = 1$ of the above algebra, corresponding to the non-existent "semidihedral group of order eight". This explains why the cohomology of this subalgebra (which is the $E_2 = E_\infty$ page of the Adams spectral sequence converging to the real Bott periodicity groups) is the same as the cohomology of the semidihedral groups.

The classification of the indecomposable modules for the generalised quaternion groups ought to be similar in nature to the case of the semidihedral groups, but no-one has succeeded in completing this case.

4.12. Almost split sequences

Almost split sequences were introduced by Auslander and Reiten [**10**]. They proved their existence for Artin algebras (a ring is an **Artin algebra** if it is finitely generated as a module over its centre, and its centre is an Artinian ring). We shall content ourselves with the case of a finite dimensional algebra, even though the proof for a general Artin algebra is not much harder. Finally we shall show how almost split sequences are related to functor categories.

DEFINITION 4.12.1. Let Λ be a ring. A short exact sequence of finitely generated Λ-modules

$$0 \to M \to E \xrightarrow{\sigma} N \to 0$$

is called an **almost split sequence** or **Auslander–Reiten sequence** if the following conditions are satisfied:

(i) M and N are indecomposable.

(ii) The map σ does not split.

(iii) Given any Λ-module N' and map $\rho : N' \to N$ which is not a split epimorphism then ρ factors through σ as in the following diagram:

$$\begin{array}{ccccccccc} & & & & & & N' & & \\ & & & & & \swarrow & \downarrow{\scriptstyle \rho} & & \\ 0 & \longrightarrow & M & \longrightarrow & E & \xrightarrow{\sigma} & N & \longrightarrow & 0 \end{array}$$

Note that if ρ is a split epimorphism, that is, N is a summand of N' and ρ is the projection, then by condition (ii) ρ cannot factor through σ.

The extraordinary thing about this definition is that such sequences should exist at all. Clearly if N is projective, there can be no almost split sequence terminating in N.

THEOREM 4.12.2 (Auslander, Reiten [**10**]). *Suppose Λ is an Artin algebra. Given any finitely generated indecomposable non-projective module N, there exists an almost split sequence terminating in N. This sequence is unique up to isomorphism of short exact sequences.*

We shall only prove this theorem for finite dimensional algebras. We first prove uniqueness, since this is a general argument and is quite easy.

LEMMA 4.12.3. *Suppose*

$$0 \to M \to E \xrightarrow{\sigma} N \to 0$$

$$0 \to M' \to E' \xrightarrow{\sigma'} N \to 0$$

are almost split sequences terminating in the same module N. Then they are isomorphic as short exact sequences.

PROOF. The definition guarantees us a commutative diagram:

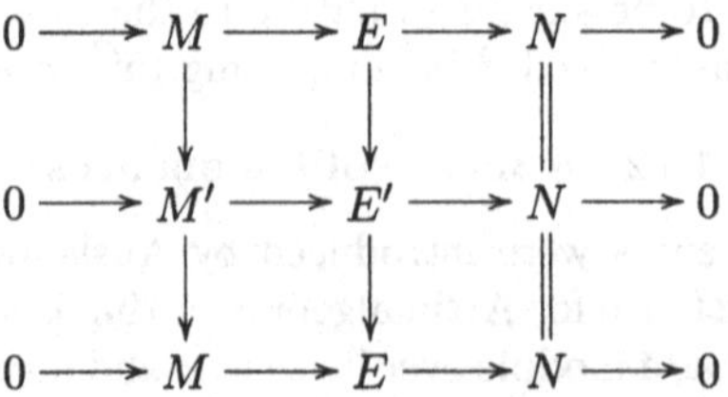

The composite map $M \to M' \to M$ is not nilpotent since otherwise the map σ would split. Thus by Fitting's Lemma 1.4.4 it is an isomorphism, and so by the five lemma the sequences are isomorphic. □

The idea of the proof of existence of almost split sequences is as follows. Given an indecomposable module N, let us suppose that we can find an indecomposable module M depending functorially on N, and such that there is a natural duality between $\mathrm{Hom}_\Lambda(N, -)$ and $\mathrm{Ext}^1_\Lambda(-, M)$. Well, we should modify this slightly since Ext does not see homomorphisms which factor through a projective module. So we work in the stable module category ${}_\Lambda\underline{\mathbf{mod}}$ (see Section 2.1), and let us suppose that $\underline{\mathrm{Hom}}_\Lambda(N, -)$ and $\mathrm{Ext}^1_\Lambda(-, M)$ are dual. Since for N non-projective $\underline{\mathrm{End}}_\Lambda(N)$ is a local ring and $\mathrm{Ext}^1_\Lambda(N, M)$ is its dual, the latter has a simple socle as an $\underline{\mathrm{End}}_\Lambda(N)$-module. We claim that any extension $0 \to M \to E \xrightarrow{\sigma} N \to 0$ representing a non-zero element of this socle is an almost split sequence. This is because

$$\begin{aligned}
&\rho : N' \to N \text{ is split epi}\\
&\Leftrightarrow \rho_* : \underline{\mathrm{Hom}}_\Lambda(N, N') \to \underline{\mathrm{End}}_\Lambda(N) \text{ has } \mathrm{id}_N \text{ in its image}\\
&\Leftrightarrow \rho_* : \underline{\mathrm{Hom}}_\Lambda(N, N') \to \underline{\mathrm{End}}_\Lambda(N) \text{ is surjective}\\
&\Leftrightarrow \rho^* : \mathrm{Ext}^1_\Lambda(N, M) \to \mathrm{Ext}^1_\Lambda(N', M) \text{ is injective}\\
&\Leftrightarrow \rho^* \text{ does not kill } \mathrm{Soc}\mathrm{Ext}^1_\Lambda(N, M)\\
&\Leftrightarrow \rho \text{ does not factor through } \sigma.
\end{aligned}$$

The required functor taking N to M is called the **Auslander–Reiten translation** DTr, and is the composite of two contravariant functors Tr and D.

The first clue as to how to obtain the required functor is the identity

$$\mathrm{Ext}^1_\Lambda(N, \mathrm{Hom}_k(N', k)) \cong \mathrm{Hom}_k(\mathrm{Tor}^\Lambda_1(N', N), k)$$

from Proposition 2.8.5. So the functor D is simply the duality functor. If M is a finitely generated module then we set

$$D(M) = M^* = \mathrm{Hom}_k(M, k) \cong \mathrm{Hom}_\Lambda(M, \mathrm{Hom}_k(\Lambda, k)) = \mathrm{Hom}_\Lambda(M, \Lambda^*),$$

as a finitely generated Λ^{op}-module. Here, Λ^* is a Λ-Λ-bimodule in the obvious way. We also write D for the corresponding functor from finitely generated Λ^{op}-modules to finitely generated Λ-modules, so that $D^2 : {}_\Lambda\mathbf{mod} \to {}_\Lambda\mathbf{mod}$ is isomorphic to the identity functor. D passes down to these quotients to give dualities $D : {}_\Lambda\underline{\mathbf{mod}} \to {}_{\Lambda^{\mathrm{op}}}\overline{\mathbf{mod}}$ and $D : {}_\Lambda\overline{\mathbf{mod}} \to {}_{\Lambda^{\mathrm{op}}}\underline{\mathbf{mod}}$. It should be pointed out that we now have three different notations for vector space duality, namely $D(M)$, M^* and $\mathrm{Hom}_k(M,k)$. We apologise for this burden on the reader, which has arisen in an attempt not to break with traditional notations.

We are thus left with the problem of finding a contravariant functor $Tr : {}_\Lambda\underline{\mathbf{mod}} \to {}_{\Lambda^{\mathrm{op}}}\underline{\mathbf{mod}}$ with the property that

$$\underline{\mathrm{Hom}}_\Lambda(N,-) \cong \mathrm{Tor}_1^\Lambda(Tr\ N,-).$$

The way to construct this functor is as follows. Form the beginning of a projective resolution

$$P_1 \xrightarrow{f} P_0 \to N \to 0$$

and let $Tr\ N$ be the cokernel of the Λ^{op}-module homomorphism

$$f^* : \mathrm{Hom}_\Lambda(P_0,\Lambda) \to \mathrm{Hom}_\Lambda(P_1,\Lambda).$$

Now Tr is a well defined contravariant functor from ${}_\Lambda\underline{\mathbf{mod}}$ to ${}_{\Lambda^{\mathrm{op}}}\underline{\mathbf{mod}}$, and it is easy to see that $Tr^2 : {}_\Lambda\underline{\mathbf{mod}} \to {}_\Lambda\underline{\mathbf{mod}}$ is isomorphic to the identity functor.

We have an exact sequence

$$0 \to \mathrm{Hom}_\Lambda(N,\Lambda) \to \mathrm{Hom}_\Lambda(P_0,\Lambda) \to \mathrm{Hom}_\Lambda(P_1,\Lambda) \to Tr\ N \to 0$$

whose last three terms form the beginning of a projective resolution of Tr N as a Λ^{op}-module, and so we may use it to calculate Tor. Namely,

$$\begin{aligned}
\mathrm{Tor}_1^\Lambda(Tr\ N, N') &= \frac{\mathrm{Ker}(\mathrm{Hom}_\Lambda(P_0,\Lambda)\otimes N' \to \mathrm{Hom}_\Lambda(P_1,\Lambda)\otimes N')}{\mathrm{Im}(\mathrm{Hom}_\Lambda(N,\Lambda)\otimes N' \to \mathrm{Hom}_\Lambda(P_0,\Lambda)\otimes N')} \\
&= \frac{\mathrm{Ker}(\mathrm{Hom}_\Lambda(P_0,N') \to \mathrm{Hom}_\Lambda(P_1,N'))}{\mathrm{Im}(\mathrm{Hom}_\Lambda(N,\Lambda)\otimes N' \to \mathrm{Hom}_\Lambda(P_0,N'))} \\
&= \frac{\mathrm{Hom}_\Lambda(N,N')}{\mathrm{Im}(\mathrm{Hom}_\Lambda(N,\Lambda)\otimes N') \to \mathrm{Hom}_\Lambda(N,N'))} \\
&= \underline{\mathrm{Hom}}_\Lambda(N,N').
\end{aligned}$$

We have thus proved that

$$\underline{\mathrm{Hom}}_\Lambda(N,N') \cong \mathrm{Tor}_1^\Lambda(Tr\ N,N')$$

so that

$$\begin{aligned}
\mathrm{Ext}^1_\Lambda(N', DTr\ N) &= \mathrm{Ext}^1_\Lambda(N', \mathrm{Hom}_k(Tr\ N,k)) \\
&\cong \mathrm{Hom}_k(\mathrm{Tor}_1^\Lambda(Tr\ N,N'),k) \\
&\cong \mathrm{Hom}_k(\underline{\mathrm{Hom}}_\Lambda(N,N'),k)
\end{aligned}$$

as required. To complete the proof of the existence of almost split sequences for finite dimensional algebras, we only need show that DTr takes indecomposables to indecomposables. But both D and Tr are self inverse, and so they must take indecomposables to indecomposables.

Note that DTr gives a bijection between the non-projective indecomposable Λ-modules and the non-injective indecomposable Λ-modules. Thus given any non-injective indecomposable Λ-module M, there is a unique almost split sequence beginning with M, and it has the form $0 \to M \to E \to TrD\,M \to 0$. This is related to the following dual property of almost split sequences.

PROPOSITION 4.12.4. *If $0 \to M \xrightarrow{\sigma'} E \to N \to 0$ is an almost split sequence and $\rho : M \to M'$ is not a split monomorphism then ρ factors through σ'.*

PROOF. Suppose ρ does not factor through σ'. Then in the pushout

$$\begin{array}{ccccccccc} 0 & \longrightarrow & M & \xrightarrow{\sigma'} & E & \longrightarrow & N & \longrightarrow & 0 \\ & & \downarrow{\scriptstyle\rho} & & \downarrow & & \| & & \\ 0 & \longrightarrow & M' & \longrightarrow & E' & \longrightarrow & N & \longrightarrow & 0 \end{array}$$

the second sequence does not split. Thus we may complete a diagram

$$\begin{array}{ccccccccc} 0 & \longrightarrow & M & \xrightarrow{\sigma'} & E & \longrightarrow & N & \longrightarrow & 0 \\ & & \downarrow{\scriptstyle\rho} & & \downarrow & & \| & & \\ 0 & \longrightarrow & M' & \longrightarrow & E' & \longrightarrow & N & \longrightarrow & 0 \\ & & \downarrow{\scriptstyle\rho'} & & \downarrow & & \| & & \\ 0 & \longrightarrow & M & \longrightarrow & E & \longrightarrow & N & \longrightarrow & 0 \end{array}$$

Since $\rho' \circ \rho$ is not nilpotent it is an isomorphism by Fitting's Lemma 1.4.4, and so ρ is a split monomorphism. □

COROLLARY 4.12.5. *If $0 \to M \to E \to N \to 0$ is an almost split sequence then so is $0 \to N^* \to E^* \to M^* \to 0$.* □

We now give an interpretation of almost split sequences in terms of functor categories. The following proposition is clear from the definition:

PROPOSITION 4.12.6. *An almost split sequence of Λ-modules*

$$0 \to M \to E \to N \to 0$$

gives rise to exact sequences

$$0 \to \mathrm{Hom}_\Lambda(N', M) \to \mathrm{Hom}_\Lambda(N', E) \to \mathrm{Hom}_\Lambda(N', N) \to 0$$

if N' has no summand isomorphic to N, and

$$0 \to \mathrm{Hom}_\Lambda(N, M) \to \mathrm{Hom}_\Lambda(N, E) \to \mathrm{End}_\Lambda(N) \to \mathrm{End}_\Lambda(N)/J\mathrm{End}_\Lambda(N) \to 0.$$

Dually, we have exact sequences

$$0 \to \mathrm{Hom}_\Lambda(N, M') \to \mathrm{Hom}_\Lambda(E, M') \to \mathrm{Hom}_\Lambda(M, M') \to 0$$

if M' has no summand isomorphic to M, and

$$0 \to \mathrm{Hom}_\Lambda(N,M) \to \mathrm{Hom}_\Lambda(E,M) \to \mathrm{End}_\Lambda(M) \to \mathrm{End}_\Lambda(M)/J\mathrm{End}_\Lambda(M) \to 0. \qquad \square$$

We may express this in terms of functors as an exact sequence of covariant functors

$$0 \to (-,M) \to (-,E) \to (-,N) \to S_N \to 0$$

where S_N is the simple functor associated to N, as in Theorem 4.8.2.

Similarly the dual property of almost split sequences shows that there is also an exact sequence of contravariant functors

$$0 \to (N,-) \to (E,-) \to (M,-) \to S_M \to 0.$$

Since M and N are indecomposable, $(M,-)$, $(N,-)$, $(-,M)$ and $(-,N)$ are projective indecomposable functors, and so the above sequences are **minimal projective resolutions** of the simple functors S_N for N non-projective indecomposable and S_M for M non-injective indecomposable in $\mathbf{Fun}^\circ(\Lambda)$ and $\mathbf{Fun}(\Lambda)$ respectively.

Similarly if N is projective indecomposable then there is a minimal projective resolution of S_N in $\mathbf{Fun}(\Lambda)$ of the form

$$0 \to (-,\mathrm{Rad}(N)) \to (-,N) \to S_N \to 0$$

while if M is injective indecomposable then there is a minimal resolution of S_M in $\mathbf{Fun}^\circ(\Lambda)$ of the form

$$0 \to (M/\mathrm{Soc}(M),-) \to (M,-) \to S_M \to 0.$$

Dually, we have injective resolutions

$$0 \to S_M \to D(M,-) \to D(E,-) \to D(N,-) \to 0$$

and

$$0 \to S_N \to D(-,N) \to D(-,E) \to D(-,M) \to 0$$

in $\mathbf{Fun}(\Lambda)$ and $\mathbf{Fun}^\circ(\Lambda)$ respectively, and similarly for the injective/projective modules.

EXERCISE. Show that if $0 \to M \to E \to N \to 0$ is a short exact sequence of finitely generated Λ-modules with the property that

$$0 \to (-,M) \to (-,E) \to (-,N) \to S_{N_0} \to 0$$

is exact, then the sequence is the sum of the almost split sequence terminating in N_0 and a split sequence. (Hint: write $N = n.N_0 \oplus N'$ and apply the sequence of functors to N)

There is one situation in which the almost split sequence is easy to write down.

PROPOSITION 4.12.7. *Suppose P is a projective and injective indecomposable module. Then there is an almost split sequence*

$$0 \to \mathrm{Rad}(P) \to P \oplus \mathrm{Rad}(P)/\mathrm{Soc}(P) \to P/\mathrm{Soc}(P) \to 0.$$

This is the only almost split sequence having P as a summand of the middle term.

PROOF. It is easy to check directly from the definitions that this is an almost split sequence. Conversely, if P is a summand of the middle term of an almost split sequence and the right-hand term is not $P/\mathrm{Soc}(P)$, then there is a map from $P/\mathrm{Soc}(P)$ to the right-hand term which is not a split epimorphism and does not lift, contradicting the definition of an almost split sequence. □

We are particularly interested in almost split sequences for group algebras of finite groups. The first observation holds for all symmetric algebras.

PROPOSITION 4.12.8. *Suppose Λ is a finite dimensional symmetric algebra. Then for any non-projective indecomposable module M we have $DTr(M) = \Omega^2(M)$, the second kernel of a minimal resolution of M.*

PROOF. Since Λ is symmetric, $\Lambda \cong \Lambda^*$ as Λ–Λ-bimodules. Hence the functors $\mathrm{Hom}_\Lambda(-,\Lambda^*)$ and $\mathrm{Hom}_\Lambda(-,\Lambda)$ coincide, and so the functor Tr applied to M yields the dual of $\Omega^2(M)$. □

Note in particular that in this situation

$$\mathrm{Ext}^1_\Lambda(N, DTr\ M) = \mathrm{Ext}^1_\Lambda(N, \Omega^2(M)) \cong \underline{\mathrm{Hom}}_\Lambda(N, \Omega(M))$$

and so the duality between $\underline{\mathrm{Hom}}_\Lambda(M,N)$ and $\mathrm{Ext}^1_\Lambda(N, DTr\ M)$ becomes a duality between the spaces $\underline{\mathrm{Hom}}_\Lambda(M,N)$ and $\underline{\mathrm{Hom}}_\Lambda(N,\Omega(M))$.

PROPOSITION 4.12.9. *Suppose Λ is a finite dimensional symmetric algebra. Then there is a natural duality between the spaces $\underline{\mathrm{Hom}}_\Lambda(M,N)$ and $\underline{\mathrm{Hom}}_\Lambda(N,\Omega(M))$.* □

REMARK. In case Λ is self injective but not necessarily symmetric, we may modify the above propositions as follows. We define ν to be the **Nakayama functor**

$$\nu = \Lambda^* \otimes_\Lambda - = D\mathrm{Hom}_\Lambda(-,\Lambda).$$

Then we have $DTr = \nu\Omega^2$, and there is a natural duality between the spaces $\underline{\mathrm{Hom}}_\Lambda(M,N)$ and $\underline{\mathrm{Hom}}_\Lambda(N,\nu\Omega(M))$.

We now specialise to group algebras.

PROPOSITION 4.12.10. *An almost split sequence $0 \to M \to E \to N \to 0$ of modules for a group algebra kG splits on restriction to a subgroup H if and only if H does not contain a vertex of N (or equivalently of M).*

PROOF. The sequence splits on restriction to H if and only if for all kH-modules N' the sequence

$$0 \to \mathrm{Hom}_{kH}(N', M\downarrow_H) \to \mathrm{Hom}_{kH}(N', E\downarrow_H) \to \mathrm{Hom}_{kH}(N', M\downarrow_H) \to 0$$

is exact. By the Nakayama relations 3.3.1 this happens if and only if the sequence

$$0 \to \mathrm{Hom}_{kG}(N'\uparrow^G, M) \to \mathrm{Hom}_{kG}(N'\uparrow^G, E) \to \mathrm{Hom}_{kG}(N'\uparrow^G, N) \to 0$$

is exact. By the defining property of almost split sequences, this happens if and only if N is not a direct summand of $N'\uparrow^G$ for any kH-module N'. □

The following theorem was proved in [**23**].

THEOREM 4.12.11. *Suppose M is an indecomposable kG-module with vertex D, and suppose H is a subgroup of G containing $N_G(D)$, so that the Green correspondent $f(M)$ is defined as a kH-module. If $0 \to \Omega^2 f(M) \to E \to f(M) \to 0$ is the almost split sequence terminating in $f(M)$, then the induced sequence $0 \to \Omega^2 f(M)\uparrow^G \to E\uparrow^G \to f(M)\uparrow^G \to 0$ is isomorphic to the direct sum of the almost split sequence terminating in M and a (possibly zero) split short exact sequence.*

PROOF. By the Nakayama Relations 3.3.1 and Proposition 4.12.6, we have a diagram

$$\begin{array}{ccccccccc}
0 \to & \mathrm{Hom}_{kG}(N,\Omega^2 f(M)\uparrow^G) & \to & \mathrm{Hom}_{kG}(N,E\uparrow^G) & \to & \mathrm{Hom}_{kG}(N,f(M)\uparrow^G) & & & \\
& \downarrow\cong & & \downarrow\cong & & \downarrow\cong & & & \\
0 \to & \mathrm{Hom}_{kH}(N\downarrow_H,\Omega^2 f(M)) & \to & \mathrm{Hom}_{kH}(N\downarrow_H,E) & \to & \mathrm{Hom}_{kH}(N\downarrow_H,f(M)) & \to & S_{f(M)}(N\downarrow_H) & \to 0.
\end{array}$$

By the Burry–Carlson–Puig Theorem 3.12.3, we have $S_{f(M)}(N\downarrow_H) \cong S_M(N)$, and so we have an exact sequence

$$0 \to \mathrm{Hom}_{kG}(N, \Omega^2 f(M)\uparrow^G) \to \mathrm{Hom}_{kG}(N, E\uparrow^G)$$
$$\to \mathrm{Hom}_{kG}(N, f(M)\uparrow^G) \to S_M(N) \to 0.$$

The result now follows from the exercise before Proposition 4.12.7. □

4.13. Irreducible maps and the Auslander–Reiten quiver

In this section, we describe a certain modulated quiver called the Auslander–Reiten quiver, associated with almost split sequences, and describe its elementary properties.

DEFINITION 4.13.1. Suppose M and N are finitely generated indecomposable Λ-modules. A map $\lambda : M \to N$ is said to be **irreducible** if λ has no left or right inverse, and whenever $\lambda = \nu \circ \mu$ is a factorisation of λ, either μ has a left inverse or ν has a right inverse.

If $M = \bigoplus M_i$ and $N = \bigoplus N_j$ with the M_i and N_j indecomposable, we denote by $\mathrm{Rad}(M, N)$ the space of maps $M \to N$ with the property that no component $M_i \to N_j$ is an isomorphism. We denote by $\mathrm{Rad}^2(M, N)$ the space spanned by the homomorphisms of the form $\nu \circ \mu$ with $\mu \in \mathrm{Rad}(M, M')$ and $\nu \in \mathrm{Rad}(M', N)$ for some M'. Then the set of irreducible maps is precisely $\mathrm{Rad}(M, N) \setminus \mathrm{Rad}^2(M, N)$. The space $\mathrm{Irr}(M, N) = \mathrm{Rad}(M, N)/\mathrm{Rad}^2(M, N)$ is an $(\mathrm{End}_\Lambda(N)/J\mathrm{End}_\Lambda(N))$–$(\mathrm{End}_\Lambda(M)/J\mathrm{End}_\Lambda(M))$-bimodule.

The **Auslander–Reiten quiver** of Λ is the modulated quiver given as follows. The vertices x_α are indexed by the finitely generated indecomposable Λ-modules M_α, with associated division ring

$$\Delta_\alpha = (\mathrm{End}_\Lambda(M_\alpha)/J\mathrm{End}_\Lambda(M_\alpha))^{\mathrm{op}}.$$

There is an arrow $x_\alpha \xrightarrow{\gamma} x_\beta$ if and only if $\mathrm{Irr}(M_\alpha, M_\beta) \neq 0$, and

$$\begin{aligned}
{}_\beta M_\alpha^\gamma &= M_{\alpha\beta} = \mathrm{Irr}(M_\alpha, M_\beta) \\
{}_\alpha M_\beta^\gamma &= M'_{\alpha\beta} = \mathrm{Hom}_k({}_\beta M_\alpha^\gamma, k) \cong \mathrm{Hom}_{\Delta_\alpha}({}_\beta M_\alpha^\gamma, \Delta_\alpha) \cong \mathrm{Hom}_{\Delta_\beta}({}_\beta M_\alpha^\gamma, \Delta_\beta) \\
{}_\beta d_\alpha^\gamma &= d_{\alpha\beta} = \dim_{\Delta_\alpha}({}_\beta M_\alpha^\gamma) \\
{}_\alpha d_\beta^\gamma &= d'_{\alpha\beta} = \dim_{\Delta_\beta}({}_\alpha M_\beta^\gamma) \quad \text{and} \\
f_\alpha &= \dim_k(\Delta_\alpha).
\end{aligned}$$

LEMMA 4.13.2. *If $\lambda : M \to N$ is irreducible, then λ is either an epimorphism whose kernel is indecomposable, or a monomorphism whose cokernel is indecomposable.*

PROOF. The factorisation $M \xrightarrow{\mu} M/\mathrm{Ker}(\lambda) \xrightarrow{\nu} N$ shows that λ is either an epimorphism or a monomorphism.

Suppose λ is an epimorphism with kernel $A \oplus B$. Then there is a factorisation

$$M \xrightarrow{\mu} M/A \xrightarrow{\nu} M/(A \oplus B) \cong N.$$

Since μ is an epimorphism it does not have a left inverse, and so ν has a right inverse $\rho : N \to M/A$. Similarly we obtain a right inverse $\rho' : N \to M/B$ for the map $\nu' : M/B \to N$. Since M is the pullback of ν and ν', ρ and ρ' determine a map $N \to M$ right inverse to λ, and so λ is not irreducible. The dual argument works for λ a monomorphism. □

PROPOSITION 4.13.3. (i) *If N is a non-projective indecomposable Λ-module, let the almost split sequence terminating in N be $0 \to DTr\,N \to E \to N \xrightarrow{\sigma} 0$. Then $\lambda : N' \to N$ is irreducible if and only if N' is a summand of E and $\lambda = \sigma \circ i$ with i an inclusion of N' as a summand of E.*

If N is projective indecomposable, $\lambda : N' \to N$ is irreducible if and only if λ is an inclusion of N' as a summand of $\mathrm{Rad}(N)$.

(ii) *If M is a non-injective indecomposable Λ-module, let the almost split sequence beginning with M be $0 \to M \xrightarrow{\sigma'} E \to TrD\,M \to 0$. Then $\lambda : M \to M'$ is irreducible if and only if M' is a summand of E and $\lambda = \pi \circ \sigma'$ with π a projection of E onto M' as a summand.*

If M is injective indecomposable, $\lambda : M \to M'$ is irreducible if and only if λ is a projection of $M/\mathrm{Soc}(M)$ onto M' as a summand.

PROOF. (i) Suppose first that N is non-projective.

$$\begin{array}{ccccccccc} & & & & & & N' & & \\ & & & & & \swarrow^{\mu} & \downarrow \lambda & & \\ 0 & \longrightarrow & DTr\,N & \longrightarrow & E & \xrightarrow{\sigma} & N & \longrightarrow & 0 \end{array}$$

Since λ is not an isomorphism, λ factors as $\sigma \circ \mu$. Since σ does not have a left inverse, μ has a right inverse. Thus we may take $i = \mu$.

Conversely if N' is a direct summand of E with inclusion i, we must show that $\sigma \circ i$ is irreducible. Suppose it can be expressed as a composite $N' \xrightarrow{\mu} N'' \xrightarrow{\nu} N$. If ν does not have a left inverse, then ν factors through σ, and so μ has a right inverse.

If N is projective then every map to N either has a right inverse or lands inside $\mathrm{Rad}(N)$. Thus an irreducible map to N must be an injection, and the inclusion into $\mathrm{Rad}(N)$ has a left inverse.

(ii) This is proved dually. □

REMARKS. (i) This proposition implies that for M_α non-injective, $d_{\alpha\beta}$ is equal to the multiplicity of M_β as a direct summand of the middle term of the almost split sequence beginning with M_α, while for M_α injective $d_{\alpha\beta}$ is equal to the multiplicity of M_β as a direct summand of $M_\alpha/\mathrm{Soc}(M_\alpha)$. Similarly for M_β non-projective $d'_{\alpha\beta}$ is the multiplicity of M_α as a direct summand of the almost split sequence terminating in M_β, while for M_β projective $d'_{\alpha\beta}$ is the multiplicity of M_α as a direct summand of $\mathrm{Rad}(M_\beta)$. It follows that the Auslander–Reiten quiver is a locally finite graph; that is, each vertex is incident with only finitely many edges.

It is conjectured that the Auslander–Reiten quiver of a finite dimensional algebra of infinite representation type always has infinitely many connected components. This has been proved by Crawley-Boevey [**63**] in the case of tame representation type over an algebraically closed field, but otherwise this conjecture is still open.

(ii) Comparing this proposition with the minimal resolutions for the simple functors found in the last section, it is not hard to see that the Auslander–Reiten quiver of Λ is the Ext-quiver of the Auslander algebra $\mathrm{Aus}(\Lambda)$.

LEMMA 4.13.4. *Suppose M and N are indecomposable Λ-modules and $f : M \to N$ is non-zero and is not an isomorphism.*

(i) *There is an irreducible map $\lambda : M \to M'$ and a map $\mu : M' \to N$ with $\mu \circ \lambda \neq 0$.*

(ii) *There is a map $\nu : M \to N'$ and an irreducible map $\lambda : N' \to N$ with $\lambda \circ \nu \neq 0$.*

PROOF. We shall prove (ii); (i) is proved dually using Proposition 4.12.4. Suppose N is not projective. Let $0 \to DTr\,N \to E \xrightarrow{\sigma} N \to 0$ be the almost split sequence terminating in N. Since f is not an isomorphism it factors through σ. Write $E = \bigoplus_i E_i$ and $f = \sigma \circ \rho = \sum_i \sigma_i \circ \rho_i$ with $\sigma_i : E_i \to N$

and $\rho_i : M \to E_i$. Since $f \neq 0$, some $\sigma_i \circ \rho_i \neq 0$, and σ_i is an irreducible map. On the other hand if N is projective then f factors through the injection $\mathrm{Rad}(N) \to N$ and we apply the same argument. □

PROPOSITION 4.13.5. *Suppose M and N are indecomposable Λ-modules, and $f : M \to N$ is non-zero and is not an isomorphism. Suppose there is no chain of irreducible maps from M to N of length less than n.*

(i) *There exists a chain of irreducible maps*

$$M = M_0 \to M_1 \to \cdots \to M_{n-1} \to M_n$$

and a map $M_n \to N$ such that the composite map from M to N is non-zero.

(ii) *There exists a chain of irreducible maps*

$$N_0 \to N_1 \to \cdots \to N_{n-1} \to N_n = N$$

and a map $M \to N_0$ such that the composite map from M to N is non-zero.

PROOF. This follows from the lemma and induction on n. □

Finite quiver components are especially easy to deal with. The idea of the proof of the following proposition will reappear in the next chapter when we deal with bilinear forms on representation rings.

PROPOSITION 4.13.6. *Suppose that a component Q of the Auslander–Reiten quiver of Λ has only finitely many vertices. Then Q consists of all the indecomposable modules in a block of Λ of finite representation type.*

PROOF. Let $\mathbb{C}^Q$ be a complex vector space whose basis elements e_x correspond to the vertices of Q. Letting M_x denote the indecomposable Λ-module corresponding to x, we impose a bilinear form on $\mathbb{C}^Q$ by setting $(e_x, e_y) = \dim_k \mathrm{Hom}_\Lambda(M_x, M_y)$. For each non-projective M_x, we have an almost split sequence

$$0 \to M_{\tau(x)} \to E_x = \bigoplus_{y \in x^-} M_y \to M_x \to 0$$

and we set

$$f_x = e_x + e_{\tau(x)} - \sum_{y \in x^-} e_y \in \mathbb{C}^Q.$$

If M_x is projective, then $\mathrm{Rad}(M_x) = -\bigoplus_{y \in x^-} M_y$ and we set

$$f_x = e_x - \sum_{y \in x^-} e_y \in \mathbb{C}^Q.$$

It follows from Proposition 4.12.6 for M_x non-projective, and is clear for M_x projective, that $(e_x, f_y) = 0$ unless $x = y$, and

$$(e_x, f_x) = \dim_k \mathrm{End}_\Lambda(M_x)/J\mathrm{End}_\Lambda(M_x) \neq 0.$$

Thus the bilinear form $(\ , \ )$ is non-singular on $\mathbb{C}^Q$.

Now if Q does not consist of all the indecomposable modules in a block of Λ, then there is an indecomposable module M not in Q and a non-zero homomorphism from M to some M_x. By the non-singularity of (,), we can find a non-zero element $\sum \lambda_x e_x \in \mathbb{C}^Q$ such that

$$\dim_k \operatorname{Hom}_\Lambda(M, M_y) = (\sum \lambda_x e_x, e_y)$$

for all $y \in Q$. Choose x with $\lambda_x \neq 0$, so that $(\sum \lambda_x e_x, f_x) \neq 0$. If M_x is not projective, then the sequence

$$0 \to \operatorname{Hom}_\Lambda(M, M_{\tau(x)}) \to \operatorname{Hom}_\Lambda(M, E_x) \to \operatorname{Hom}_\Lambda(M, M_x)$$

is not exact on the right. Hence using Proposition 4.12.6 again, $M \cong M_x$, contradicting the fact that M is not in Q. On the other hand, if M_x is projective, then

$$\operatorname{Hom}_\Lambda(M, \operatorname{Rad}(M_x)) \to \operatorname{Hom}_\Lambda(M, M_x)$$

is not surjective, and so there is a surjective map $M \to M_x$. Since M_x is projective, this forces $M \cong M_x$. □

The projective and injective modules often get in the way when we are looking at the Auslander–Reiten quiver. Indeed, if we remove all the modules of the form $(DTr)^n(I)$ for I injective and $(TrD)^n(P)$ for P projective, we obtain the largest subquiver of the Auslander–Reiten quiver for which DTr is an automorphism. This is called the **stable quiver**. In Section 4.15 we investigate the possible structure of a connected component of the stable quiver.

EXERCISE. Show that any short exact sequence $0 \to M \to E \to N \to 0$ with $M \cong DTr(N)$ indecomposable and both maps irreducible is an almost split sequence.

4.14. Rojter's theorem

The main theorem of this section (4.14.3) was conjectured by Brauer and Thrall, and first proved by Rojter. The proof we give is due to Auslander. We refer to Ringel [**176**], Bautista [**13**], and Nazarova and Rojter [**155**] for more information on this and another conjecture of Brauer and Thrall, solved by Nazarova and Rojter.

We begin with a lemma.

LEMMA 4.14.1 (Harada, Sai [**123**]). *Let $M_0, \dots, M_{2^n-1}$ be indecomposable modules, each having at most n composition factors, and suppose $f_i : M_{i-1} \to M_i$ is a homomorphism which is not an isomorphism. Then $f_{2^n-1} \circ \cdots \circ f_2 \circ f_1 = 0$.*

PROOF. We show by induction on m that the image of $f_{2^m-1} \circ \cdots \circ f_1$ has at most $n - m$ composition factors. The assertion is clear for $m = 1$, since f_1 is not an isomorphism. Let $f = f_{2^{m-1}-1} \circ \cdots \circ f_1$, $g = f_{2^{m-1}}$ and $h = f_{2^m-1} \circ \cdots \circ f_{2^{m-1}+1}$. By the inductive hypothesis, the images of f and h each have at most $n-m+1$ composition factors. If either has strictly less, we are done, so suppose the images of f, h and $h \circ g \circ f$ each has exactly $n-m+1$

composition factors. Then $\text{Ker}(h \circ g) \cap \text{Im}(f) = 0$ and $\text{Ker}(h) \cap \text{Im}(g \circ f) = 0$, and so by counting composition factors we have $M_{2^m-1} \cong \text{Ker}(h \circ g) \oplus \text{Im}(f)$ and $M_{2^m} \cong \text{Ker}(h) \oplus \text{Im}(g \circ f)$. Since each is indecomposable, $h \circ g$ is injective and $g \circ f$ is surjective. Thus g is an isomorphism, contrary to hypothesis. □

THEOREM 4.14.2 (Auslander [**7**]). *Suppose Λ is a finite dimensional algebra and Q is an infinite connected component of the Auslander–Reiten quiver of Λ-modules. Then Q has modules with an arbitrarily large number of composition factors.*

PROOF. Suppose to the contrary that all modules in Q have at most n composition factors. Suppose M is an indecomposable Λ-module in Q. For some projective indecomposable Λ-module P we have a non-zero homomorphism $\phi : P \to M$ (for example, take for P the projective cover of a simple submodule of M). If P is not in the component Q, then by Proposition 4.13.5 there is a chain of irreducible maps

$$M_0 \xrightarrow{f_1} M_1 \xrightarrow{f_2} \cdots \xrightarrow{f_{2^n-1}} M_{2^n-1} = M$$

and $f : P \to M_0$ such that $f_{2^n-1} \circ \cdots \circ f_2 \circ f_1 \circ f = \phi$, so that by the lemma $\phi = 0$. It follows that P is in Q, and P is connected to M by a chain of irreducible maps of length at most $2^n - 1$. Since Q has finite valence, and there are only finitely many projective Λ-modules, this forces Q to have only finitely many vertices, contrary to assumption. □

THEOREM 4.14.3 (Rojter [**181**]). *Suppose Λ is a finite dimensional algebra of infinite representation type. Then there are finitely generated indecomposable Λ-modules with an arbitrarily large number of composition factors.*

PROOF. (Auslander) By Proposition 4.13.6 the Auslander–Reiten quiver of finitely generated Λ-modules has an infinite connected component. Now apply the above theorem. □

REMARK. Almost split sequences exist for an arbitrary Artin algebra, and so Auslander's proof of Rojter's theorem works in this generality.

4.15. The Riedtmann structure theorem

We now describe the Riedtmann structure theorem, which describes the structure of an abstract **stable representation quiver**, of which the stable quivers described in the last section are examples. The necessary terminology is given in the following definitions. The proof of the structure theorem involves a variant of the classical universal cover construction.

DEFINITION 4.15.1. A **morphism of quivers** $\phi : Q \to Q'$ assigns to each vertex x of Q a vertex $\phi(x)$ of Q' and to each arrow $x \xrightarrow{\gamma} y$ in Q an arrow $\phi(x) \xrightarrow{\phi(\gamma)} \phi(y)$ in Q'.

If x is a vertex in a quiver Q we write x^- for the set of vertices y in Q such that there is an arrow $y \to x$ in Q, and we write x^+ for the set of vertices y in Q such that there is an arrow $x \to y$ in Q.

A quiver is **locally finite** if the sets x^+ and x^- are finite sets for each x in Q. A **loop** in Q is an arrow from a vertex to itself. A **multiple arrow** in Q is a set of at least two arrows from a given vertex to another given vertex. To a quiver Q without loops or multiple arrows, we associate an undirected graph $\bar{Q}$ whose vertices are the same as the vertices of Q, and where two vertices x and y are joined by an edge in $\bar{Q}$ if there is an arrow $x \to y$ or $y \to x$ in Q.

A **stable representation quiver** or **translation quiver** is a quiver Q together with an automorphism τ called the **translation** such that the following conditions are satisfied.

(i) Q contains no loops or multiple arrows.

(ii) For all vertices x in Q, $x^- = \tau(x)^+$.

A morphism of stable representation quivers is a morphism of quivers commuting with the translation.

A stable representation quiver is **connected** if it is non-empty and cannot be written as a disjoint union of two subquivers each stable under the translation. Note that this does not imply that the underlying quiver is connected.

The **reduced graph** or **orbit graph** of a stable representation quiver Q is the graph obtained from Q by identifying each vertex x with $\tau(x)$ and then replacing each pair of arrows $x \to y$ and $y \to x$ by an undirected edge x—y

Thus for example the stable quiver of finitely generated Λ-modules is a stable representation quiver with translation DTr.

To a directed tree B we associate a stable representation quiver $\mathbb{Z}B$ as follows. The vertices of $\mathbb{Z}B$ are the pairs (n, x) with $n \in \mathbb{Z}$ and x a vertex of B. For each arrow $x \to y$ in B and each $n \in \mathbb{Z}$ we have two arrows $(n, x) \to (n, y)$ and $(n, y) \to (n-1, x)$. The translation is defined via $\tau(n, x) = (n+1, x)$. We regard B as embedded in $\mathbb{Z}B$ as the vertices $(0, x)$ and the arrows connecting them.

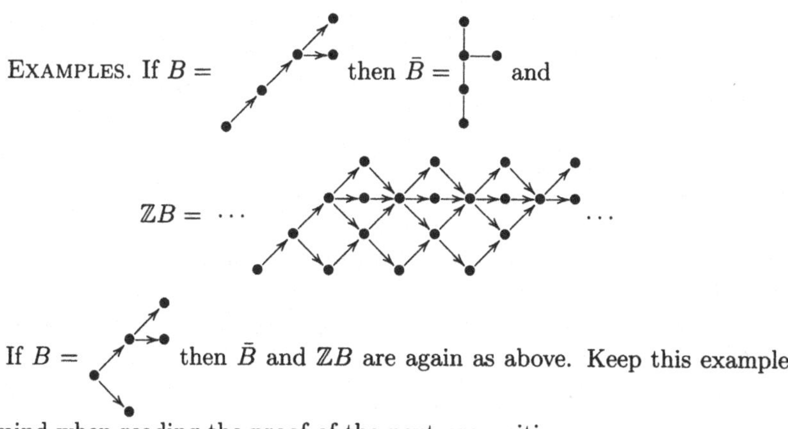
EXAMPLES. If $B =$
then $\bar{B} =$
and
$\mathbb{Z}B = \cdots$
$\cdots$
If $B =$
then $\bar{B}$ and $\mathbb{Z}B$ are again as above. Keep this example

in mind when reading the proof of the next proposition.

LEMMA 4.15.2. *Let B be a directed tree and Q a stable representation quiver. Given a quiver morphism $\phi : B \to Q$ and an integer n, there is a unique morphism of stable representation quivers $f : \mathbb{Z}B \to Q$ such that $f(n, x) = \phi(x)$.*

PROOF. $f(m, x) = \tau^{m-n}\phi(x)$ is clearly the unique such morphism. □

PROPOSITION 4.15.3. *Let B and B' be directed trees. Then $\mathbb{Z}B \cong \mathbb{Z}B'$ as stable representation quivers if and only if $\bar{B} \cong \bar{B}'$.*

PROOF. Since $\bar{B}$ is the reduced graph of $\mathbb{Z}B$, if $\mathbb{Z}B \cong \mathbb{Z}B'$ then $\bar{B} \cong \bar{B}'$. Conversely, suppose $\phi : \bar{B} \cong \bar{B}'$. Choose a vertex x of B, and send it to $(0, \phi(x))$ in $\mathbb{Z}B'$. Since B is connected we may extend this uniquely to a morphism of quivers $B \to \mathbb{Z}B'$ in such a way that each x in B is sent to some $(a_x, \phi(x))$ in B'. Now by the lemma, we obtain a morphism of stable representation quivers $\mathbb{Z}B \to \mathbb{Z}B'$ sending (n, x) to $(n + a_x, \phi(x))$. This is an isomorphism with inverse sending $(n, \phi(x))$ to $(n - a_x, x)$. □

DEFINITION 4.15.4. A group Π of automorphisms of a stable representation quiver Q is said to be **admissible** if no orbit of Π on the vertices of Q intersects a set of the form $\{x\} \cup x^+$ or $\{x\} \cup x^-$ in more than one point. The quotient quiver Q/Π, defined in the obvious way, is then a stable representation quiver.

A morphism of stable representation quivers $\phi : Q \to Q'$ is called a **covering** if it is surjective, and for each vertex x of Q the induced maps $x^- \to \phi(x)^-$ and $x^+ \to \phi(x)^+$ are bijective. It is clearly enough to check that $x^+ \to \phi(x)^+$ is bijective for each vertex x of Q.

Thus for example the canonical projection $Q \to Q/\Pi$, for Π an admissible group of automorphisms of Q, is a covering.

EXAMPLE. Taking $B = A_\infty$, we obtain a stable representation quiver

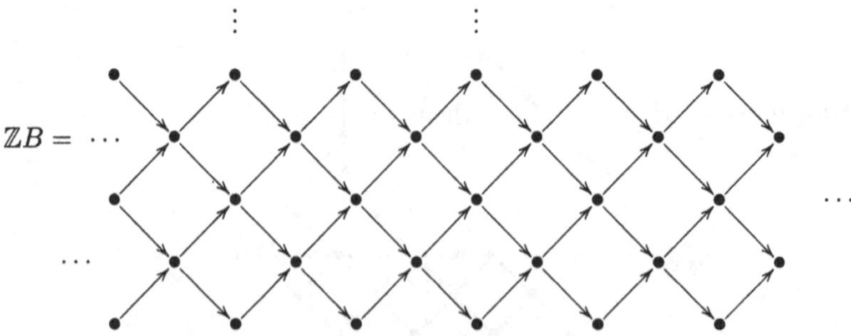

Let Π be the group of automorphisms generated by the nth power of the translation, τ^n. Then Π is admissible, and the resulting stable representation quiver $\mathbb{Z}A_\infty/(n)$ is called an **tube**; see Section 4.16.

LEMMA 4.15.5. *Let B be a directed tree, $\pi : \mathbb{Z}B \to Q$ a morphism of stable representation quivers, $\phi : Q' \to Q$ a covering, and (n, x) a vertex of*

$\mathbb{Z}B$. Then for each vertex y of Q' with $\phi(y) = \pi(n, x)$, there is a unique morphism $\psi : \mathbb{Z}B \to Q'$ with $\phi \circ \psi = \pi$ and $y = \psi(n, x)$.

$$\begin{array}{ccc} & & \mathbb{Z}B \\ & \overset{\psi}{\swarrow} & \downarrow \pi \\ Q' & \xrightarrow{\phi} & Q \end{array}$$

PROOF. The map $\psi(n, x) = y$ clearly extends uniquely to a map from the copy of B consisting of the elements $(n, -)$ to Q whose composite with ϕ is π. The lemma now follows from Lemma 4.15.2. □

THEOREM 4.15.6 (Riedtmann structure theorem).
Given a connected stable representation quiver Q, there is a directed tree B and an admissible group of automorphisms $\Pi \subseteq \mathrm{Aut}(\mathbb{Z}B)$ such that $Q \cong \mathbb{Z}B/\Pi$. The graph $\bar{B}$ associated to B is determined by Q uniquely up to canonical isomorphism, and Π is uniquely defined up to conjugation in $\mathrm{Aut}(\mathbb{Z}B)$.

PROOF. Given Q, we construct B as follows. Choose a vertex x of Q, and let B have as vertices the paths

$$(x = y_0 \to y_1 \to \cdots \to y_n) \qquad (n \geq 0)$$

for which no $y_i = \tau(y_{i+2})$. The arrows of B are

$$(x = y_0 \to \cdots \to y_{n-1}) \longrightarrow (x = y_0 \to \cdots \to y_{n-1} \to y_n).$$

Clearly B is a directed tree.

The quiver morphism $B \to Q$ given by

$$(x = y_0 \to \cdots \to y_n) \mapsto y_n$$

extends uniquely, by Lemma 4.15.2, to a morphism $\phi : \mathbb{Z}B \to Q$. We check that $\phi : \mathbb{Z}B \to Q$ is a covering morphism. If

$$u = (x = y_0 \to \cdots \to y_n)$$

is a vertex of B, then u^+ is the set of vertices of the form

$$(x = y_0 \to \cdots \to y_n \to z)$$

for which $\tau(z) \neq y_{n-1}$ if $n \geq 1$, while u^- consists of the single vertex $(x = y_0 \to \cdots \to y_{n-1})$. Thus

$$(m, u)^+ = \{(m, v),\ v \in u^+\} \cup \{(m-1, v),\ v \in u^-\}$$

has image $\{z \in y_n^+ \mid \tau(z) \neq y_{n-1}\} \cup \{\tau^{-1}(y_{n-1})\} = y_n^+$ in Q. Thus $(m, u)^+$ is in bijection with $\tau(y_n)^+$ as desired.

Now let Π be the **fundamental group** of Q at x, namely the group of morphisms of stable representation quivers $\rho : \mathbb{Z}B \to \mathbb{Z}B$ with $\phi \circ \rho = \phi$. Since Q has no loops, Π is admissible. It follows from Lemma 4.15.5 that Π is transitive on the vertices of $\mathbb{Z}B$ whose image is a given vertex of Q, and so $Q \cong \mathbb{Z}B/\Pi$.

Also by Lemma 4.15.5, we see that if $\mathbb{Z}B \to Q$ and $\mathbb{Z}B' \to Q$ are two such covers, then we obtain inverse isomorphisms $\mathbb{Z}B \xrightarrow{g} \mathbb{Z}B'$ and $\mathbb{Z}B' \xrightarrow{g^{-1}} \mathbb{Z}B$. Hence $\Pi' = g\Pi g^{-1}$, and so by Proposition 4.15.3, $\bar{B} \cong \bar{B}'$. □

The stable representation quiver $\mathbb{Z}B$ is called the **universal cover** of Q, and the isomorphism type of $\bar{B}$ is called the **tree class** of Q.

LEMMA 4.15.7. *There is a natural map κ from the tree $\bar{B}$ associated to Q to the reduced graph of Q, which is surjective and does not identify adjacent points in $\bar{B}$.*

PROOF. The composite map from $\mathbb{Z}B$ to Q and then to the reduced graph of Q is surjective, and has the property that (n, x) and $(n+1, x)$ have the same image. Thus we have a well defined surjective map from $\bar{B}$ to the reduced graph of Q. Since $Q \cong \mathbb{Z}B/\Pi$ with Π an admissible group of automorphisms, this map does not identify adjacent points. □

Finally, we note that a connected component Q of the stable quiver of Λ-modules comes with a modulation which is invariant under the translation DTr. Thus the tree class and reduced graph of Q are modulated graphs, and in particular have labelled graphs associated to them.

We shall see that for group algebras the tree class and reduced graph are very restricted in possible shape.

4.16. Tubes

In this section we investigate a special type of stable quiver component called tubes.

DEFINITION 4.16.1. An infinite **n-tube** is a stable representation quiver of the form $(\mathbb{Z}/n)A_\infty$. A finite n-tube of length q is a stable representation quiver of the form $(\mathbb{Z}/n)A_q$.

A module M is said to be **DTr-periodic** (or just periodic) if $(DTr)^n M \cong M$ for some $n \geq 1$.

Note that in case Λ is symmetric we have $DTr = \Omega^2$ by Proposition 4.12.8, so that a DTr-periodic module is the same as an Ω-periodic module.

THEOREM 4.16.2 (Happel, Preiser and Ringel [**122**]). *Suppose that Q is a connected component of the stable quiver of finitely generated Λ-modules which contains some periodic module. Then every module in Q is periodic. If Q is infinite then it is a tube. If Q is finite then the tree class is a finite Dynkin diagram, and is equal to the reduced graph (but Q need not be a tube).*

PROOF. Suppose $(DTr)^n M = M$, and let x be the vertex of Q corresponding to M. Then $(DTr)^n$ induces a permutation on x^-, which is a finite set by Proposition 4.13.3, and so some power of $(DTr)^n$ stabilises x^- pointwise. The same is true of x^+, so arguing by induction we see that every module in Q is periodic.

Now the function on the reduced graph or tree class of Q which assigns to each vertex the average number of composition factors in the (finite) DTr-orbit in Q corresponding to the vertex is easily seen to be subadditive. So by Theorem 4.5.8 (i) this graph is either a Dynkin diagram or a Euclidean diagram. If Q is infinite and this subadditive function is additive, then Q is a component of the Auslander-Reiten quiver, and so by Theorem 4.14.2 this additive function is unbounded. By Theorem 4.5.8 (ii) and (iv) this cannot happen. If Q is infinite and this subadditive function is not additive, then by Theorem 4.14.2 (iv) the tree class is A_∞. Since the only automorphisms of $\mathbb{Z}A_\infty$ are the translations, by Theorem 4.15.6 it follows from the fact that every module is periodic that for some $n \geq 1$, $Q \cong (\mathbb{Z}/n)A_\infty$.

If Q is finite, then it is connected to at least one projective or injective module, since otherwise Q is a component of the Auslander-Reiten quiver, and we deduce from Proposition 4.13.6 that Q consists of all the indecomposable modules in a block of Λ of finite representation type. Thus the above subadditive function is not additive. So by Theorem 4.5.8 (ii) the tree class and reduced graph are both finite Dynkin diagrams. Finally by Lemma 4.15.7 and the fact that finite Dynkin diagrams are trees, we see that the tree class and reduced graph are equal. □

4.17. Webb's Theorem

Webb [**204**] constructed a subadditive function on the labelled tree associated by the Riedtmann structure theorem to a connected component of the stable quiver of kG-modules. It then follows from Theorem 4.5.8 that this labelled tree is either a Dynkin diagram (finite or infinite) or a Euclidean diagram. He then went on to examine each possibility in detail. His construction used the finite generation of group cohomology. We shall present Okuyama's approach [**156**] to this theorem, in which it is only necessary to understand the cohomology of cyclic groups of prime order.

LEMMA 4.17.1. *Let M_1 and M_2 be indecomposable kG-modules in the same connected component of the stable quiver, and let N be a kG-module. Then $M_1 \otimes N$ is projective if and only if $M_2 \otimes N$ is projective.*

PROOF. It suffices to prove this in case there is an irreducible map $M_1 \to M_2$. If $M_2 \otimes N$ is projective, then so is $\Omega^2(M_2) \otimes N$ by Corollary 3.1.6. Thus if we tensor the almost split sequence terminating in M_2 with N we obtain a sequence with both ends projective, and hence the middle is projective. But $M_1 \otimes N$ is a summand of this middle term by Proposition 4.13.3, and is hence projective. The dual argument shows that if $M_1 \otimes N$ is projective then so is $M_2 \otimes N$. □

Now given a connected component Q of the stable quiver of kG-modules, we construct a subadditive function as follows. Choose a fixed indecomposable kG-module M_0 in Q. Let P be a minimal p-subgroup of G such that the restriction $M_0 \downarrow_P$ is not projective. If P' is a maximal subgroup of P, then $M_0 \downarrow_{P'}$ is projective, P' is normal in P and P/P' is a cyclic group of order p. Thus by Corollary 3.5.3, if X is an indecomposable summand of $M_0 \downarrow_P$ then $X \cong \Omega^2(X)$. By Proposition 3.1.10, $X^* \otimes X$ is not projective, and hence neither are $X^* \otimes M_0 \downarrow_P$ nor $(X^* \otimes M_0 \downarrow_P)\uparrow^G \cong X^* \uparrow^G \otimes M_0$. Thus by the lemma, for any indecomposable kG-module M in Q, $X^* \uparrow^G \otimes M$ is not projective.

We now define a function f from the vertices of Q to the natural numbers via

$$f(M) = \dim_k \operatorname{Ext}^1_{kG}(X\uparrow^G, M),$$

which is equal to $\dim_k \operatorname{Ext}^1_{kP}(X, M\downarrow_P)$ by the Eckmann–Shapiro lemma.

LEMMA 4.17.2. *The above constructed function f from the vertices of Q to the natural numbers has the following properties:*

(i) *$f(M) > 0$ for every M in Q*

(ii) *$f(\Omega^2(M)) = f(M)$*

(iii) *If $0 \to \Omega^2(M) \to E \to M \to 0$ is the almost split sequence terminating in M then*

$$f(E) \le f(M) + f(\Omega^2(M)) = 2f(M).$$

If $f(E) < 2f(M)$ then M is periodic (i.e., $\Omega^n M \cong M$ for some $n \ge 1$).

PROOF. (i) This follows from Corollary 3.14.5.

(ii) By Proposition 2.5.7, we have

$$\begin{aligned} f(\Omega^2(M)) &= \dim_k \operatorname{Ext}^1_{kG}(X\uparrow^G, \Omega^2(M)) \\ &= \dim_k \operatorname{Ext}^1_{kG}(\Omega^2(X\uparrow^G), \Omega^2(M)) \\ &= \dim_k \operatorname{Ext}^1_{kG}(X\uparrow^G, M) = f(M). \end{aligned}$$

(iii) This inequality follows from the long exact Ext sequence

$$\cdots \to \operatorname{Ext}^1_{kP}(X, \Omega^2 M\downarrow_P) \to \operatorname{Ext}^1_{kP}(X, E\downarrow_P) \to \operatorname{Ext}^1_{kP}(X, M\downarrow_P) \to \cdots.$$

If M is not periodic then the restriction of M to a vertex is still not periodic, and so the subgroup P used in the construction of f is a proper subgroup of the vertex. But then by Proposition 4.12.7 the sequence $0 \to \Omega^2(M) \to E \to M \to 0$ splits on restriction to P and so the long exact Ext sequence reduces to a short exact sequence and $f(E) = 2f(M)$. □

THEOREM 4.17.3. (i) *(Webb* **[204]***) Let T be the tree class of a connected component Q of the stable quiver of kG-modules. Then T is either a Dynkin diagram (finite or infinite) or a Euclidean diagram (apart from $\tilde{A}_n$).*

(ii) *The reduced graph $\bar{T}$ of Q is also either a Dynkin diagram (finite or infinite) or a Euclidean diagram (this time $\tilde{A}_n$ is allowed).*

PROOF. (Okuyama) By the lemma, the function f commutes with the translation Ω^2 of Q, and satisfies $2f(x) \ge \sum_{y\in x^-} f(y)$. Thus f passes to a subadditive function on both T and $\bar{T}$. The result now follows from Theorem 4.5.8. □

REMARK. Webb's original subadditive function was constructed as follows [**204**] (see also the treatment given in [**17**]). Let

$$\eta_M(t) = \sum_{n=0}^{\infty} t^n \dim_k(P_n)$$

where

$$\cdots \to P_2 \to P_1 \to P_0$$

is a minimal resolution of M. Then it follows from the finite generation of cohomology that $\eta_M(t)$ is a rational function of t of the form $p(t)/\prod_{i=1}^{r}(1-t^{k_i})$. If the pole at $t = 1$ of $\eta_M(t)$ has order c (this is the same as the **complexity** of M, see Chapter 5 of Volume II) then let $\eta(M)$ be the value of the rational function $(\prod_{i=1}^{r} k_i)\eta_M(t)(1-t)^c$ at $t = 1$. Then $\eta(\Omega^2(M)) = \eta(M)$, and if $0 \to \Omega^2(M) \to E \to M \to 0$ is the almost split sequence terminating in M then $\eta(E) \le 2\eta(M)$. Thus η passes to a subadditive function on T and $\bar{T}$ as above.

Following Webb, we now investigate the possibilities allowed by Theorem 4.17.3.

THEOREM 4.17.4. *Let Q be a connected component of the stable quiver of kG-modules. Then one of the following occurs.*

(i) *Q consists of all the non-projective modules in a block of kG of finite representation type. (We investigate this situation in Corollary 6.3.5 and Theorem 6.5.5, where we see that Q is a finite tube $(\mathbb{Z}/e)A_{p^n-1}$.)*

(ii) *Q is isomorphic to $\mathbb{Z}T$ for some infinite Dynkin diagram $T = A_\infty$, B_∞, C_∞, D_∞ or A_∞^∞. (Components of type $\mathbb{Z}A_\infty$ are in fact by far the most common. To the best of my knowledge nobody knows of an example where $\mathbb{Z}B_\infty$ or $\mathbb{Z}C_\infty$ occurs. $\mathbb{Z}A_\infty^\infty$ occurs for infinitely many components in the case of a block with dihedral defect group in characteristic two, and $\mathbb{Z}D_\infty$ occurs for infinitely many components in the case of a block with semidihedral defect group in characteristic two.)*

(iii) *Q contains a periodic module. In this case Q is an infinite n-tube $(\mathbb{Z}/n)A_\infty$ for some $n \ge 1$.*

(iv) *The reduced graph is a Euclidean diagram. (This possibility has been investigated by Okuyama [**156**] and Bessenrodt [**32**], who have shown that this only occurs in characteristic two for blocks whose defect group is a Klein four group. In this case the reduced graph is $\tilde{A}_{12}$, $\tilde{B}_3$ or $\tilde{A}_5$.)*

PROOF. By Theorem 4.17.3, the reduced graph is either a Dynkin diagram (finite or infinite) or a Euclidean diagram. The latter is covered by case (iv), so we concentrate on the former.

It the reduced graph is a finite Dynkin diagram then by Theorem 4.5.8 (iii) the subadditive function f used in the proof of Theorem 4.17.3 is not additive. So by Lemma 4.17.2 (iii) there is a periodic module in Q. By Theorem 4.16.2 every module in Q is periodic, so Q has only finitely many vertices. Now by

Proposition 4.13.6, Q consists of all the modules in a block of kG of finite representation type, and we are in case (i).

If the reduced graph is an infinite Dynkin diagram, i.e., one of A_∞, B_∞, C_∞, D_∞, A_∞^∞, then the tree class T is also an infinite Dynkin diagram by Lemma 4.15.7, and by the Riedtmann structure theorem $Q \cong \mathbb{Z}T/\Pi$ for some admissible group of automorphisms Π. If Π is trivial then $Q \cong \mathbb{Z}T$ and we are in case (ii). If T is one of A_∞, B_∞, C_∞ or D_∞, then every non-trivial admissible group of automorphisms will identify two vertices in the same DTr-orbit. In this case Q contains a periodic module and so by Theorem 4.16.2, Q is an infinite tube, and we are in case (iii). If $T \cong A_\infty^\infty$ then every non-trivial admissible group of automorphisms will either identify two vertices in the same DTr-orbit so that again using Theorem 4.16.2 we are in case (iii), or produce a quotient with only finitely many DTr-orbits so that the reduced graph is of type $\tilde{A}_n$ and we are in case (iv). □

Examples. (i) In Section 4.3 we described the classification of modules for the Klein four group. It is not hard to see that for the indecomposable modules M of even dimension we have $\Omega(M) \cong M$, so that M lies in a 1-tube. The indecomposable rational canonical forms corresponding to the modules going up this 1-tube correspond to the ascending powers of a given irreducible polynomial. For example, if k is algebraically closed, the 1-tubes take the form

$$V_{1,a} \rightleftarrows V_{2,a} \rightleftarrows V_{3,a} \rightleftarrows \cdots$$

and so the corresponding almost split sequences are

$$0 \to V_{n,a} \to V_{n-1,a} \oplus V_{n+1,a} \to V_{n,a} \to 0.$$

So these modules form a family of 1-tubes parametrised by $\mathbb{P}^1(k)$. The remaining modules are the projective indecomposable and the modules of the form $\Omega^{\pm n}(k)$. These form a quiver component of type $\mathbb{Z}\tilde{A}_{12}$.

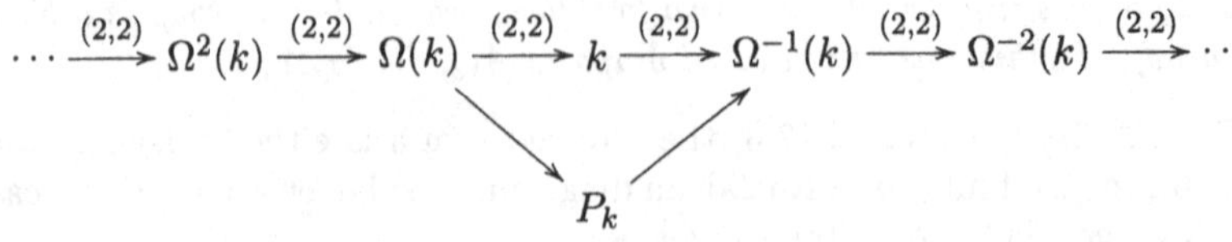

$$\cdots \xrightarrow{(2,2)} \Omega^2(k) \xrightarrow{(2,2)} \Omega(k) \xrightarrow{(2,2)} k \xrightarrow{(2,2)} \Omega^{-1}(k) \xrightarrow{(2,2)} \Omega^{-2}(k) \xrightarrow{(2,2)} \cdots$$
$$\Omega(k) \searrow P_k \nearrow \Omega^{-1}(k)$$

(ii) We also described in Section 4.3 the classification of modules for the alternating group A_4 in characteristic two. The modules $W_{n,\alpha}$ for $\alpha \notin \{\omega, \bar{\omega}\}$ are periodic of period one and hence lie in 1-tubes. For $\alpha \in \{\omega, \bar{\omega}\}$ the modules $W_{n,\alpha}(1)$, $W_{n,\alpha}(\omega)$ and $W_{n,\alpha}(\bar{\omega})$ form a 3-tube as follows:

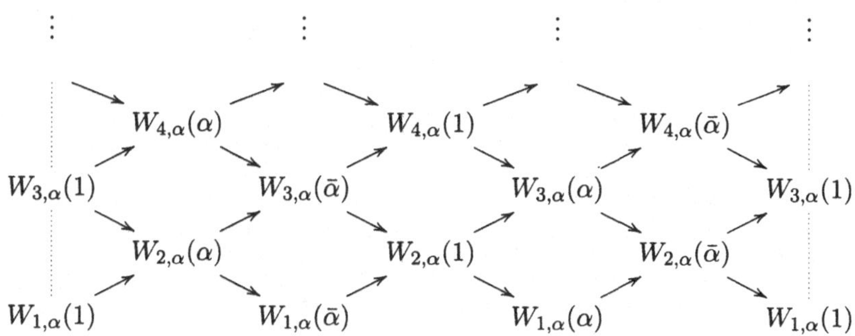

(identify left and right edges to make a 3-tube)
The modules $\Omega^{\pm n}$ of simples all lie in a single quiver component isomorphic to $\mathbb{Z}\tilde{A}_5$ as follows:

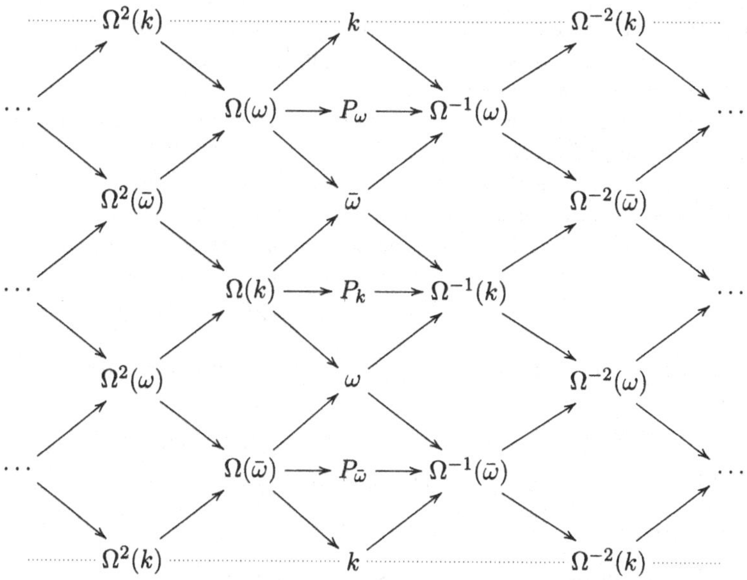

(identify top and bottom edges to make a $\mathbb{Z}\tilde{A}_5$).

The same story holds over any field containing three cube roots of unity. If k does not have three cube roots of unity the the simple modules are k and a two dimensional module S which when tensored with the field extension obtained by adjoining cube roots of unity gives $\omega \oplus \bar{\omega}$. In this case the only difference is that the modules $\Omega^{\pm n}$ of simples form a quiver component isomorphic to $\mathbb{Z}\tilde{B}_3$ as follows:

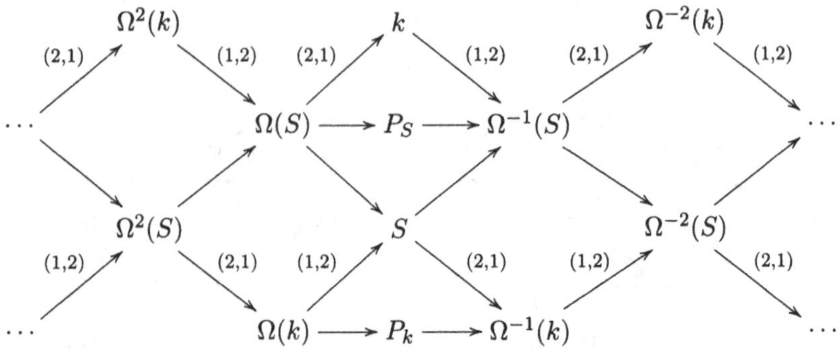

(No identifications).

(iii) In Section 4.11 we described the classification of modules for the dihedral group D_{4q} in characteristic two, where q is a power of two. The almost split sequences and Auslander–Reiten quiver are described as follows. We define two functions L_q and R_q from the set $\mathcal{W}$ of words to itself as follows. Let $A = (ab)^{q-1}a$ and $B = (ba)^{q-1}b$. If a word C starts with Ab^{-1} or Ba^{-1} then $L_q(C)$ is obtained by removing that part; otherwise $L_q(C) = A^{-1}bC$ or $B^{-1}aC$, whichever is a word. Similarly if C ends in aB^{-1} or bA^{-1} then $R_q(C)$ is obtained by removing that part; otherwise $R_q(C) = Ca^{-1}B$ or $Cb^{-1}A$, whichever is a word. The maps L_q and R_q are bijections from $\mathcal{W}$ to itself, and we have $L_q \circ R_q = R_q \circ L_q$.

The Auslander–Reiten translate on modules for D_{4q} is given on modules of the first kind by $\Omega^2 M(C) = M(L_qR_qC)$, and on modules of the second kind by $\Omega^2 M(C,\phi) \cong M(C,\phi)$. The almost split sequence terminating in $M(C)$ is

$$0 \to M(L_qR_qC) \to M(L_qC) \oplus M(R_qC) \to M(C) \to 0$$

unless C or C^{-1} is A, B or AB^{-1}, in which case the almost split sequences are

$$0 \to M(A) \to M(Ab^{-1}A) \to M(A) \to 0$$
$$0 \to M(B) \to M(Ba^{-1}B) \to M(B) \to 0$$
$$0 \to M(A^{-1}B) \to M((ab)^{q-1}) \oplus M((ba)^{q-1}) \oplus P \to M(AB^{-1}) \to 0$$

where P is the projective indecomposable of dimension $4q$.

If $p(x)$ is an irreducible polynomial in $k[x]$, let $\phi_{n,p}$ denote an indecomposable automorphism of a finite dimensional vector space, with rational canonical form associated to the polynomial $p(x)^n$. Then the almost split sequence terminating in the module of the second kind $M(C,\phi_{n,p})$ is

$$0 \to M(C,\phi_{n,p}) \to M(C,\phi_{n+1,p}) \oplus M(C,\phi_{n-1,p}) \to M(C,\phi_{n,p}) \to 0$$

(where the term $M(C,\phi_{n-1,p})$ is absent if $n = 1$).

The modules of the first kind form an infinite set of quiver components of type $\mathbb{Z}A_\infty^\infty$

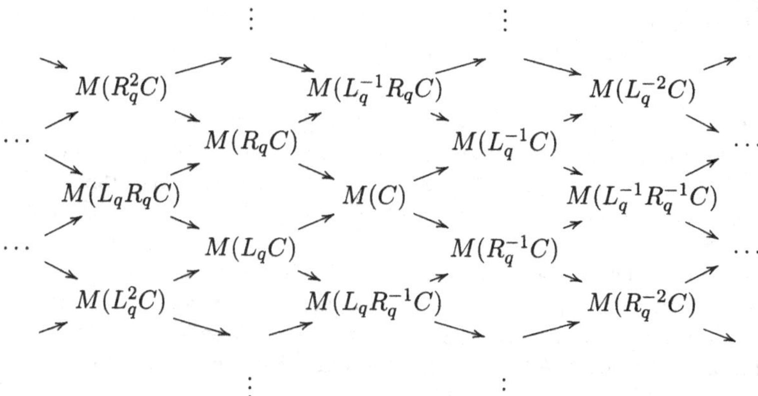

One of these has P attached to it:

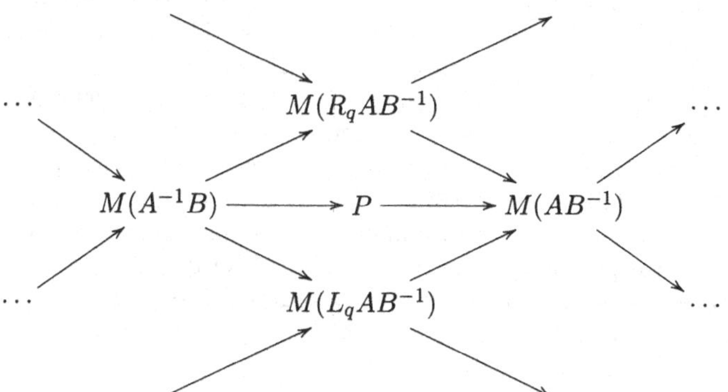

There are also two 1-tubes of modules of the first kind:

$$M(A) \rightleftarrows M(R_qA) \rightleftarrows M(R_q^2A) \rightleftarrows \cdots$$

$$M(B) \rightleftarrows M(R_qB) \rightleftarrows M(R_q^2B) \rightleftarrows \cdots$$

All the modules of the second kind lie in 1-tubes:

$$M(C, \phi_{1,p}) \rightleftarrows M(C, \phi_{2,p}) \rightleftarrows M(C, \phi_{3,p}) \rightleftarrows \cdots$$

4.18. Brauer graph algebras

In this section we describe some finite dimensional algebras which arise in the representation theory of finite groups, namely the class of Brauer graph algebras. We show that there is a simple condition given in terms of almost split sequences which guarantees that a finite dimensional algebra is a Brauer graph algebra. Among the Brauer graph algebras, we identify the ones of finite representation type as being the Brauer tree algebras. In Section 6.5, we shall see that blocks of cyclic defect of a finite group are Brauer tree algebras. The ideas involved in this section are an adaptation of the ideas in Erdmann **[101]**. Throughout this section we work over an algebraically closed field k of coefficients.

DEFINITION 4.18.1. A **Brauer graph** consists of a finite undirected connected graph (possibly with loops and multiple edges), together with the following data. To each vertex we assign a cyclic ordering of the edges incident to it, and an integer greater than or equal to one, called the **multiplicity** of the vertex.

A **Brauer tree** is a Brauer graph which is a tree, and having at most one vertex with multiplicity greater than one. If there is such a vertex, it is called the **exceptional vertex**, and its multiplicity is called the **exceptional multiplicity**; otherwise the exceptional multiplicity is defined to be one.

Note that at least in the case of a tree, the cyclic ordering on the edges around a vertex is usually indicated by drawing the tree in such a way that the ordering is anticlockwise around each vertex. Thus the cyclic orderings are sometimes thought of as being given by a "planar embedding".

We say a finite dimensional algebra Λ is a Brauer graph algebra for a given Brauer graph, if there is a one-one correspondence between the edges j of the graph and the simple Λ-modules S_j in such a way that the projective cover P_j of S_j has the following description. We have $P_j/\mathrm{Rad}(P_j) \cong \mathrm{Soc}(P_j) \cong S_j$, and $\mathrm{Rad}(P_j)/\mathrm{Soc}(P_j)$ is a direct sum of two (possibly zero) uniserial modules U_j and V_j corresponding to the two vertices u and v at the end of the edge j. If the edges around u are cyclically ordered $j, j_1, j_2, \dots, j_r, j$ and the multiplicity of the vertex u is e_u, then the corresponding uniserial module U_j has composition factors (from the top)

$$S_{j_1}, S_{j_2}, \dots, S_{j_r}, S_j, S_{j_1}, \dots, S_{j_r}, S_j, \dots, \dots, S_{j_r}$$

so that $S_{j_1}, \dots, S_{j_r}$ appear e_u times and S_j appears $e_u - 1$ times.

For example, a Brauer graph algebra for the Brauer graph

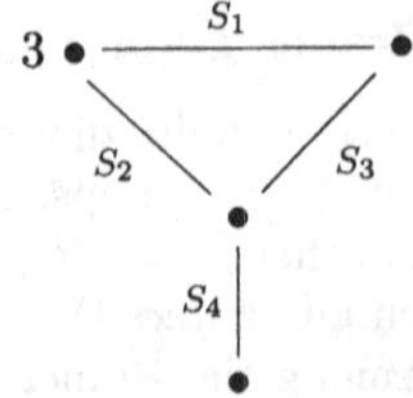

(unmarked vertices have multiplicity one, and the cyclic ordering is anticlockwise around the vertices) has projective modules as follows:

$$\begin{array}{ccc} & S_1 & \\ S_2 & & \\ S_1 & & \\ S_2 & & S_3 \\ S_1 & & \\ S_2 & & \\ & S_1 & \end{array} \qquad \begin{array}{ccc} & S_2 & \\ S_1 & & \\ S_2 & & S_4 \\ S_1 & & \\ S_2 & & S_3 \\ S_1 & & \\ & S_2 & \end{array} \qquad \begin{array}{ccc} & S_3 & \\ & & S_2 \\ S_1 & & \\ & & S_4 \\ & S_3 & \end{array} \qquad \begin{array}{c} S_4 \\ S_3 \\ S_2 \\ S_4 \end{array}$$

Note that by the methods of Section 4.1, at least if there are no loops, the basic algebra of a Brauer graph algebra over an algebraically closed field is almost determined by the Brauer graph. Namely, the Ext-quiver consists of one vertex for each edge of the Brauer graph, and directed edges going in oriented cycles corresponding to the vertices of the Brauer graph (or rather, those with either valency or multiplicity greater than one). For example, the Ext-quiver in the above example is

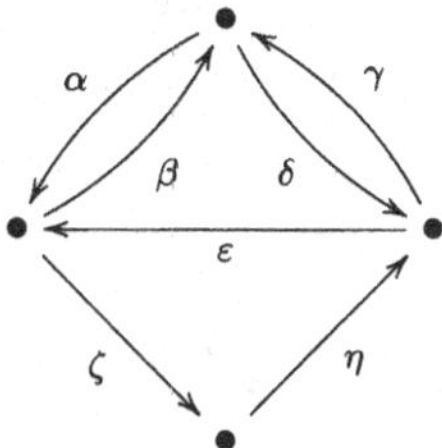

There is one relation corresponding to each edge, and it says that going round the cycle corresponding to the vertex at one end, a number of times equal to the multiplicity, is equal to some non-zero multiple of doing the same for the vertex at the other end. In the above example the relations say that there are constants λ_1, λ_2, λ_3 with

$$\beta\alpha\beta\alpha\beta\alpha = \lambda_1.\gamma\delta, \quad \alpha\beta\alpha\beta\alpha\beta = \lambda_2.\varepsilon\eta\zeta, \quad \delta\gamma = \lambda_3.\eta\zeta\varepsilon, \quad \zeta\varepsilon\eta = 0.$$

By replacing the generators by multiples of the same generators, some of the parameters can be set equal to one. If the graph is a tree, then all the parameters can be set equal to one by this method, but in general the number of remaining parameters will equal the number of edges minus the number of vertices plus one (this is H^1 of the graph!). However, even these parameters are forced to be equal to one if the algebra is assumed to be symmetric. Thus in the above example λ_1 and λ_2 may be set equal to one by replacing γ and ε by suitable multiples, but then λ_3 is fixed. But if λ is a linear map on Λ as in the definition of a symmetric algebra, then

$$\begin{aligned} \lambda(\eta\zeta\varepsilon) &= \lambda(\varepsilon\eta\zeta) = \lambda(\alpha\beta\alpha\beta\alpha\beta) = \lambda(\beta\alpha\beta\alpha\beta\alpha) \\ &= \lambda(\gamma\delta) = \lambda(\delta\gamma) = \lambda_3.\lambda(\eta\zeta\varepsilon) \end{aligned}$$

and so since λ may not vanish on the one dimensional ideal defined by this element, we have $\lambda_3 = 1$.

We now investigate some conditions which force a finite dimensional symmetric algebra to be a Brauer graph algebra. Suppose Λ is a finite dimensional symmetric algebra. If S is a simple Λ-module with projective cover P_S, and U is a summand of $\mathrm{Rad}(P_S)/\mathrm{Soc}(P_S)$, then we write $\overline{U}$ for the extension of $S = P_S/\mathrm{Rad}(P_S)$ by U

$$0 \to U \to \overline{U} \to S \to 0$$

and $\underline{U}$ for the extension of U by $S = \mathrm{Soc}(P_S)$

$$0 \to S \to \underline{U} \to U \to 0.$$

LEMMA 4.18.2 (Erdmann [**101**]). *Suppose Λ, S, P_S, U and $\overline{U}$ are as above. If $0 \neq U \neq \mathrm{Rad}(P_S)/\mathrm{Soc}(P_S)$ then in the almost split sequence ending in $\overline{U}$, the middle term is indecomposable.*

PROOF. Suppose $\mathrm{Rad}(P_S)/\mathrm{Soc}(P_S)$ decomposes as $U \oplus V$ with U and V non-zero. Note that $\underline{V} \cong \Omega(\overline{U})$. Let m be the number of simple summands in $\mathrm{Soc}(U) \cong \mathrm{Soc}(\overline{U})$ and n be the number of simple summands in

$$V/\mathrm{Rad}(V) \cong \underline{V}/\mathrm{Rad}(\underline{V}) \cong \mathrm{Soc}(\Omega\underline{V}) \cong \mathrm{Soc}(\Omega^2\overline{U}).$$

If the almost split sequence ending in $\overline{U}$ is

$$0 \to \Omega^2\overline{U} \to M \xrightarrow{\alpha} \overline{U} \to 0$$

then since $\overline{U} \not\cong P_S/\mathrm{Soc}(P_S)$, M has no projective summands. The almost split sequence beginning with $\underline{V} = \Omega\overline{U}$ is

$$0 \to \underline{V} \to \Omega^{-1}M \to \Omega^{-2}\underline{V} \to 0$$

and so by dualising and reversing the rôles of U and V if necessary, we may assume that $n \leq m$. Since $\mathrm{Soc}(\overline{U})$ is a proper submodule of $\overline{U}$, the inclusion $\mathrm{Soc}(\overline{U}) \hookrightarrow \overline{U}$ factors through α and so $\mathrm{Soc}(M)$ has $m + n \leq 2m$ simple summands.

If $M = \bigoplus_{i=1}^k M_i$ then the components $\alpha_i : M_i \to \overline{U}$ of α are irreducible maps by Proposition 4.13.3. Since $\overline{U}$ has a unique maximal submodule, if α_i is injective, then M_i is a summand of U. But then α_i has the non-trivial factorisation $M_i \hookrightarrow P_S/\mathrm{Soc}(P_S) \twoheadrightarrow \overline{U}$. So each α_i is surjective, and the map $P_S \twoheadrightarrow \overline{U}$ lifts to $P_S \to M_i$. This lift has $\mathrm{Soc}(P_S)$ in its kernel, and $P_S/\mathrm{Soc}(P_S) \twoheadrightarrow \overline{U}$ is surjective on socles, so $\alpha_i(\mathrm{Soc}(M_i)) = \mathrm{Soc}(\overline{U})$. If α_i is injective on socles then it is injective, contradicting the fact that it is an irreducible map. Thus $\mathrm{Soc}(M_i)$ has strictly greater than m summands. Since $\mathrm{Soc}(M)$ has at most $2m$ summands, this forces M to be indecomposable. □

THEOREM 4.18.3. *Suppose Λ is a finite dimensional symmetric algebra with e simple modules, none of which is projective or periodic of period one. Then there are at least $2e$ almost split sequences whose middle terms have only one non-projective summand. If there are exactly $2e$ such almost split sequences then Λ is a Brauer graph algebra.*

PROOF. For each simple module S we produce at least two almost split sequences satisfying the given condition. Let P_S be the projective cover of S. Then $\mathrm{Soc}(P_S) \leq \mathrm{Rad}(P_S)$ and $\mathrm{Rad}(P_S)/\mathrm{Soc}(P_S)$ is non-zero, since S is not projective or periodic of period one. If it is indecomposable then the almost split sequences terminating in S and $P_S/\mathrm{Rad}(P_S)$ satisfy the given condition (see Proposition 4.12.7). If $\mathrm{Rad}(P_S)/\mathrm{Soc}(P_S) = U \oplus V$ with U and V non-zero then by the lemma the almost split sequence terminating in $\overline{U}$ and $\overline{V}$ satisfy the given condition. It is easy to see that the $2e$ almost split sequences produced in this way are distinct.

Now suppose there are e simple modules $S_1, \ldots, S_e$ and exactly $2e$ such almost split sequences. Let P_i be the projective cover of S_i. Then the quotient $\mathrm{Rad}(P_i)/\mathrm{Soc}(P_i)$ has at most two indecomposable summands, since otherwise we can produce more almost split sequences by the above process. If there are exactly two summands, we call them U_i and V_i. If $\mathrm{Rad}(P_i)/\mathrm{Soc}(P_i)$ is indecomposable we write U_i for it, and set $V_i = 0$ and $\overline{V}_i = \underline{V}_i = S_i$. Now the $2e$ almost split sequences ending in the $\overline{U}_i$ and $\overline{V}_i$ have to equal, as a set, the $2e$ almost split sequences ending in $\underline{U}_i$ and $\underline{V}_i$. Thus each $\overline{U}_i$ and $\overline{V}_i$ is one of the $\underline{U}_j$ or $\underline{V}_j$. In particular there is a permutation ρ of the set $\{\overline{U}_i, \overline{V}_i,\ 1 \leq i \leq e\}$ with $\rho(\overline{U}_i) = \underline{U}_i$ and $\rho(\overline{V}_i) = \underline{V}_i$. Thus $\mathrm{Soc}(\overline{U}_i)$ and $\mathrm{Soc}(\overline{V}_i)$ are simple. Arguing by induction using the maps $\underline{U}_i \to \overline{U}_i$ and $\underline{V}_i \to \overline{V}_i$ which have kernel and cokernel S_i, we see that each $\overline{U}_i$ and $\overline{V}_i$ is uniserial and of the same length as $\rho(\overline{U}_i)$, resp. $\rho(\overline{V}_i)$. The composition factors, from the top, of $\overline{U}_i$ are given by repeated application of ρ:

$$\overline{U}_i/\mathrm{Rad}\overline{U}_i,\ \rho\overline{U}_i/\mathrm{Rad}\rho\overline{U}_i,\ \rho^2\overline{U}_i/\mathrm{Rad}\rho^2\overline{U}_i,\ \ldots$$

We build the Brauer graph corresponding to Λ by taking an edge for each S_i. The two ends of this edge correspond to $\overline{U}_i$ and $\overline{V}_i$. The edges in cyclic order around the vertex at the end at $\overline{U}_i$ correspond to the simple modules in the list above. The length of $\overline{U}_i$ is a multiple of the number of distinct modules in this list, namely the number of images of $\overline{U}_i$ under ρ, and this multiple is used as the multiplicity at the vertex. □

PROPOSITION 4.18.4. *If Λ is a Brauer graph algebra of finite representation type then the Brauer graph is a Brauer tree.*

PROOF. We first show that if the graph has a cycle then the algebra has infinite representation type. If $S_1, \ldots, S_r$ are simple modules which form a cycle in the graph of minimal length, then the projective cover P_i of S_i has a quotient M_i with $M_i/\mathrm{Rad}(M_i) \cong S_i$ and $\mathrm{Soc}(M_i) \cong S_{i-1} \oplus S_{i+1}$, where the subscripts are taken modulo r, and $\mathrm{Rad}(M_i)/\mathrm{Soc}(M_i)$ does not involve any of $S_1, \ldots, S_r$ as subquotients. So $M_1 \oplus M_3$ has a diagonal copy of S_2 as a submodule, which we may quotient out. Similarly $M_1 \oplus M_3 \oplus M_5 \oplus \cdots \oplus M_{r-1}$ (going once round the cycle if r is even and twice if r is odd) has a diagonally embedded submodule $S_2 \oplus S_4 \oplus \cdots \oplus S_{r-2}$ which we quotient out to make a module $\hat{M}$, which still has two copies of S_r in its socle. Since k is infinite, there are infinitely many isomorphisms λ from the copy of S_r in

$\mathrm{Soc}(M_1)$ and the copy in $\mathrm{Soc}(M_{r-1})$. Each such isomorphism defines a diagonally embedded copy of S_1 in $\hat{M}$, and we write $M(\lambda)$ for the quotient. We claim that the $M(\lambda)$ are non-isomorphic indecomposable Λ-modules. First we show that $M(\lambda)$ is indecomposable. If $r = 1$ or 2 then $M(\lambda)$ has a unique minimal submodule and is hence indecomposable, so suppose $r > 2$. Note that $M(\lambda)/\mathrm{Rad}(M(\lambda))$ has no composition factors in common with $\mathrm{Rad}(M(\lambda))/\mathrm{Soc}(M(\lambda))$. Denote by $\phi_i(\lambda)$ the map $M_i \to M(\lambda)$ defined by the construction. Suppose $M(\lambda) = M' \oplus M''$. Since $M(\lambda)/\mathrm{Rad}(M(\lambda))$ is multiplicity free, the image of $\phi_i(\lambda)$ composed with projection onto one of the summands, say M', has non-zero image in $M'/\mathrm{Rad}(M')$ and the composite with the other projection has zero image in $M''/\mathrm{Rad}(M'')$. Since S_i does not appear as a composition factor of $\mathrm{Rad}(M'')/\mathrm{Soc}(M'')$ this means that the image of $M_i \xrightarrow{\phi_i(\lambda)} M(\lambda) \to M''$ lies in $\mathrm{Soc}(M'')$ and so the image of $\mathrm{Soc}(M_i)$ in M'' is zero. Thus both the simple summands of $\mathrm{Soc}(M_i)$ lie in M'. One of these simple summands also lies in the image of $\mathrm{Soc}(M_{i+1})$, and so arguing by induction we see that every simple module in the socle of M lies in the socle of M' and so $M'' = 0$.

Next we show that if $M(\lambda) \cong M(\mu)$ then $\lambda = \mu$. Suppose that $\psi : M(\lambda) \to M(\mu)$ is an isomorphism. Then $\psi \circ \phi_i(\lambda)$ agrees on $\mathrm{Soc}(M_i)$ with some non-zero multiple $\alpha_i\phi_i(\mu)$ of $\phi_i(\mu)$. Comparing the socles of M_i and M_{i+1}, we see that $\alpha_i = \alpha_{i+1}$, so that by induction $\alpha_1 = \alpha_r$. Then comparing the socles of M_1 and M_r we see that $\lambda = \mu$. We have assumed that $r > 2$ in this argument, but a similar argument works for $r \leq 2$.

We have now shown that if Λ is of finite representation type then the graph has no cycles, so it remains to show that there is at most one vertex with multiplicity greater than one. If there is more than one, we may choose a path with no repetitions, where the two end vertices have multiplicity greater than one and none of the remaining vertices do. Let $S_1, \dots, S_r$ be the simple modules corresponding to the edges along this path. As before, the projective cover P_i of S_i has a quotient M_i with $M_i/\mathrm{Rad}(M_i) \cong S_i$, $\mathrm{Rad}(M_i)/\mathrm{Soc}(M_i)$ does not involve any of $S_1, \dots, S_r$, $\mathrm{Soc}(M_i) \cong S_{i-1} \oplus S_{i+1}$ for $2 \leq i \leq r-1$, $\mathrm{Soc}(M_1) \cong S_1 \oplus S_2$ and $\mathrm{Soc}(M_r) \cong S_{r-1} \oplus S_r$. So we may apply exactly the same procedure as before to obtain quotients $M(\lambda)$ of $M_1 \oplus M_3 \oplus \cdots \oplus M_{r-1} \oplus M_r \oplus M_{r-2} \oplus \cdots \oplus M_2$ if r is even and $M_1 \oplus M_3 \oplus \cdots \oplus M_{r-2} \oplus M_r \oplus M_{r-1} \oplus M_{r-3} \oplus \cdots \oplus M_2$ if r is odd. Exactly the same arguments show that the $M(\lambda)$ are distinct and indecomposable. □

Recent Progress: Since the first edition of this book was published, Erdmann [**226**] has proved that for a block of a group algebra having wild representation type, all connected components of the stable quiver have type A_∞.

CHAPTER 5

Representation rings and Burnside rings

Representation rings are a convenient way of organising information about direct sums and tensor products of modules. J. A. Green was the first person to make a systematic investigation of representation rings in the 1960's [**117**]. For this reason representation rings are also known as Green rings. In this chapter we investigate representation theory from the point of view of the structure of representation rings.

In the study of representation theory in characteristic zero, it is customary to work in terms of the character table, namely the square table whose rows are indexed by the ordinary irreducible representations, whose columns are indexed by the conjugacy classes of group elements, and where a typical entry gives the *trace* of the group element on the representation. Why do we use the trace function? This is because the maps $M \mapsto \mathrm{tr}(g, M)$ are precisely the ring homomorphisms from the representation ring to $\mathbb{C}$, and these homomorphisms separate representations. In particular, in this case the representation ring is semisimple. This has the effect that we can compute with representations easily and effectively in terms of their characters; representations are distinguished by their characters, direct sum corresponds to addition and tensor product corresponds to multiplication. The orthogonality relations state that we may determine the dimension of the space of homomorphisms from one representation to another by taking the inner product of their characters.

How much of this carries over to characteristic p, where $p \mid |G|$? The first problem is that Maschke's theorem no longer holds; a representation may be indecomposable without being irreducible. Thus the concepts of representation ring $a(G)$ and Grothendieck ring do not coincide. The latter is a quotient of the former by the "ideal of short exact sequences" $a_0(G, 1)$. Brauer discovered the remarkable fact (he did not state it in this language) that the Grothendieck ring $\mathcal{R}(G) = a(G)/a_0(G, 1)$ is *semisimple*, and found the set of algebra homomorphisms from this to $\mathbb{C}$, in terms of lifting eigenvalues. Thus he gets a square character table, giving information about composition factors of modules, but saying nothing about how they are glued together. For some time, it was conjectured that $a(G)$ has no nilpotent elements in general. However, it is now known that $a(G)$ has no nilpotent elements whenever the Sylow p-subgroups of G are cyclic (p is the characteristic of k), as well as a few other cases in characteristic two, whilst in general there are nilpotent elements (Green [**117**], O'Reilly [**158**], Zemanek [**208, 210**], Benson and

Carlson [**20**]). It is still not known whether there are nilpotent elements in the representation ring of an elementary abelian 2-group of order at least eight.

The next feature of ordinary character theory which we may wish to mimic is the bilinear form and orthogonality relations. There are two sensible bilinear forms to use here, which both agree with the usual inner product in the case of characteristic zero. These are

$$(M,N) = \dim_k \operatorname{Hom}_{kG}(M,N), \quad \langle M,N\rangle = \text{rank of} \sum_{g\in G} g \text{ on } \operatorname{Hom}_k(M,N).$$

There are elements u and v of $a(G)$ with $uv = 1$, $(M,N) = \langle v.M,N\rangle = \langle M,u.N\rangle$ and $\langle M,N\rangle = (u.M,N) = (M,v.N)$. It is thus easy to pass back and forth between these two bilinear forms, and the second has the advantage that it is symmetric. The non-singularity of these bilinear forms follows from the fact that the almost split sequences provide dual elements to the indecomposable modules. These dual elements correspond to the simple modules and the almost split sequences.

5.1. Representation rings and Grothendieck rings

Suppose R is a commutative ring of coefficients. We define the **representation ring** $a(G) = a(RG)$ to be the ring with generators the isomorphism classes $[M]$ of RG-lattices (finitely generated R-projective RG-modules), and relations

$$[M] + [N] = [M \oplus N], \quad [M].[N] = [M \otimes N].$$

The identity and zero elements of this ring are given by

$$1 = [R], \qquad 0 = [0]$$

where R is the trivial RG-lattice and 0 is the zero lattice.

The usual properties of direct sum and tensor product show that $a(G)$ is a commutative associative ring with identity. In the presence of the Krull–Schmidt theorem, the additive structure of $a(G)$ is clear. Each element can be written uniquely as a finite sum $\sum_i n_i[M_i]$ with the M_i indecomposable. Thus the additive group of $a(G)$ is free abelian on generators $[M_\alpha]$, one for each isomorphism class of indecomposable lattice M_α. Note that there are usually infinitely many of these, so that $a(G)$ is quite a large object. It is, for example, usually not a Noetherian ring.

It is often convenient to introduce various rings of coefficients (not to be confused with the ring R) into the representation ring. We set

$$A(G) = \mathbb{C} \otimes_{\mathbb{Z}} a(G), \quad a(G)_{\mathbb{Q}} = \mathbb{Q} \otimes_{\mathbb{Z}} a(G), \quad a(G)_p = \mathbb{Z}[1/p] \otimes_{\mathbb{Z}} a(G).$$

More generally, if S is a set of primes, we can look at the localised representation ring $a(G)_{(S)}$, obtained by allowing denominators coprime to S. Thus for example if $S = \{p\}$, we write $a(G)_{(p)}$ for the p-local representation ring, and if S is the set of all primes other than p, we obtain $a(G)_p$ as above.

We also introduce various ideals and subrings of $a(G)$ as follows. If H is a subgroup of G, we denote by $a(G,H)$ the ideal (cf. Corollary 3.6.7) in $a(G)$ spanned by the relatively H-projective RG-lattices. If X is a permutation representation of G, we write $a(G,X)$ for the ideal spanned by the relatively X-projective RG-lattices (see Definition 3.6.13). Similarly, if $\mathcal{X}$ is a collection of subgroups closed under conjugation and intersections, we denote by $a(G,\mathcal{X})$ the ideal (cf. Corollary 3.6.8) in $a(G)$ spanned by summands of sums of relatively H-projective RG-lattices for $H \in \mathcal{X}$. We similarly write $A(G,H)$, $A(G,X)$, $A(G,\mathcal{X})$, $a(G,H)_{\mathbb{Q}}$, and so on, for the corresponding notions in $A(G)$, $a(G)_{\mathbb{Q}}$, etc.

If H is a subgroup of G, we denote by $a_0(G,H)$ the ideal spanned by the difference elements of the form $M_2 - M_1 - M_3$, where $0 \to M_1 \to M_2 \to M_3 \to 0$ is a short exact sequence of RG-lattices which splits on restriction to H. If X is a permutation representation of G, we write $a_0(G,X)$ for the ideal spanned by the difference elements of the above form for X-split sequences. If $\mathcal{X}$ is a collection of subgroups closed under conjugation and intersections, we write $a_0(G,\mathcal{X})$ for the intersection of the $a_0(G,H)$ for $H \in \mathcal{X}$. The **Grothendieck ring** of RG-lattices is by definition the quotient $\mathcal{R}(G) = a(G)/a_0(G,1)$. Note that all short exact sequences of RG-lattices split on restriction to the trivial subgroup, since RG-lattices are by definition projective as R-modules.

We write $K_0(RG)$ for the ring with generators the isomorphism classes $[P]$ of finitely generated projective RG-modules, with the same relations

$$[P] + [Q] = [P \oplus Q], \quad [P].[Q] = [P \otimes Q]$$

as before. This is called the Grothendieck ring of projective RG-modules. The same definition may be made with RG replaced by any ring Λ.

Now in case $H = 1$, an RG-lattice which is relatively 1-projective is projective. It follows that $a(RG,1)$ is the image in $a(RG)$ of the natural map $K_0(RG) \to a(RG)$. If the Krull–Schmidt theorem holds for RG-lattices, this map is injective and we have $K_0(RG) \cong a(RG,1)$. It is unclear whether this is true in the absence of the Krull–Schmidt theorem.

If H is a subgroup of G, we have restriction and induction maps on representation rings and Grothendieck rings given by restriction and induction of RG-lattices and RH-lattices

$$\begin{aligned} &\mathrm{res}_{G,H} : a(RG) \to a(RH) && \mathrm{res}_{G,H} : K_0(RG) \to K_0(RH) \\ &\mathrm{ind}_{H,G} : a(RH) \to a(RG) && \mathrm{ind}_{H,G} : K_0(RH) \to K_0(RG). \end{aligned}$$

Note that $\mathrm{res}_{G,H}$ is a ring homomorphism, while $\mathrm{ind}_{H,G}$ is just an additive map, whose image is an ideal by Proposition 3.3.3 (i).

Tensor induction is not additive. However, by Proposition 3.15.2 (i), (iii) and (iv), it induces a well defined ring homomorphism

$$\mathrm{ind}^{\otimes}_{H,G} : a(RH)/\sum_{K<H} \mathrm{Im}(\mathrm{ind}_{K,H}) \to a(RG)/\sum_{K<G} \mathrm{Im}(\mathrm{ind}_{K,G}).$$

5.2. Ordinary character theory

The easiest case in which to understand $a(G)$ is the case $R = \mathbb{C}$. In a sense, this is the model for all further developments. In this case, we have ring homomorphisms

$$\begin{aligned} t_g : a(G) &\to \mathbb{C} \\ M &\mapsto \mathrm{tr}(g, M), \end{aligned}$$

one for each conjugacy class of elements $g \in G$. Using modules induced from cyclic subgroups, it is easy to see that these homomorphisms are distinct (the argument is spelled out in detail in Lemma 5.3.1).

LEMMA 5.2.1. *The $\mathbb{Z}$-rank of $a(\mathbb{C}G)$ (i.e., the number of isomorphism types of ordinary irreducible representations of G) is equal to the number of conjugacy classes of elements of G.*

PROOF. By Maschke's theorem, $a(\mathbb{C}G) = \mathcal{R}(\mathbb{C}G)$ has a $\mathbb{Z}$-basis consisting of the irreducible $\mathbb{C}G$-modules. Using the Wedderburn structure theorem (Section 1.3), $\mathbb{C}G$ is a sum of matrix algebras over $\mathbb{C}$. Each matrix component corresponds to an isomorphism class of simple module, and has a one dimensional centre given by scalar matrices. Thus the $\mathbb{Z}$-rank of $a(\mathbb{C}G)$ is equal to the dimension of $Z(\mathbb{C}G)$. The lemma now follows from the fact that the conjugacy class sums form a basis for $Z(\mathbb{C}G)$. □

LEMMA 5.2.2. *Suppose R_1 is a commutative ring, and R_2 is an integral domain. Then any set of distinct ring homomorphisms $\lambda_i : R_1 \to R_2$ is linearly independent over R_2.*

PROOF. Suppose $\sum_{i=1}^n \alpha_i \lambda_i = 0$ is a linear relation between such ring homomorphisms, with $\alpha_i \in R_2$ and n minimal. There is no linear relation with $n = 1$. Choose $y \in R_1$ such that $\lambda_1(y) \neq \lambda_2(y)$. Then for all $x \in R_1$ we have

$$\sum_{i=1}^n \alpha_i \lambda_i(x)\lambda_i(y) = \sum_{i=1}^n \alpha_i \lambda_i(xy) = 0$$

and so

$$\sum_{i=2}^n \alpha_i(\lambda_i(y) - \lambda_1(y))\lambda_i(x) = 0.$$

Since R_2 is an integral domain, the coefficient of λ_2 is non-zero, and so this is a shorter linear relation among the λ_i, contradicting the minimality of n. □

PROPOSITION 5.2.3. *Every ring homomorphism $\mathcal{R}(\mathbb{C}G) = a(\mathbb{C}G) \to \mathbb{C}$ is of the form t_g for some $g \in G$. The sum of these maps is an isomorphism after tensoring with $\mathbb{C}$:*

$$\sum t_g : \mathbb{C} \otimes_{\mathbb{Z}} \mathcal{R}(G) = A(G) \xrightarrow{\cong} \bigoplus_{\substack{\text{ccl's} \\ \text{of } g \in G}} \mathbb{C}.$$

PROOF. Using the above two lemmas, with $R_1 = a(\mathbb{C}G)$ and $R_2 = \mathbb{C}$, we see that there can be at most as many ring homomorphisms $a(\mathbb{C}G) \to \mathbb{C}$ as there are conjugacy classes of elements in G. Since the t_g are such ring homomorphisms, we have equality. □

The **character** of a $\mathbb{C}G$-module is its image under the above map $\sum t_g$.

5.3. Brauer character theory

If $R = k$ is a field of characteristic p, we have to modify the procedure given in the last section for obtaining ring homomorphisms $a(kG) \to \mathbb{C}$. Let $\hat{k}$ be an algebraic closure of k, and let γ be the p'-part of the exponent of G. Then the γth roots of unity in $\hat{k}$ and in $\mathbb{C}$ both form a cyclic group of order γ. We choose once and for all an isomorphism between these cyclic groups. Let g be a p'-element of G. Given a kG-module M, we restrict it to $\langle g \rangle$ and extend the field to $\hat{k}$. Then the representation breaks up as a direct sum of eigenspaces of g, and each eigenvalue of g is a γth root of unity in $\hat{k}$. We define $t_g(M)$ to be the sum of the corresponding roots of unity in $\mathbb{C}$. It is clear that $t_g : a(G) \to \mathbb{C}$ is a ring homomorphism. We also write t_g for the corresponding algebra homomorphism $A(G) \to \mathbb{C}$.

LEMMA 5.3.1. *Suppose k contains the primitive γth roots of unity, where γ is the p'-part of the exponent of G. Let t_g be as above. Then there is an element $x \in A(G,1)$ with $t_g(x) \neq 0$ and $t_{g'}(x) = 0$ for g' a p'-element not conjugate to g in G.*

PROOF. Suppose g has order n, and let ε be the primitive nth root of unity in k corresponding to $e^{2\pi i/n} \in \mathbb{C}$. Let M_j be the one dimensional representation $g \mapsto (\varepsilon^j)$ of $\langle g \rangle$ and take

$$x = \sum_{j=1}^{n} e^{-2\pi ij/n}[M_j \uparrow^G] \in A(G).$$

If g' is not conjugate to an element of $\langle g \rangle$ then $M_j \uparrow^G \downarrow_{\langle g' \rangle}$ is a free module (for example by the Mackey decomposition theorem) and so $t_{g'}(x) = 0$. We also have

$$M_j \uparrow^G \downarrow_{\langle g \rangle} = \bigoplus_{h \in N_G\langle g \rangle / \langle g \rangle} {}^hM_j \oplus (\text{free})$$

so that since

$$\sum_{j=1}^{n} e^{-2\pi ij/n} t_{g^m}({}^hM_j) = \begin{cases} n & \text{if } hgh^{-1} = g^m \\ 0 & \text{otherwise} \end{cases}$$

it follows that $t_{g^m}(x) \neq 0$ if and only if g^m is conjugate to g in $N_G\langle g \rangle$. □

PROPOSITION 5.3.2. *Suppose M and M' are kG-modules. Then the following are equivalent:*

(i) *$t_g(M) = t_g(M')$ for all p'-elements $g \in G$.*

(ii) *M and M' have the same composition factors.*

PROOF. Since all short exact sequences split on restriction to $\langle g\rangle$ for g a p'-element, it is clear that (ii) implies (i). Conversely, suppose $t_g(M) = t_g(M')$ for all p'-elements $g \in G$. We may replace M and M' with completely reducible modules with the same composition factors, without affecting the values of $t_g(M)$ and $t_g(M')$. Let the irreducible kG-modules be $M_1, \dots, M_r$, and let the multiplicity of M_i in M be a_i and in M' be b_i.

If k' is a finite extension of k, then the map $\mathcal{R}(kG) \to \mathcal{R}(k'G)$ given by extension of scalars is injective, since composing with the map $\mathcal{R}(k'G) \to \mathcal{R}(kG)$ given by restriction of scalars gives multiplication by $|k' : k|$. We may thus assume that k is a splitting field for G.

By the Wedderburn structure theorem, we may choose elements $x_i \in kG$ with trace 1 on M_i and zero on the other M_j. Namely, regarding $kG/J(kG)$ as a sum of matrix algebras over k, we choose a pre-image x_i in kG of the element consisting of a 1 in the top left entry of the ith matrix component and zeros everywhere else.

Since the trace of an element of G is equal to the trace of its p'-part, the hypothesis tells us that every element of kG has the same trace on M as on M'. In particular, the elements x_i do, and so $a_i \equiv b_i \bmod p$. So stripping off some common direct summands, we may assume that $a_i \equiv b_i \equiv 0 \bmod p$. If we divide these multiplicities by p, the values of t_g are also divided by p, and so we may continue by induction to show that the multiplicities are equal. $\square$

THEOREM 5.3.3. *Every ring homomorphism $\mathcal{R}(G) = a(kG)/a_0(kG,1) \to \mathbb{C}$ is of the form t_g for some p'-element $g \in G$. If k contains the γth roots of unity, then the sum of these maps is an isomorphism after tensoring with $\mathbb{C}$:*

$$\sum t_g : \mathbb{C} \otimes_{\mathbb{Z}} \mathcal{R}(G) = A(G)/A_0(G,1) \xrightarrow{\cong} \bigoplus_{\substack{\text{ccl's of} \\ p'\text{-elements } g\in G}} \mathbb{C}.$$

PROOF. By the proposition, the map $\sum t_g$ is injective. By Lemmas 5.3.1 and 5.2.2, the t_g are linearly independent, and so $\sum t_g$ is surjective, and there are no more ring homomorphisms $a(kG)/a_0(kG,1) \to \mathbb{C}$. $\square$

The **Brauer character** of a kG-module is its image under the above map $\sum t_g$. In contrast to the situation in characteristic zero, it turns out that in characteristic p the field of definition of a representation may be deduced from the character values, in the following sense.

PROPOSITION 5.3.4. *Let k be an algebraically closed field of characteristic p, and let $\phi : G \to GL_r(k)$ be a map affording a kG-module M. Denote by $F^n : GL_r(k) \to GL_r(k)$ the* **Frobenius map**, *which replaces matrix entries by their p^nth powers. If the representation $F^n(M)$ afforded by the map $F^n \circ \phi$ is isomorphic to M, then there is an $\mathbb{F}_{p^n}G$-module M_0 with*

$$k \otimes_{\mathbb{F}_{p^n}} M_0 \cong M.$$

Note that the character of $F^n(M)$ is given by $t_g(F^n(M)) = t_{g^{p^n}}(M)$.

PROOF. Since $M \cong F^n(M)$, there is a matrix $X \in GL_r(k)$ such that for all $g \in G$, $X\phi(g)X^{-1} = F^n\phi(g)$. By a theorem of Lang (see for example Srinivasan [**191**], p. 11), the map $Y \mapsto F^n(Y)^{-1}Y$ on $GL_r(k)$ is surjective, and so we may write $X = F^n(Y)^{-1}Y$ for some $Y \in GL_r(k)$. Then for all $g \in G$ we have $Y\phi(g)Y^{-1} = F^n(Y\phi(g)Y^{-1})$. Thus changing basis by means of Y, we see that the image of $G \to GL_r(k)$ lies in $GL_r(\mathbb{F}_{p^n})$ as required. □

COROLLARY 5.3.5. *Suppose k contains the primitive γth roots of unity, where γ is the p'-part of the exponent of G. Then the $\mathbb{Z}$-rank of $\mathcal{R}(kG)$ (i.e., the number of isomorphism types of p-modular irreducible representations of G) is equal to the number of conjugacy classes of p'-elements of G.* □

EXERCISE. Use the above arguments to count the number of isomorphism types of p-modular irreducible representations of G in case k does not contain the γth roots of unity.

COROLLARY 5.3.6. *The representation ring $A(G)$ decomposes as a direct sum of ideals*

$$A(G) = A(G,1) \oplus A_0(G,1).$$

The **Cartan homomorphism**

$$c : A(G,1) \hookrightarrow A(G) \twoheadrightarrow A(G)/A_0(G,1)$$

is an isomorphism. In particular, the Cartan matrix is non-singular. Thus projective kG-modules with the same Brauer character are isomorphic.

PROOF. By Lemmas 5.3.1 and 5.2.2, the t_g are linearly independent on $A(G,1)$, and so c is injective. But the number of isomorphism types of projective indecomposables is equal to the number of isomorphism types of irreducibles, so the dimensions of $A(G,1)$ and $A(G)/A_0(G,1)$ are equal, so c is an isomorphism. Letting $e = c^{-1}(1)$, we see that e is an idempotent, $A(G,1) = e.A(G)$, and $A_0(G,1) = (1-e).A(G)$. □

5.4. G-sets and the Burnside ring

We now go through the same process with permutation representations as we went through in the last two sections with linear representations. The result is called the Burnside ring. There is a natural homomorphism from the Burnside ring to the representation ring of RG for any coefficient ring R, and we shall use this fact to obtain information about representation rings (namely various "induction theorems") in Section 5.6.

We define the **Burnside ring** $b(G)$ to be the ring with generators the isomorphism classes $[X]$ of permutation representations of G on finite sets, and relations

$$[X] + [Y] = [X \dot{\cup} Y], \quad [X].[Y] = [X \times Y]$$

giving the addition and multiplication in terms of disjoint union and Cartesian product. The identity element of this ring corresponds to the one point set with trivial action, and the zero element corresponds to the empty set.

Now any permutation representation X of G may be expressed uniquely as a disjoint union of orbits. The isomorphism classes of transitive permutation representations are in one-one correspondence with the conjugacy classes of subgroups in such a way that the permutation representation G/H corresponds to the conjugacy class of H, which is characterised as the stabiliser of a point. So the additive structure of $b(G)$ is easy to describe. It is the free abelian group, with basis corresponding to the transitive permutation representations $[G/H]$, one for each conjugacy class of subgroups $H \leq G$.

EXAMPLE. Suppose G is the symmetric group S_3. Then we shall denote the transitive permutation representations of G by 1, a, b and c, on 1, 2, 3 and 6 objects respectively. The multiplication table of $b(G)$ is as follows:

$\times$	1	a	b	c
1	1	a	b	c
a	a	$2a$	c	$2c$
b	b	c	$b+c$	$3c$
c	c	$2c$	$3c$	$6c$

What are the ring homomorphisms $f : b(G) \to \mathbb{C}$? Clearly for any such ring homomorphism we have $f(1) = 1$. Since $f(c)^2 = 6f(c)$, either $f(c) = 0$ or $f(c) = 6$.

Case (i): $f(c) = 6$. In this case $f(a)f(c) = 2f(c)$ implies that $f(a) = 2$, while $f(b)f(c) = 3f(c)$ implies that $f(b) = 3$.

Case (ii) : $f(c) = 0$ implies that $f(a)f(b) = 0$. In this case we have $f(a)^2 = 2f(a)$ and $f(b)^2 = f(b)$, and so we have either $f(a) = 0$, $f(b) = 1$, or $f(a) = 2$, $f(b) = 0$, or $f(a) = 0$, $f(b) = 0$.

We may summarise this information in the following table:

c	6	0	0	0
b	3	1	0	0
a	2	0	2	0
1	1	1	1	1

We can interpret this table in terms of the numbers of fixed points on subgroups as follows. If $H \leq G$, the map

$$f_H : b(G) \to \mathbb{Z} \subseteq C$$

sending a permutation representation X to $|X^H|$ is a ring homomorphism, since

$$|(X \dot{\cup} Y)^H| = |X^H| + |Y^H|, \quad |(X \times Y)^H| = |X^H||Y^H|.$$

Clearly if H is not conjugate to K then $f_H \neq f_K$ (evaluate on G/H and G/K).

LEMMA 5.4.1. *We have $f_H(G/K) \neq 0$ if and only if H is conjugate to a subgroup of K.* □

Let $B(G) = \mathbb{C} \otimes_{\mathbb{Z}} b(G)$. Then f_H extends in an obvious way to a $\mathbb{C}$-linear ring homomorphism

$$f_H : B(G) \to \mathbb{C}$$

THEOREM 5.4.2. *Every ring homomorphism $b(G) \to \mathbb{C}$ is of the form f_H for some $H \leq G$. The sum of these maps is an isomorphism after tensoring with $\mathbb{C}$:*

$$\sum f_H : B(G) \to \bigoplus_{H \leq_G G} \mathbb{C}.$$

PROOF. By Lemma 5.2.2, the f_H are linearly independent, and so the above map $\sum f_H$ is surjective. Since $\dim_{\mathbb{C}} B(G)$ is equal to the number of conjugacy classes of subgroups, it follows that it is an isomorphism. □

We write ε_H for the primitive idempotent corresponding to H in the right hand side of the above isomorphism, and e_H for the corresponding element of $B(G)$.

It follows from the above theorem that

$$\sum f_H : b(G) \to \bigoplus_{H \leq_G G} \mathbb{Z}$$

is injective with finite cokernel. How big is this cokernel? Choosing bases by listing subgroups in non-decreasing order of size, the matrix of $\sum f_H$ is

$$\begin{pmatrix} * & & 0 \\ & \ddots & \\ * & & * \end{pmatrix}.$$

The diagonal entries are $f_H(G/H) = |N_G(H) : H|$, and so the size of the cokernel, which is the determinant of this matrix, is equal to

$$\prod_{H \leq_G G} |N_G(H) : H|.$$

REMARK. Burnside calls the above matrix the **table of marks**. In section 185 of his book [**44**], you will find the following table of marks for the alternating group A_4.

	1	C_2	C_3	V_4	A_4
1	12	0	0	0	0
C_2	6	2	0	0	0
C_3	4	0	1	0	0
V_4	3	3	0	3	0
A_4	1	1	1	1	1

CONGRUENCES AND IDEMPOTENTS IN $b(G)$. The idempotents in $B(G)$ are the elements

$$\sum_{H \in \mathcal{H}} e_H$$

where $\mathcal{H}$ is a collection of representatives of distinct conjugacy classes of subgroups.

QUESTION 5.4.3. *When is* $\sum_{H\in\mathcal{H}} e_H$ *in* $b(G)$*?*

More generally, if S is a set of primes, we can look at the localised Burnside ring $b(G)_{(S)}$, obtained by allowing denominators coprime to S. Thus for example if $S = \{p\}$, we write $b(G)_{(p)}$ for the p-local Burnside ring, and if S is the set of all primes other than p, we write $b(G)_p$ for $\mathbb{Z}[1/p] \otimes_{\mathbb{Z}} b(G)$. All these rings may be thought of as subrings of $B(G)$.

QUESTION 5.4.4. *When is* $\sum_{H\in\mathcal{H}} e_H$ *in* $b(G)_{(S)}$*?*

LEMMA 5.4.5 (Burnside). *The number of orbits of* G *on* X *is equal to*

$$\frac{1}{|G|}\sum_{g\in G}|X^{\langle g\rangle}|$$

and in particular

$$\sum_{g\in G}|X^{\langle g\rangle}| \equiv 0 \pmod{|G|}.$$

PROOF. Count $\{(x,g) \mid xg = x\}$ in two ways. □

Hence, if $H \trianglelefteq K \leq G$, then the quotient group K/H acts on X^H, and we have

$$\sum_{\bar{k}\in K/H}|X^{\langle H,k\rangle}| \equiv 0 \pmod{|K/H|}.$$

Here, k denotes any pre-image of $\bar{k}$ in K, and the group $\langle H, k\rangle$ is clearly independent of this choice.

We can obtain a complete set of congruences (i.e., characterising the image of $b(G)$ under the map $\sum f_H$) by taking $K = N_G(H)$ for each conjugacy class of subgroups H of G.

These congruences are independent since they form a lower triangular matrix of congruences with ones on the diagonal. For example, in the case of the example above from Burnside's book, the congruences say that

$$\begin{pmatrix}12&0&0&0&0\\6&2&0&0&0\\4&0&1&0&0\\3&3&0&3&0\\1&1&1&1&1\end{pmatrix}\begin{pmatrix}1&0&0&0&0\\3&1&0&0&0\\8&0&1&0&0\\0&1&0&1&0\\0&0&0&2&1\end{pmatrix} \equiv 0 \pmod{\begin{pmatrix}12&2&1&3&1\end{pmatrix}}.$$

The second matrix in the above equation has its rows and columns labelled by the conjugacy classes of subgroups of G, and has a non-zero entry if and only if the subgroup H corresponding to the column is contained normally in a conjugate of the subgroup K corresponding to the row, with cyclic quotient. The entry is the number of times conjugates of K appear in the above congruence for $N_G(H)/H$; namely the sum, over the subgroups K in the given G-conjugacy class that contain H normally with cyclic quotient, of the number of generators of the cyclic group K/H.

These congruences therefore define an additive subgroup of $\sum_{H\leq_G G}\mathbb{Z}$ of the same index as $b(G)$, which therefore *is* $b(G)$. We have therefore proved the following theorem of Dress [**86**] (our presentation follows tom Dieck [**72**]).

THEOREM 5.4.6. *The image of the map*

$$\sum f_H : b(G) \to \bigoplus_{H\leq_G G} \mathbb{Z}$$

is given by the congruences

$$\sum_k |\{\text{generators of } K/H\}|.f_H(x) \equiv 0 \pmod{|N_G(H):H|}$$

where the sum runs over the subgroups $K \leq G$ with $H \trianglelefteq K$ and K/H cyclic. □

Note that we can separate these congruences into p-primary components by using the pairs of groups $H \trianglelefteq N \leq G$ with $N/H \in \mathrm{Syl}_p(N_G(H)/H)$.

THEOREM 5.4.7 (Dress). (i) *An idempotent $\sum_{H\in\mathcal{H}} e_H \in B(G)$ lies in $b(G)$ if and only if, whenever $H \trianglelefteq H'$ with cyclic quotient, $H \in \mathcal{H} \Leftrightarrow H' \in \mathcal{H}$.*

(ii) *An idempotent $\sum_{H\in\mathcal{H}} e_H \in B(G)$ lies in $b(G)_{(S)}$ if and only if, whenever $H \trianglelefteq H'$ of index $p \in S$, H is conjugate to a subgroup in $\mathcal{H}$ if and only if H' is.*

PROOF. Suppose $\sum_{H\in\mathcal{H}} e_H \in b(G)_{(S)}$ and $H \trianglelefteq H'$ with cyclic quotient of order $p \in S$. Then the congruence $|X^H| \equiv |X^{H'}| \bmod |H' : H|$ implies that for all $x \in b(G)_{(S)}$, $f_H(x) \equiv f_{H'}(x)$. Since $0 \not\equiv 1 \bmod p$, $H \in \mathcal{H} \Leftrightarrow H' \in \mathcal{H}$. Conversely, by the above theorem, if these congruences are satisfied then $\sum_{H\in\mathcal{H}} e_H \in b(G)_{(S)}$. □

COROLLARY 5.4.8 (Dress). (i) *The primitive idempotents in $b(G)$ are of the form*

$$\sum_{H^{(\infty)}=H_0} e_H$$

where H runs through representatives of conjugacy classes of subgroups of G for which $H^{(\infty)}$ is a given perfect subgroup H_0 of G.

In particular, G is soluble if and only if the only idempotents in $b(G)$ are 0 and 1.

(ii) *The primitive idempotents in $b(G)_{(p)}$ are of the form*

$$\sum_{O^p(H)=H_0} e_H$$

where H runs through representatives of conjugacy classes of subgroups of G for which $O^p(H)$ is a given p-perfect subgroup H_0 of G (i.e., subgroup with no normal subgroup of index p). □

INDUCTION AND RESTRICTION. If $H \leq G$, **restriction** is a ring homomorphism

$$\mathrm{res}_{G,H} : b(G) \to b(H).$$

Induction is defined on permutation representations as follows. If X is an H-set, define

$$G \times_H X = (G \times X)/\sim$$

where the equivalence relation $\sim$ is given by $(gh, x) \sim (g, hx)$ for all $h \in H$. The map

$$\begin{aligned} \mathrm{ind}_{H,G} : b(H) &\to b(G) \\ X &\mapsto G \times_H X \end{aligned}$$

is a homomorphism of additive groups, but not of rings. The following lemma summarises some properties of induction and restriction.

LEMMA 5.4.9. (i) $(G \times_H X)^G = \emptyset$ *if* $H < G$.
(ii) $\mathrm{ind}_{H,G}(a.\mathrm{res}_{G,H}(b)) = \mathrm{ind}_{H,G}(a).b$.
(iii) *(Mackey formula):*

$$\mathrm{res}_{G,H}\mathrm{ind}_{K,G}(a) = \sum_{HgK} \mathrm{ind}_{H\cap {}^gK,H}\mathrm{res}_{{}^gK,H\cap {}^gK}({}^ga).$$

PROOF. This is left as an easy exercise. □

THEOREM 5.4.10. *Suppose H is a subgroup of G. Then*
(i) $\mathrm{ind}_{H,G}(e_H) = |N_G(H) : H|e_H$, *where the first e_H is in $B(H)$ and the second is in $B(G)$. In particular, as an element of $B(G)$, we have*

$$e_H \in \mathrm{Im}\,(\mathrm{ind}_{H,G})\,.$$

(ii) $B(G) = \mathrm{Im}(\mathrm{ind}_{H,G}) \oplus \mathrm{Ker}(\mathrm{res}_{G,H})$ *as a direct sum of ideals.*

PROOF. The restriction of e_H to any proper subgroup of H is zero, since all f_k vanish on it. So by the Mackey formula, for $K \leq G$ we have

$$\mathrm{res}_{G,K}\mathrm{ind}_{H,G}(e_H) = \sum_{KgH} \mathrm{ind}_{K\cap {}^gH,K}\mathrm{res}_{{}^gH,K\cap {}^gH}({}^ge_H) = \sum_{\substack{KgH \\ K \geq {}^gH}} \mathrm{ind}_{{}^gH,K}({}^ge_H).$$

The value of f_k on an element induced from a proper subgroup of K is zero, and so we have

$$f_k(\mathrm{ind}_{H,G}(e_H)) = f_k(\mathrm{res}_{G,K}\mathrm{ind}_{H,G}(e_H)) = \begin{cases} |N_G(H) : H| & H \sim K \\ 0 & H \not\sim K. \end{cases}$$

It follows that $f_k(\mathrm{ind}_{H,G}(e_H) = f_k(|N_G(H) : H|e_H)$ for all $K \leq G$, and so by Theorem 5.4.2 we have $\mathrm{ind}_{H,G}(e_H) = |N_G(H) : H|e_H$.

It now follows that the $e_{H'}$ with $H' \leq H$ lie in $\mathrm{Im}(\mathrm{ind}_{H,G})$ while the $e_{H'}$ with H' not conjugate to a subgroup of H lie in $\mathrm{Ker}(\mathrm{res}_{G,H})$, which proves the given direct sum decomposition. □

The second part of the above theorem has the following implication for representation rings [**23**].

COROLLARY 5.4.11. *For any coefficient ring R we have*

$$A(RG) = \text{Im}(\text{ind}_{H,G}) \oplus \text{Ker}(\text{res}_{G,H})$$

as a direct sum of ideals.

PROOF. There is a natural map $\phi : b(G) \to a(RG)$, which takes a permutation representation to the matrix representation with the permuted points as basis. We also write ϕ for the corresponding map $B(G) \to A(RG)$. Writing $1 = e + e'$ with $e \in \text{Im}(\text{ind}_{H,G})$ and $e' \in \text{Ker}(\text{res}_{G,H})$ as in the theorem, we see that $1 = \phi(e) + \phi(e')$ in $A(RG)$ so that

$$A(RG) = \text{Im}(\text{ind}_{H,G}) + \text{Ker}(\text{res}_{G,H}).$$

If $x \in \text{Im}(\text{ind}_{H,G}) \cap \text{Ker}(\text{res}_{G,H})$ then $x = x.1 = x.\phi(e) + x.\phi(e') = 0$ since the product of elements of $\text{Im}(\text{ind}_{H,G})$ and $\text{Ker}(\text{res}_{G,H})$ is zero. □

5.5. The trivial source ring

At the end of the last section, we made use of the natural map $\phi : b(G) \to a(RG)$, which takes a permutation representation to the matrix representation with the permuted points as basis. In other words, if X is a permutation representation, then the free R-module RX is an RG-module with the obvious action. The map ϕ is usually not injective, but we wish to study its image. For many purposes, it is better to work with subrings of $a(RG)$ which are closed under taking direct summands, so we make the following definition.

DEFINITION 5.5.1. We write $a(G, \text{Triv})$ or $a(RG, \text{Triv})$ for the subring of $a(RG)$ consisting of linear combinations of trivial source modules (i.e., direct summands of permutation modules). As usual, we write $A(G, \text{Triv})$ for $\mathbb{C} \otimes_{\mathbb{Z}} a(G, \text{Triv})$, and so on.

Let $(K, \mathcal{O}, k)$ be a splitting p-modular system for G and all its subgroups. We investigated trivial source modules in Section 3.11, and found that a trivial source kG-module has a unique lift to a trivial source $\mathcal{O}G$-module. Thus we have the following.

LEMMA 5.5.2. *The natural map*

$$a(\mathcal{O}G, \text{Triv}) \to a(kG, \text{Triv}),$$

given by reducing trivial source $\mathcal{O}G$-modules modulo $\mathfrak{p}$, is an isomorphism of rings. □

We now investigate the ring homomorphisms

$$a(kG, \text{Triv}) \cong a(\mathcal{O}G, \text{Triv}) \to \mathbb{C}.$$

DEFINITION 5.5.3. A group H is said to be **p-hypo-elementary** if the quotient $H/O_p(H)$ is cyclic; in other words, if H has a normal p-subgroup for which the quotient is a cyclic p'-group.

If H is a p-hypo-elementary subgroup of G, and $g \in H/O_p(H)$, we define a ring homomorphism

$$s_{H,g} : a(kG, \text{Triv}) \to \mathbb{C}$$

as follows. If M is a trivial source kG-module, let $M\downarrow_H = M_1 \oplus M_2$, where M_1 is a direct sum of indecomposable modules with vertex $O_p(H)$, and M_2 is a direct sum of indecomposable modules with vertex properly contained in $O_p(H)$. Since M is a trivial source module, $O_p(H)$ acts trivially on M_1, and so M_1 is a module for $H/O_p(H)$. We define

$$s_{H,g}(M) = t_g(M_1).$$

Since the tensor product of any module with a module whose vertex is properly contained in $O_p(H)$ is a direct sum of such modules (Corollary 3.6.7), it is clear that $s_{H,g}$ is a ring homomorphism.

PROPOSITION 5.5.4 (Conlon). *Suppose M_1 and M_2 are trivial source kG-modules and $s_{H,g}(M_1) = s_{H,g}(M_2)$ for all pairs (H, g). Then $M_1 \cong M_2$.*

PROOF. By stripping off common direct summands, we may suppose that no direct summand of M_1 is isomorphic to a direct summand of M_2. Let D be a maximal element of the set of vertices of summands of M_1 and M_2. Suppose $M_1\downarrow_{N_G(D)} = M_1' \oplus M_1''$ and $M_2\downarrow_{N_G(D)} = M_2' \oplus M_2''$, where M_1' and M_2' are sums of modules with vertex D, and M_1'', M_2'' are sums of modules whose vertex does not contain D. Thus M_1' and M_2' are projective $N_G(D)/D$-modules. Since $t_g(M_1') = t_g(M_2')$ for all p'-elements $g \in N_G(D)/D$, by Corollary 5.3.6 we have $M_1' \cong M_2'$ as modules for $N_G(D)/D$. By Theorem 3.12.2, the Green correspondents of summands of M_1' and M_2' are summands of M_1 and M_2, contradicting the fact that M_1 and M_2 have no isomorphic summands. □

COROLLARY 5.5.5. *Every ring homomorphism $a(G, \text{Triv}) \to \mathbb{C}$ is of the form $s_{H,g}$ for some pair (H, g) with H p-hypo-elementary and g a generator of $H/O_p(H)$. The sum of these maps is an isomorphism after tensoring with $\mathbb{C}$:*

$$\sum s_{H,g} : A(G, \text{Triv}) \to \bigoplus_{\substack{\text{ccl's of} \\ \text{pairs } (H,g)}} \mathbb{C}.$$

PROOF. By the proposition, the map $\sum s_{H,g}$ is injective. To show that the $s_{H,g}$ are distinct, we evaluate them on Green correspondents of projective $N_G(D)/D$-modules (viewed as $N_G(D)$-modules) as in the proof of the proposition. By Lemma 5.2.2 it follows that $\sum s_{H,g}$ is an isomorphism, and that there are no more ring homomorphisms from $a(G, \text{Triv})$ to $\mathbb{C}$. □

5.6. Induction theorems

As an application of the theory of Burnside rings and trivial source rings, we shall prove some induction theorems in representation theory. Each of these theorems says that a representation ring can be written as a sum of the images of induction from a certain class of subgroups. Which class of subgroups is involved depends on whether we look at the full representation ring or just the Grothendieck ring, and what coefficients we allow in the representation ring. The four induction theorems we have in mind are:

A) Artin's induction theorem,
B) Brauer's induction theorem,
C) Conlon's induction theorem,
D) Dress' induction theorem.

The first two of these are theorems about Grothendieck rings, while the last two are about representation rings.

The approach we shall be taking was communicated to me by Ken Brown, to whom I am very grateful for permission to use this material here. It is a distillation of ideas of many people, and I am sure they'll not feel too hurt if they are not mentioned here.

THEOREM 5.6.1 (Artin's induction theorem). *Suppose k is a field. Then*

$$\mathbb{C} \otimes_{\mathbb{Z}} \mathcal{R}(kG) = \sum_{\langle g \rangle \leq G} \mathrm{Im}(\mathrm{ind}_{\langle g \rangle, G}).$$

Here, $\langle g \rangle$ runs over the cyclic subgroups of G if k has characteristic zero, and the cyclic subgroups of order prime to the characteristic otherwise.

PROOF. The proof of this theorem is easy, and will be our model for proving the other induction theorems. We start off by writing down a commutative diagram:

$$\begin{array}{ccc} B(G) & \longrightarrow & \bigoplus_{H \leq_G G} \mathbb{C} \\ \downarrow \phi & & \downarrow \lambda \\ \mathbb{C} \otimes_{\mathbb{Z}} \mathcal{R}(G) & \longrightarrow & \bigoplus_{g \in_G G} \mathbb{C} \end{array}$$

In this diagram, ϕ is the map which takes a permutation representation to the matrix representation with the permuted points as basis. The elements g run through a set of representatives of the conjugacy classes of elements of G if k has characteristic zero, and the conjugacy classes of elements of order prime to the characteristic otherwise. The horizontal maps are the ones given in Proposition 5.2.3, Theorem 5.3.3 and Theorem 5.4.2. Write ε_H for the primitive idempotents in the top right hand corner of this diagram, and ε_g for the primitive idempotents in the bottom right. Recall that e_H is the preimage in $B(G)$ or ε_H. The map λ takes a primitive idempotent ε_H to $\sum_{\langle g \rangle = H} \varepsilon_g$ if H is cyclic, and 0 otherwise. It is an easy exercise to check that this diagram commutes. One proceeds coordinate at a time, and one uses the fact that the trace of a group element on a permutation representation is equal to the number of fixed points.

It follows that $\phi(e_H) = 0$ unless H is cyclic. But in Section 5.4 we showed that $e_H \in \mathrm{Im}(\mathrm{ind}_{H,G})$. So applying ϕ to the equation

$$1 = \sum_{H \leq_G G} e_H$$

in $B(G)$, we obtain the equation

$$1 = \sum_{\substack{H \leq_G G \\ \text{cyclic}}} \phi(e_H)$$

in $\mathbb{C} \otimes_{\mathbb{Z}} \mathcal{R}(G)$. In particular, we have

$$1 \in \sum_{\langle g \rangle \leq G} \mathrm{Im}(\mathrm{ind}_{\langle g \rangle, G}).$$

However, by the identity

$$\mathrm{ind}_{H,G}(a.\mathrm{res}_{G,H}(b)) = \mathrm{ind}_{H,G}(a).b$$

this sum of images of induction maps is an ideal in $\mathbb{C} \otimes_{\mathbb{Z}} \mathcal{R}(G)$. Artin's induction theorem now follows from the fact that an ideal containing the identity element must be the whole ring. □

We now turn our attention to Brauer's induction theorem. This time, we work with integer, rather than complex coefficients in our representation rings. However, since $b(G)$ does not have many idempotents (see Section 5.4), we must work with q-local coefficients as an intermediate step, where q is a prime which may or may not equal the characteristic of k.

$$\begin{array}{ccc} b(G)_{(q)} & \longrightarrow & \bigoplus_{H \leq_G G} \mathbb{C} \\ \downarrow \phi & & \downarrow \lambda \\ \mathcal{R}(G)_{(q)} & \longrightarrow & \bigoplus_{g \in_G G} \mathbb{C} \end{array}$$

Recall from Section 5.4 that the primitive idempotents in $b(G)_{(q)}$ are of the form

$$\sum_{O^q(H) = H_0} e_H$$

where H runs through representatives of conjugacy classes of subgroups of G for which $O^q(H)$ is a given q-perfect subgroup H_0 of G. Now ϕ kills such an idempotent if and only if H_0 is non-cyclic. So in $\mathcal{R}(G)_{(q)}$ we have

$$1 = \sum_{H_0 \text{ cyclic}} \left(\sum_{O^q(H) = H_0} \phi(e_H) \right).$$

DEFINITION 5.6.2. A group H is **q-hyperelementary** if $O^q(H)$ is cyclic. In other words, H has a normal cyclic q'-subgroup for which the quotient is a q-group.

PROPOSITION 5.6.3.

$$\mathcal{R}(kG)_{(q)} = \sum_{\substack{H\ q\text{-hyper-}\\ \text{elementary}}} \mathrm{Im}(\mathrm{ind}_{H,G})$$

Similarly, if S is the set of all primes not dividing $|G|$ then

$$\mathcal{R}(G)_{(S)} = \mathcal{R}(G)[|G|^{-1}] = \sum_{H\ \text{cyclic}} \mathrm{Im}(\mathrm{ind}_{H,G}).$$

Combining these statements using the Chinese remainder theorem, we obtain the first form of the Brauer induction theorem.

THEOREM 5.6.4 (Brauer's induction theorem).

$$\mathcal{R}(kG) = \sum_{\substack{H\ q\text{-hyper-}\\ \text{elementary}\\ \text{for some } q}} \mathrm{Im}(\mathrm{ind}_{H,G}). \qquad \square$$

COROLLARY 5.6.5. *Suppose $(K, \mathcal{O}, k)$ is a p-modular system. Then the decomposition map*

$$d : \mathcal{R}(KG) \to \mathcal{R}(kG)$$

(see Section 1.9) is surjective.

PROOF. Since $\mathrm{ind}_{H,G}$ commutes with d, it follows from Brauer's induction theorem that it suffices to prove that d is surjective in case G is q-hyperelementary for some prime q. Since $\mathcal{R}(kG) = \mathcal{R}(kG/O_p(G))$, we may also assume that $O_p(G) = 1$. If $q \neq p$ it follows that G has order prime to p so that d is an isomorphism. If $q = p$ then G is a split extension of a cyclic p'-group by a p-group acting faithfully, and one can check using Clifford theory that in this case the simple kG-modules lift to $\mathcal{O}G$-lattices so that d is onto in this case. $\square$

Examination of the representations of a q-hyperelementary group shows the following, which we shall not prove.

DEFINITION 5.6.6. If H is a q-hyperelementary group, let ε be a primitive $|H|$th root of unity, and let $O^q(H) = \langle a \rangle$. Then H is **k-elementary** if for all Galois automorphisms $\varepsilon \mapsto \varepsilon^t$ of $k(\varepsilon)$ over k, there exists an element $b \in H$ such that $bab^{-1} = a^t$.

THEOREM 5.6.7 (Witt–Berman). *We have*

$$\mathcal{R}(kG) = \sum_{\substack{H \leq G\\ k\text{-elementary}}} \mathrm{Im}(\mathrm{ind}_{H,G}).$$

In particular, if $k = \mathbb{C}$, the **elementary** subgroups are the subgroups which are a direct product of a q-group and a cyclic q'-group. The full form of Brauer's induction theorem is the statement that $\mathcal{R}(\mathbb{C}G)$ is the sum of the images of induction from the elementary subgroups.

Next, we turn our attention to Conlon's induction theorem. This time, we are working with the full ring of all representations in characteristic p, and there are usually more of those than we can reasonably handle. The usual way to get around this is to choose particular kinds of modules which are suited to the problem in hand. In our case, we are trying to prove induction theorems, which means that we are trying to prove that the identity element of the representation ring is in a certain sum of ideals. This means that we need to look at a suitable subring of the representation ring containing at least the image of the natural map from the Burnside ring. It turns out that the trivial source subring is the right one to look at.

Recall from Proposition 5.5.4 and its corollary that every ring homomorphism from $a(G, \mathrm{Triv})$ to $\mathbb{C}$ is of the form $s_{H,g}$ for some pair (H, g) with H p-hypo-elementary and g an element of $H/O_p(H)$, and that we have an isomorphism

$$A(G, \mathrm{Triv}) \cong \bigoplus_{\substack{\text{ccl's of} \\ \text{pairs } (H,g)}} \mathbb{C}.$$

Just as before we draw a commutative diagram:

$$\begin{array}{ccc} B(G) & \longrightarrow & \bigoplus_{H \leq_G G} \mathbb{C} \\ \Big\downarrow \phi & & \Big\downarrow \lambda \\ A(G, \mathrm{Triv}) & \longrightarrow & \bigoplus_{\substack{\text{ccl's of} \\ \text{pairs } (H,g)}} \mathbb{C} \end{array}$$

Write ε_H for the primitive idempotents in the top right hand corner of this diagram, and $\varepsilon_{H,g}$ for the primitive idempotents in the bottom right. The map λ sends ε_H to $\sum_g \varepsilon_{(H,g)}$ if H is p-hypo-elementary, and zero otherwise. We now deduce in the usual way that

$$1_{A(G,\mathrm{Triv})} \in \sum_{\substack{H\ p\text{-hypo-} \\ \text{elementary}}} \mathrm{Im}(\mathrm{ind}_{H,G}).$$

We have thus proved the following theorem.

THEOREM 5.6.8 (Conlon's induction theorem).
Suppose k is a field of characteristic p. Then

$$A(kG) = \sum_{\substack{H\ p\text{-hypo-} \\ \text{elementary}}} \mathrm{Im}(\mathrm{ind}_{H,G}). \qquad \square$$

In contrast to the situation for the Brauer induction theorem, however, none of these subgroups goes away when the field becomes large. One can show that all the conjugacy classes of maximal p-hypo-elementary subgroups are really necessary in the above theorem.

COROLLARY 5.6.9. *The sum of the restriction maps*

$$\sum \mathrm{res}_{G,H} : A(kG) \to \bigoplus_{\substack{\text{ccl's of } H \\ p\text{-hypo-elementary}}} A(kH)$$

is injective.

PROOF. We express the identity element $1 \in A(kG)$ as a sum of elements $\mathrm{ind}_{H,G}(x_H)$ with $x_H \in A(kH)$ and H p-hypo-elementary. If $y \in A(kG)$, then

$$y = y.1 = y.\sum \mathrm{ind}_{H,G}(x_H) = \sum \mathrm{ind}_{H,G}(\mathrm{res}_{G,H}(y).x_H).$$

It follows that if $\mathrm{res}_{G,H}(y)$ is zero for all H p-hypo-elementary then $y = 0$. □

Finally, Dress' induction theorem consists of doing all of the above at once. Let q be a prime, which may equal p or not. Then we have as usual a commutative diagram:

$$\begin{array}{ccc} b(G)_{(q)} & \longrightarrow & \bigoplus_{H \leq_G G} \mathbb{C} \\ \downarrow \phi & & \downarrow \lambda \\ a(G,\mathrm{Triv})_{(q)} & \longrightarrow & \bigoplus_{\substack{\text{ccl's of} \\ \text{pairs } (H,g)}} \mathbb{C} \end{array}$$

The primitive idempotent

$$\sum_{O^q(H)=H_0} e_H \in b(G)_{(q)}$$

goes to zero in $a(G,\mathrm{Triv})_{(q)}$ unless H_0 is p-hypo-elementary, and so $a(G)$ is the sum of the images of induction from the so-called "Dress subgroups".

DEFINITION 5.6.10. If q is a prime (not necessarily different from p), a (p,q)**-Dress subgroup** is a subgroup H such that $H/O_p(H)$ is q-hyper-elementary. A k**-Dress subgroup** is a subgroup H such that $H/O_p(H)$ is k-elementary for the field k of characteristic p.

THEOREM 5.6.11 (Dress' induction theorem). *Suppose that k is a field of characteristic p and S is a set of primes. Then*

$$a(kG)_{(S)} = \sum_{\substack{H\ (p,q)\text{-Dress} \\ \text{subgroup, } q \in S}} \mathrm{Im}(\mathrm{ind}_{H,G}).$$

In particular, taking for S the set of all primes,

$$a(kG) = \sum_{\substack{H\ (p,q)\text{-Dress} \\ \text{subgroup, any } q}} \operatorname{Im}(\operatorname{ind}_{H,G}). \qquad \square$$

As with the Witt–Berman theorem, one can show that over a given field k, the necessary subgroups are the k-Dress subgroups. Thus for a large enough field k, the subgroups needed are the subgroups H for which $H/O_p(H)$ is a direct product of a q-group and a cyclic group of order coprime to p and q. This time, we may assume $p \neq q$, since otherwise the O_p just grows.

5.7. Relatively projective and relatively split ideals

Suppose k is a field of characteristic p. The main theorem in this section is a theorem of Dress [**89**], which states that for any subgroup $H \leq G$, the representation ring with p inverted

$$a(G)_p = \mathbb{Z}[1/p] \otimes_{\mathbb{Z}} a(G)$$

decomposes as a direct sum of the ideal of relatively H-projective modules $a(G,H)_p$ and the ideal of H-split sequences $a_0(G,H)_p$. More generally, for any permutation representation X of G, $a(G)_p$ decomposes as a direct sum of $a(G,X)_p$ and $a_0(G,X)_p$. This latter statement is more amenable to proof by induction. The case $H = 1$ (or equivalently X is the trivial permutation representation on one point) will tell us that the determinant of the Cartan matrix is a power of p. We give another proof of this using psi operations in Section 5.9. Brauer's original proof used his characterisation of characters.

THEOREM 5.7.1 (Dress). *Suppose X is a permutation representation of G, and k is a field of characteristic p. Then*

$$a(kG)_p = a(G,X)_p \oplus a_0(G,X)_p.$$

PROOF. We first observe that it suffices to show that the identity element $1 \in a(G)_p$ lies in $a(G,X)_p + a_0(G,X)_p$. For since $a(G,X)_p$ and $a_0(G,X)_p$ are ideals, we then have $a(G)_p = a(G,X)_p + a_0(G,X)_p$. Elements of $a(G,X)_p$ and $a_0(G,X)_p$ have zero product, so if $1 = \alpha + \beta$ with $\alpha \in a(G,X)_p$ and $\beta \in a_0(G,X)_p$ then for any $x \in a(G,X)_p \cap a_0(G,X)_p$ we have

$$x = x.1 = x.\alpha + x.\beta = 0,$$

and so $a(G,X)_p \cap a_0(G,X)_p = 0$.

We prove the theorem by induction on the order of G. For any subgroup H of G we have

$$\operatorname{ind}_{H,G}(a(H,X)_p) \subseteq a(G,X)_p, \quad \operatorname{ind}_{H,G}(a_0(H,X)_p) \subseteq a_0(G,X)_p.$$

If G is not a (p,q)-Dress subgroup for some $q \neq p$, then Dress' induction theorem 5.6.11 (with S equal to the collection of all primes other then p) says that $a(G)_p$ is the sum of the images of induction from proper subgroups. We may thus suppose that G is a (p,q)-Dress subgroup for some prime $q \neq p$. In particular, setting $P = O_p(G)$, we see that G/P is a p'-group. If P stabilises any point in X, then by Corollary 3.6.9, $a(G)_p = a(G,X)_p$, and we are done. So we may suppose P does not stabilise any point in X.

Recall that the tensor induction map

$$\mathrm{ind}^{\otimes}_{P,G} : a(P)_p / \sum_{K<P} \mathrm{Im}(\mathrm{ind}_{K,P}) \to a(G)_p / \sum_{K<G} \mathrm{Im}(\mathrm{ind}_{K,G}).$$

is a ring homomorphism. By Proposition 3.15.2 (vi), the image of $a_0(P,X)_p$ under this map lies in $a_0(G,X)_p$. It thus suffices to show that the identity element of $a(P)_p$ lies in $a_0(P,X)_p + \sum_{K<P} \mathrm{Im}(\mathrm{ind}_{K,P})$. Now if $X = X_1 \dot\cup X_2$ then $a_0(P,X)_p = a_0(P,X_1)_p \cap a_0(P,X_2)_p$, so since P does not stabilise any point in X, we may suppose that X is a non-trivial transitive permutation representation of P. Let P' be a maximal subgroup of P containing the stabiliser of a point in X, so that $a_0(P,X)_p \geq a_0(P,P')_p$. Since P' is normal in P, and P/P' is a cyclic group of order p, $a(P/P',1) = \mathrm{Im}(\mathrm{ind}_{1,P/P'})$ is spanned by the regular representation. The composition factors of this are exactly p copies of the trivial module, and so we have

$$a(P/P')_p = a(P/P',1)_p \oplus a_0(P/P',1)_p = \mathrm{Im}(\mathrm{ind}_{1,P/P'}) \oplus a_0(P/P',1)_p$$

(cf. Corollary 5.3.6). Inflating to P, we see that the identity element of $a(P)_p$ lies in $\mathrm{Im}(\mathrm{ind}_{P',P}) + a_0(P,P')_p$ as required. □

REMARK. The statement of the theorem is equivalent to the statement that the quotient $a(G)/(a(G,X) + a_0(G,X))$ is entirely p-torsion.

COROLLARY 5.7.2 (Brauer). *Suppose that k is a field of characteristic p. Then the determinant of the Cartan matrix of kG is a power of p.*

PROOF. This is the case of the theorem in which X is a single point (cf. Corollary 5.3.6). For another proof of this theorem using psi operations, see Theorem 5.9.3. □

5.8. A quotient without nilpotent elements

In this section, we prove that if k is a field of characteristic p, then $A(kG)$ has a fairly large quotient $A(G)/A(G;p)$ with no non-zero nilpotent elements. As an application of this, we show that if G has cyclic Sylow p-subgroups then $A(G)$ has no non-zero nilpotent elements. This was first proved by Green and O'Reilly [**118, 159**] using some heavy computations for metacyclic groups. The proof we present here is taken from [**20**].

PROPOSITION 5.8.1. *Let k be an algebraically closed field of characteristic p. Suppose M is a finitely generated kG-module with $p \mid \dim_k M$. Then for any kG-module N and any summand U of $M \otimes N$ we have $p \mid \dim_k U$.*

PROOF. Suppose $\dim_k U$ is not divisible by p. Then by Theorem 3.1.9, the trivial module k is a summand of $U \otimes U^*$ and hence of $(M \otimes N) \otimes U^* = M \otimes (N \otimes U^*)$. But again applying Theorem 3.1.9, we see that this implies that $\dim_k M$ is not divisible by p, contradicting the hypothesis. □

DEFINITION 5.8.2. Suppose k is a field of characteristic p. We write $a(G;p)$ for the additive span in $a(kG)$ of the elements $[M]$ with the property that for any extension field $\hat{k} \supseteq k$, every summand of $\hat{k} \otimes_k M$ has dimension divisible by p. We write $A(G;p)$ for $\mathbb{C} \otimes_{\mathbb{Z}} a(G;p) \subseteq A(G)$.

LEMMA 5.8.3. *(i) $a(G;p)$ is an ideal in $a(G)$.*
(ii) $A(G;p)$ is an ideal in $A(G)$.

PROOF. This follows immediately from Proposition 5.8.1. □

LEMMA 5.8.4. *If $x = \sum a_i[M_i] \in A(G)$, write x^* for $\sum \bar{a}_i[M_i^*]$ where $\bar{a}_i$ is the complex conjugate of a_i. If $xx^* \in A(G;p)$ then $x \in A(G;p)$.*

PROOF. Without loss of generality, k is algebraically closed. If the trivial module $[k]$ does not appear with positive multiplicity in

$$xx^* = \sum_i |a_i|^2[M_i \otimes M_i^*] + \sum_{i \neq j} a_i\bar{a}_j[M_i \otimes M_j^*]$$

then by Theorem 3.1.9, each $[M_i]$ lies in $A(G;p)$. □

THEOREM 5.8.5. *The quotient ring $A(G)/A(G;p)$ has no non-zero nilpotent elements.*

PROOF. If $A(G)/A(G;p)$ has a non-zero nilpotent element, then there is a non-zero element $x \in A(G)$, not in $A(G;p)$, but with $x^2 \in A(G;p)$. Let $y = xx^*$. Then $yy^* = (xx^*)^2 \in A(G;p)$. Applying the lemma twice, we deduce first that $y \in A(G;p)$, and then that $x \in A(G;p)$. □

LEMMA 5.8.6. *Suppose k is algebraically closed. Suppose H is p-hypo-elementary with $1 \neq D = O_p(H)$ cyclic. Let H_1 be a subgroup of H of index p (it is easy to see that there is one, but it need not be normal). Then*

$$A(H;p) = \mathrm{Im}(\mathrm{ind}_{H_1,H}).$$

PROOF. It follows from Jordan normal form that the indecomposable kD-modules are just the Jordan blocks of size at most $|D|$ with eigenvalue one, so that they are uniserial modules.

Since $kH/J(kH) = k(H/D)$ is a direct sum of distinct one dimensional modules, the Idempotent Refinement Theorem 1.7.3 shows that each projective indecomposable module for kH restricts to the regular representation of kD, which is uniserial of length $|D|$. It follows that if M is any indecomposable kH-module then M is uniserial and $M\downarrow_D$ is indecomposable.

Let D_1 be the subgroup of D of index p. If M_1 is an indecomposable kD_1-module then $\mathrm{Hom}_{kD}(k, M_1\uparrow^D) \cong \mathrm{Hom}_{kD_1}(k, M_1)$ is one dimensional, and so $M_1\uparrow^D$ is indecomposable. It follows that a kD-module has dimension divisible by p if and only if it is induced from a kD_1-module. Thus an indecomposable kH-module M has dimension divisible by p if and only if it is projective relative to H_1.

It only remains to show that if N is an indecomposable kH_1-module then $N\uparrow^H$ is indecomposable. But this is clear since $N\uparrow^H\downarrow_D \cong N\downarrow_{D_1}\uparrow^D$ by the Mackey decomposition theorem. □

THEOREM 5.8.7 (Green, O'Reilly). *Suppose that the group G has cyclic Sylow p-subgroups, and k is a field of characteristic p. Then $A(kG)$ has no nilpotent elements.*

PROOF. Without loss of generality k is algebraically closed. By Corollary 5.6.9, the sum of the restriction maps to $A(kH)$ with H p-hypo-elementary is injective, so it suffices to prove the theorem in case $G = H$ is p-hypo-elementary with $D = O_p(H)$ cyclic. We prove the theorem by induction on $|D|$. If $|D| = 1$ then H has order coprime to p, and the theorem follows from Proposition 5.2.3. If $|D| > 1$, let H_1 be a subgroup of H of index p. Then by Corollary 5.4.11 and the lemma, we have

$$A(H) = \mathrm{Im}(\mathrm{ind}_{H_1,H}) \oplus \mathrm{Ker}(\mathrm{res}_{H,H_1}) = A(H;p) \oplus \mathrm{Ker}(\mathrm{res}_{H,H_1})$$

as a direct sum of ideals. Thus res_{H,H_1} maps $A(H;p)$ injectively into $A(H_1)$, and so $A(H;p)$ has no nilpotent elements by the inductive hypothesis. Since $A(H)/A(H;p)$ has no nilpotent elements by Theorem 5.8.5, this completes the proof of the theorem. □

REMARK. It turns out that if the Sylow p-subgroups of G are not cyclic, or elementary abelian 2-groups (with $p = 2$), then there are nilpotent elements in $A(G)$ (Zemanek [**208, 210**]). In case $p = 2$ and the Sylow 2-subgroups of G are elementary abelian of order four, Conlon [**55**] has shown that there are no nilpotent elements in $A(G)$. For larger elementary abelian 2-groups, this question is still open. See also Section 5.9 of Volume II for a cohomological method for producing nilpotent elements in $A(G)$.

5.9. Psi operations

In this section we construct **psi operations** $\psi^n : a(kG) \to a(kG)$. These are the representation theoretic version of raising group elements to the nth power, in the sense that the effect on Brauer characters is given by $t_g(\psi^n(x)) = t_{g^n}(x)$. They were first introduced in characteristic zero by Frobenius, who used them to study the number of solutions of $x^n = 1$ in a group. As an application of these operations, we give Kervaire's proof [**136**] that the determinant of the Cartan matrix is a power of p. In Chapter 2 of Volume II we shall study the analogous operations in topological K-theory, and explain how these are related to the cohomology of the finite general linear groups.

We first construct the operations ψ^n in the case where n is coprime to p. Let k be a field of characteristic p and let $k[\varepsilon]$ be the field obtained from k by adjoining a primitive nth root of unity ε to k. Let

$$T = \langle t \mid t^n = 1 \rangle$$

be a cyclic group of order n. If M is a kG-module then $M^{\otimes n}$ is a $k(T \times G)$-module with T permuting the tensor multiplicands (this is just the tensor induced module $M{\otimes\!\!\uparrow}^{T\times G}$). Since n is coprime to p, after tensoring with $k[\varepsilon]$, this breaks up as a direct sum of eigenspaces on which t acts with eigenvalues

ε^j. Each eigenspace is a $k[\varepsilon]G$-module, and whenever $\langle t^j \rangle = \langle t^{j'} \rangle$ the ε^j eigenspace is isomorphic to the $\varepsilon^{j'}$ eigenspace. By considering the action of the Galois group $\mathrm{Gal}(k[\varepsilon]/k)$, it follows that these eigenspaces are defined over k. We write $[M^{\otimes n}]_{\varepsilon^j} = [M^{\otimes n}]_d$ for the eigenspace corresponding to a primitive dth root of unity ε^j with $d|n$, considered as a kG-module. We define

$$\psi^n(M) = \sum_{j=1}^{n} e^{2\pi i j/n} [M^{\otimes n}]_{\varepsilon^j} \in A(G).$$

Now $\sum_{\substack{1 \le j \le d \\ (j,d)=1}} e^{2\pi i j/d}$ is equal to the Möbius function $\mu(d)$, which takes values 0 or ± 1. It follows that

$$\psi^n(M) = \sum_{d|n} \mu(d) [M^{\otimes n}]_d$$

is an element of $a(G)$.

PROPOSITION 5.9.1. *If M_1 and M_2 are kG-modules then*
(i) $\psi^n(M_1 \oplus M_2) = \psi^n(M_1) + \psi^n(M_2)$.
(ii) $\psi^n(M_1 \otimes M_2) = \psi^n(M_1)\psi^n(M_2)$.

PROOF. (i) As kG-modules, we have

$$(M_1 \oplus M_2)^{\otimes n} = \bigoplus_{\substack{j_1=1,2, \\ \cdots \\ j_n=1,2}} (M_{j_1} \otimes \cdots \otimes M_{j_n}).$$

Under the action of T, there are two fixed summands, namely $M_1^{\otimes n}$ and $M_2^{\otimes n}$. Apart from these, each orbit forms a $k(T \times G)$-module induced from a proper subgroup of the form $T' \times G$. It is easy to check that for such a module, the sum of $e^{2\pi i j/n}$ times the ε^j eigenspace is zero.

(ii) $(M_1 \otimes M_2)^{\otimes n} \cong M_1^{\otimes n} \otimes M_2^{\otimes n}$, and so

$$[(M_1 \otimes M_2)^{\otimes n}]_{\varepsilon^j} \cong \bigoplus_{m=1}^{n} [M_1^{\otimes n}]_{\varepsilon^m} [M_2^{\otimes n}]_{\varepsilon^{j-m}}.$$

Thus we have

$$\begin{aligned}\psi^n(M_1 \otimes M_2) &= \sum_{j=1}^{n} e^{2\pi i j/n} [(M_1 \otimes M_2)^{\otimes n}]_{\varepsilon^j} \\ &= \sum_{j,m=1}^{n} e^{2\pi i m/n} [M_1^{\otimes n}]_{\varepsilon^m} e^{2\pi i (j-m)/n} [M_2^{\otimes n}]_{\varepsilon^{j-m}} = \psi^n(M_1)\psi^n(M_2). \qquad \square\end{aligned}$$

It follows from this proposition that we may extend ψ^n linearly to give a ring homomorphism

$$\psi^n : a(G) \to a(G).$$

Note that since ψ^n is given in terms of direct summands of tensor powers, the image of $a(G,H)$ is contained in $a(G,H)$, and the image of $a(G,\mathrm{Triv})$ is contained in $a(G,\mathrm{Triv})$.

PROPOSITION 5.9.2. *Let t_g be the ring homomorphism $a(G)/a_0(G,1) \to \mathbb{C}$ defined in Section 5.3. Then*

$$t_g(\psi^n(x)) = t_{g^n}(x).$$

PROOF. Suppose M is a kG-module, and suppose without loss of generality that k is algebraically closed. Then we may choose a basis $m_1, \dots, m_r$ of M consisting of eigenvectors of g. Let $gm_j = \lambda_j m_j$. Then

$$\begin{aligned} t_g(\psi^n(M)) &= t_g(\mathrm{res}_{G,\langle g\rangle}\psi^n(M)) = t_g(\psi^n(M\downarrow_{\langle g\rangle})) = t_g(\psi^n(\bigoplus_{j=1}^{r}\langle m_j\rangle)) \\ &= \sum_{j=1}^{r} t_g\psi^n(\langle m_j\rangle) = \sum_{j=1}^{r}\lambda_j^n = t_{g^n}(M). \end{aligned}$$ □

THEOREM 5.9.3 (Brauer). *Suppose k is a field of characteristic p. Then the determinant of the Cartan matrix is a power of p.*

PROOF. (Kervaire; for another proof see Corollary 5.7.2). This is the same as saying that the cokernel of the Cartan homomorphism

$$c : a(G,1) \to a(G) \to a(G)/a_0(G,1) = \mathcal{R}(G)$$

is a p-group.

Let m be the p'-part of the exponent of G. If $x \in a(G,1)$ then $\psi^m(x)$ is again an element of $a(G,1)$. By Proposition 5.9.2, each t_g has value

$$t_g(\psi^m(x)) = t_1(x)$$

equal to the "dimension" of x. Thus by Proposition 5.3.2, if P is a projective kG-module then $(\dim_k P).1$ is in the image of c.

For each prime $q \neq p$ dividing $|G|$, let Q be a Sylow q-subgroup of G. Then $k_Q\uparrow^G$ is a projective kG-module since it is induced from a projective kQ-module. Thus by the above, $|G:Q|.1 \in \mathrm{Im}(c)$. It now follows from the Chinese remainder theorem that $|G|_p.1 \in \mathrm{Im}(c)$, where $|G|_p$ is the p-part of the order of G. Since $\mathrm{Im}(c)$ is an ideal, it follows that $|G|_p$ annihilates the cokernel of c, and the theorem is proved. □

Finally, we show how Corollary 5.6.5 leads to an explicit description of $\mathcal{R}(kG)$ in case k is a field of p^n elements, in terms of the operation ψ^{p^n}.

PROPOSITION 5.9.4 (The Brauer Lift). *Let $(K,\mathcal{O},k)$ be a p-modular system. Then for every n the decomposition map*

$$d : \mathcal{R}(KG) \to \mathcal{R}(kG)$$

induces an isomorphism

$$\mathcal{R}(KG)^{\psi^{p^n}} \cong \mathcal{R}(kG)^{\psi^{p^n}} = \mathcal{R}(k_0G)$$

where $k_0 = \{\lambda \in k \mid \lambda^{p^n} = \lambda\}$.

PROOF. By Theorem 5.3.3 we have a diagram

$$\begin{array}{ccc} \mathcal{R}(KG) & \xrightarrow{\sum t_g} & \bigoplus_{H \leq_G G} \mathbb{C} \\ \downarrow d & & \downarrow \\ \mathcal{R}(kG) & \xrightarrow{\sum t_g} & \bigoplus_{\substack{p'\text{-elements} \\ g \in_G G}} \mathbb{C} \end{array}$$

so that by Proposition 5.9.2, d commutes with the action of the ψ operations. It also follows from Proposition 5.9.2 that for m large enough, ψ^m kills the kernel of d. Suppose $x \in \mathcal{R}(kG)$ is fixed by ψ^{p^n}. Choose m as above to be a multiple of n. By Corollary 5.6.5, x has a pre-image y in $\mathcal{R}(KG)$. Since $\psi^{p^n}(y) - y$ is killed by d, it is killed by ψ^{p^m}, and so $\psi^{p^{m+n}}(y) = \psi^{p^m}(y)$ is a pre-image of x which is fixed by ψ^{p^n}. Thus $d : \mathcal{R}(KG)^{\psi^{p^n}} \to \mathcal{R}(kG)^{\psi^{p^n}}$ is an isomorphism.

Finally, by Proposition 5.3.4 we have $\mathcal{R}(kG)^{\psi^{p^n}} = \mathcal{R}(k_0G)$. □

The way the above proposition is usually stated is that given a modular representation M, the class function which assigns to each element $g \in G$ the value of $t_{g'}(M)$, where g' is the p'-part of g, is a generalised ordinary character (i.e., the character of a difference of two characteristic zero representations).

5.10. Bilinear forms on representation rings

The material in this and the next section comes from **[23]**. We define two different bilinear forms $(\ ,\ )$ and $\langle\ ,\ \rangle$ on $a(kG)$ and $A(kG)$ as follows. If M and N are finitely generated kG-modules, we let

$$([M],[N]) = \dim_k \operatorname{Hom}_{kG}(M,N) = \dim_k(M^* \otimes N)^G.$$

We extend $(\ ,\ )$ bilinearly to give a bilinear form on $a(G)$ and $A(G)$.

Now the bilinear form $(\ ,\ )$ is usually not symmetric, but it is very closely related to the symmetric bilinear form $\langle\ ,\ \rangle$ defined by bilinearly extending

$$\langle [M],[N] \rangle = \dim_k(M,N)_1^G,$$

the dimension of the space of homomorphisms from M to N which factor through some projective module (see Definition 3.6.2).

DEFINITION 5.10.1. We define elements u and v of $a(G)$ via

$$u = u_{kG} = [P_k] - [\Omega^{-1}(k)], \quad v = v_{kG} = [P_k] - [\Omega(k)]$$

where P_k is the projective cover (= injective hull) of the trivial kG-module k, $\Omega(k)$ is the kernel of $P_k \twoheadrightarrow k$ and $\Omega^{-1}(k)$ is the cokernel of $k \hookrightarrow P_k$. Note that $P_k \cong P_k^*$ and $\Omega^{-1}(k) \cong \Omega(k)^*$, so that v is the 'dual' of u.

LEMMA 5.10.2. *The following expressions are equal:*
(i) $\langle [M], [N] \rangle$,
(ii) *The multiplicity of* P_k *as a summand of* $\mathrm{Hom}_k(M, N) = M^* \otimes N$,
(iii) $(u, [\mathrm{Hom}_k(M, N)]) = (u.[M], [N])$,
(iv) *The rank of the element* $\sum_{g \in G} g$ *of the group algebra* kG *in its matrix representation on* $\mathrm{Hom}_k(M, N)$.

In particular, since P_k *is self-dual,* (ii) *shows that* $\langle\ ,\ \rangle$ *is symmetric.*

PROOF. Since each of these expressions is unaffected if we replace M by the trivial module k and N by $\mathrm{Hom}_{kG}(M, N)$, we may assume that $M = k$. Also, since each expression is additive in N, we may suppose that N is indecomposable. We shall show that each of these expressions is equal to 1 when $N \cong P_k$ and 0 otherwise.

(i) If $\langle [k], [N] \rangle \neq 0$ then there is a non-zero map $k \to N$ which is a transfer from the trivial subgroup. The image of 1 under such a map is of the form $\sum_{g \in G} g(x)$ for some $x \in N$. Let $\lambda : P \twoheadrightarrow N$ be the projective cover of N, and choose $y \in P$ with $\lambda(y) = x$. Then $\sum_{g \in G} g(y)$ is a non-zero G-invariant element of P which is sent to the non-zero element $\sum_{g \in G} g(x)$ by λ. Since projective kG-modules are injective (Proposition 3.1.2) and $\mathrm{Soc}(P_k) \cong k$ (Theorem 1.6.3), the injective map $k \to P$ taking 1 to $\sum_{g \in G} g(y)$ extends to an inclusion of P_k as a direct summand of P. The map λ does not kill the socle of this copy of P_k, so since P_k is injective, it is a summand of N. Since N is indecomposable this forces $N \cong P_k$. Clearly $\langle [k], [P_k] \rangle = 1$.

(ii) The multiplicity of P_k as a summand of an indecomposable module N is clearly equal to 1 when $N \cong P_k$ and zero otherwise.

(iii) A homomorphism from P_k to N factors through $\Omega^{-1}(k) = P_k/k$ unless $N \cong P_k$, since P_k is injective. Thus if $N \not\cong P_k$ we have $(u, [N]) = 0$. On the other hand, if $N \cong P_k$, then any homomorphism $P_k \to N$ is a multiple of this isomorphism plus a homomorphism factoring through $\Omega^{-1}(k)$, and so $(u, [P_k]) = 1$.

(iv) This is clearly equal to (i). □

LEMMA 5.10.3. *In* $a(G)$ *we have* $uv = 1$.

PROOF. Tensoring the short exact sequence $0 \to \Omega(k) \to P_k \to k \to 0$ with $\Omega^{-1}(k)$, we obtain a short exact sequence

$$0 \to \Omega(k) \otimes \Omega^{-1}(k) \to P_k \otimes \Omega^{-1}(k) \to \Omega^{-1}(k) \to 0.$$

We also have a short exact sequence $0 \to k \to P_k \to \Omega^{-1}(k) \to 0$ so that by Schanuel's lemma we obtain

$$P_k \oplus \Omega(k) \otimes \Omega^{-1}(k) \cong k \oplus P_k \otimes \Omega^{-1}(k).$$

Since tensoring with P_k splits short exact sequences, we also have

$$P_k \otimes P_k \cong P_k \otimes \Omega(k) \oplus P_k$$

and so

$$P_k \otimes P_k \oplus \Omega(k) \otimes \Omega^{-1}(k) \cong k \oplus P_k \otimes \Omega^{-1}(k) \oplus P_k \otimes \Omega(k).$$

It follows that

$$([P_k] - [\Omega(k)])([P_k] - [\Omega^{-1}(k)]) = [k]$$

in $a(G)$. □

PROPOSITION 5.10.4. *If x and y are elements of $a(G)$ or $A(G)$, then*
(i) $(x, y) = \langle vx, y\rangle = \langle x, uy\rangle$
(ii) $\langle x, y\rangle = (ux, y) = (x, vy)$
(iii) $(x, y) = (y, v^2x) = (u^2y, x)$.

PROOF. It follows from Lemma 5.10.2 that $\langle x, y\rangle = (ux, y)$. The rest follows from the identity $uv = 1$ and the fact that u is obtained from v by dualising. □

LEMMA 5.10.5. *If H is a subgroup of G then* $\mathrm{res}_{G,H}(u_{kG}) = u_{kH}$ *and* $\mathrm{res}_{H,G}(v_{kG}) = v_{kH}$.

PROOF. The first of these follows by applying Schanuel's lemma to the sequences

$$0 \to \Omega(k)_{kG}\downarrow_H \to (P_k)_{kG}\downarrow_H \to k \to 0$$

$$0 \to \Omega(k)_{kH} \to (P_k)_{kH} \to k \to 0$$

and the second follows by duality. □

PROPOSITION 5.10.6. *If $x \in a(H)$ and $y \in a(G)$ then*
(i) $(x, \mathrm{res}_{G,H}(y)) = (\mathrm{ind}_{H,G}(x), y)$ *and* $(\mathrm{res}_{G,H}(y), x) = (y, \mathrm{ind}_{H,G}(x))$.
(ii) $\langle x, \mathrm{res}_{G,H}(y)\rangle = \langle \mathrm{ind}_{H,G}(x), y\rangle$.

PROOF. The first of these follows from the Nakayama relations, and the second follows since

$$\begin{aligned}\langle x, \mathrm{res}_{G,H}(y)\rangle &= (u_{kH}x, \mathrm{res}_{G,H}(y)) = (\mathrm{ind}_{H,G}(u_{kH}x), y)\\ &= (\mathrm{ind}_{H,G}(\mathrm{res}_{G,H}(u_{kG})x), y) = (u_{kG}\mathrm{ind}_{H,G}(x), y) = \langle \mathrm{ind}_{H,G}(x), y\rangle.\end{aligned}$$ □

5.11. Non-singularity

In this section we use the almost split sequences, discussed in Chapter 4, to prove that the bilinear forms $(\ ,\ )$ and $\langle\ ,\ \rangle$ on $a(G)$ introduced in the last section are non-singular. The material in this section comes from [**23**].

Recall from Theorem 4.12.2 that given any non-projective module N, there is an almost split sequence terminating in N

$$0 \to M \to E \to N \to 0$$

unique up to isomorphism of short exact sequences, and by Proposition 4.12.7 we have $M \cong \Omega^2(N)$.

Recall also from Proposition 4.12.6 that an almost split sequence

$$0 \to M \to E \to N \to 0$$

gives rise to an exact sequence

$$0 \to \mathrm{Hom}_\Lambda(N', M) \to \mathrm{Hom}_\Lambda(N', E) \to \mathrm{Hom}_\Lambda(N', N) \to 0$$

if N' has no summand isomorphic to N, and

$$0 \to \mathrm{Hom}_\Lambda(N,M) \to \mathrm{Hom}_\Lambda(N,E) \to \mathrm{End}_\Lambda(N) \to \mathrm{End}_\Lambda(N)/J\mathrm{End}_\Lambda(N) \to 0.$$

It follows that if we set $\tau_0[N] = [N] + [M] - [E]$ in $a(G)$ then we have

$$([N'], \tau_0[N]) = \begin{cases} d(N) & \text{if } N \cong N' \\ 0 & \text{otherwise} \end{cases}$$

(where $d(N) = \dim_k \mathrm{End}_{kG}(N)/J\mathrm{End}_{kG}(N)$, which is always 1 if k is algebraically closed), so that the $\tau_0[N]$ form a sort of "dual basis" to the basis of indecomposable modules $[N]$ for $a(G)$.

There are two problems with this statement. The first is that we have only defined $\tau_0[N]$ for N a non-projective indecomposable. However, it is easy to see that the right definition for N projective is $\tau_0[N] = [N] - \mathrm{Rad}(N)$. This is because any homomorphism to N whose image does not lie in $\mathrm{Rad}(N)$ is surjective and hence splits.

Having now defined τ_0 on the basis elements $[N]$, we extend antilinearly to define

$$\tau_0 \sum_i a_i[N_i] = \sum_i \bar{a}_i \tau_0[N_i]$$

in $A(G)$, where $\bar{a}_i$ denotes the complex conjugate of a_i.

The second problem is that for infinite dimensional vector spaces, duality does not work very well, and it turns out that the $\tau_0[N]$ do not form a basis of $A(G)$ unless $A(G)$ is finite dimensional (which happens if and only if the Sylow p-subgroups of G are cyclic; see Theorem 4.4.4). But in any case, we have now proved the following theorem, which may be thought of as a non-singularity statement for the bilinear form (,).

THEOREM 5.11.1. *Suppose N and N' are indecomposable kG-modules, and τ_0 is as defined above. Then*

$$([N'], \tau_0[N]) = \begin{cases} d(N) & \text{if } N \cong N' \\ 0 & \text{otherwise} \end{cases}$$

where $d(N) = \dim_k \mathrm{End}_{kG}(N)/J\mathrm{End}_{kG}(N)$.

Thus for any $x = \sum_i a_i[M_i] \in A(G)$ we have $(x, \tau_0(x)) = \sum_i |a_i|^2 \geq 0$ with equality if and only if $x = 0$. □

The following corollary may also be proved directly without using almost split sequences.

COROLLARY 5.11.2. *Suppose N_1 and N_2 are two kG-modules, such that for every kG-module M we have*

$$\dim_k \mathrm{Hom}_{kG}(M, N_1) = \dim_k \mathrm{Hom}_{kG}(M, N_2).$$

Then $N_1 \cong N_2$. □

So far, everything we have done works just as well for an arbitrary finite dimensional algebra. However, we now wish to do the same for the inner product $\langle\ ,\ \rangle$, and for this we need to use the multiplication in $A(G)$, as well as the fact that kG is a symmetric algebra.

In the last section, we saw that it is very easy to pass between the $(\ ,\ )$ and $\langle\ ,\ \rangle$ using the elements u and v. We take

$$\tau_1[N] = u.\tau_0[N],$$

so that $\langle[N'], \tau_1[N]\rangle = ([N'], \tau_0[N])$ is equal to $d(N)$ if $N \cong N'$ and zero otherwise.

Now suppose N is a non-projective indecomposable. By Corollary 3.1.6 we have $N \otimes \Omega(k) \cong \Omega(N) \oplus (\text{projective})$, and so modulo projectives we have

$$u.\tau_0[N] = -\Omega^{-1}(\tau_0[N]) = -[\Omega^{-1}(N)] - [\Omega^{-1}(M)] + [\Omega^{-1}(E)].$$

But recall from Corollary 5.3.6 that $A(G) = A(G,1) \oplus A_0(G,1)$ as a direct sum of ideals. Since $\tau_0[N] \in A_0(G,1)$, this shows that the above equation holds without working modulo projectives.

On the other hand, if N is projective indecomposable, then $N = P_S$ is the projective cover of a simple module S. Schanuel's lemma shows that $v.[S] = [P_S] - [\Omega(S)] = \tau_0[N]$, and so $\tau_1[N] = u.v.[S] = [S]$.

We record what we have proved in the following theorem, which is a non-singularity statement for $\langle\ ,\ \rangle$.

THEOREM 5.11.3. *Suppose N and N' are indecomposable kG-modules, and define $\tau_1[N]$ to be $N/\text{Rad}(N) \cong \text{Soc}(N)$ if N is projective, and*

$$[X] - [\Omega^{-1}(N)] - [\Omega(N)]$$

if N is not projective, where

$$0 \to \Omega^{-1}(N) \to X \to \Omega(N) \to 0$$

is the almost split sequence terminating in $\Omega(N)$. Then

$$\langle[N'], \tau_1[N]\rangle = \begin{cases} d(N) & \text{if } N \cong N' \\ 0 & \text{otherwise} \end{cases}$$

where $d(N) = \dim_k \text{End}_{kG}(N)/J\text{End}_{kG}(N)$. □

CHAPTER 6

Block theory

There are now quite a few decent expositions of block theory available. The reader is advised to consult Alperin [**3**], Curtis and Reiner [**66**], Dornhoff [**84**], Feit [**107**], Landrock [**148**], Nagao and Tsushima [**153**], and of course the collected works of Brauer [**36**]. For this reason, we shall not attempt an encyclopædic treatment, but we shall rather try to concentrate on aspects of the theory which are closely related to other topics discussed in this book.

The approach we shall take is the module theoretic approach initiated by Green [**119**]. In this approach, the group algebra kG is regarded as a module for $k(G \times G)$, and the indecomposable direct summands of this module are the blocks. A vertex of a block as a $k(G \times G)$-module is always conjugate to a diagonally embedded subgroup $\mathrm{diag}(D)$, $D \leq G$, and D is called a defect group of the block. It turns out that the group D in some sense determines how complicated the block is.

Throughout this chapter, $(K, \mathcal{O}, k)$ will denote a p-modular system. Thus $\mathcal{O}$ is a complete rank one discrete valuation ring with field of fractions K of characteristic zero, maximal ideal $\mathfrak{p}$, and quotient field k of characteristic p. When we write R for the coefficient ring, we shall assume that $R \in \{\mathcal{O}, k\}$.

6.1. Blocks and defect groups

Recall from Section 1.8 that a block of kG is an indecomposable two-sided ideal direct summand. A decomposition of kG into blocks

$$kG = B_1 \oplus \cdots \oplus B_s$$

corresponds to a decomposition of the identity element

$$1 = e_1 + \cdots + e_s$$

as a sum of orthogonal primitive central idempotents. The correspondence is given by $B_i = e_i.kG$.

Since both $Z(\mathcal{O}G)$ and $Z(kG)$ have a basis consisting of the conjugacy class sums in G, it follows that reduction modulo $\mathfrak{p}$ is a surjective map $Z(\mathcal{O}G) \to Z(kG)$, and so by Theorem 1.9.4 (iii), the idempotents $e_i \in kG$ may be lifted to orthogonal primitive central idempotents $f_i \in \mathcal{O}G$. We thus have

$$\mathcal{O}G = \hat{B}_1 \oplus \cdots \oplus \hat{B}_s$$

with $\hat{B}_i = f_i.\mathcal{O}G$ and $B_i = k \otimes_{\mathcal{O}} \hat{B}_i$.

Now if M is an indecomposable kG-module, the equation

$$M = e_1 M \oplus \cdots \oplus e_s M$$

shows that $M = e_i M$ for a unique i, and $e_j M = 0$ for $j \neq i$. Similarly if M is an indecomposable $\mathcal{O}G$-lattice, then $f_i M = M$ and $f_j M = 0$ for $j \neq i$. Finally, if M is an irreducible KG-module (recall that all indecomposable KG-modules are irreducible by Maschke's theorem) then $f_i M = M$ for a unique i, and $f_j M = 0$ for $j \neq i$.

We thus think of a block as a sort of receptacle, into which are thrown indecomposable summands of the algebras kG and $\mathcal{O}G$, primitive idempotents in $Z(kG)$ and $Z(\mathcal{O}G)$, indecomposable modules for kG, $\mathcal{O}G$ and KG, and so on.

Now the two sided ideal direct summands of RG (recall $R \in \{\mathcal{O}, k\}$) are the same as the direct summands of RG as an $R(G \times G)$-module, where the action of $R(G \times G)$ on RG is via left and right multiplication

$$(g_1, g_2) : g \to g_1 g g_2^{-1}.$$

With this action, RG is equal to the permutation module of $R(G \times G)$ on the cosets of the diagonal

$$\Delta(G) = \{(g, g), g \in G\} \subseteq G \times G.$$

In other words, $RG \cong R_{\Delta(G)} \uparrow^{G \times G}$. It follows that the $R(G \times G)$-module RG is projective relative to $\Delta(G)$, and so the vertices of any indecomposable summand $B = e.RG$ are conjugate to some subgroup of the form $\Delta(D) \subseteq \Delta(G)$, where D is a p-subgroup of G determined up to conjugacy by the block B. The group $D = D(B)$ is called the **defect group** of the block B. If $|D| = p^a$, we say that B is a block of **defect** a.

PROPOSITION 6.1.1 (J. A. Green). *The defect group D of any block B of RG is expressible as an intersection $S \cap {}^g S$ of two Sylow p-subgroups of G.*

PROOF. By the Mackey decomposition theorem, the restriction of RG to an $R(S \times S)$-module is

$$R_{\Delta(G)} \uparrow^{G \times G} \downarrow_{S \times S} = \bigoplus_{(1,g) \in S \times S \backslash G \times G / \Delta(G)} R_{(S \times S) \cap {}^{(1,g)}\Delta(G)} \uparrow^{S \times S}.$$

The double coset representatives may be chosen to be of the form $(1, g)$ by adjusting by elements of $\Delta(G)$. But

$$(S \times S) \cap {}^{(1,g)}\Delta(G) = {}^{(1,g)}\Delta(S \cap {}^{g^{-1}} S)$$

and so the restriction of RG to $R(S \times S)$ is a direct sum of permutation modules on cosets of subgroups of the form ${}^{(1,g)}\Delta(S \cap {}^{g^{-1}} S)$. Each transitive permutation module for $S \times S$ is indecomposable, since for $R = k$ it has a one dimensional socle (see Section 3.14). So the vertex of $R_{(1,g)\Delta(S \cap g^{-1} S)} \uparrow^{S \times S}$

is exactly ${}^{(1,g)}\Delta(S \cap {}^{g^{-1}}S)$. Since RG is projective relative to $S \times S$ (Corollary 3.6.9) it follows that each indecomposable summand of RG as an $R(G \times G)$-module has a subgroup of the form $\Delta(S \cap {}^gS)$ as a vertex. □

PROPOSITION 6.1.2 (J. A. Green). *Suppose $B = e.RG$ is a block of RG with defect group D. Then e lies in $RG^{\Delta(G)}_{\Delta(H)} \subseteq Z(RG)$ if and only if H contains a conjugate of D (see Definition 3.6.2 for the notation). Furthermore, every RG-module in the block B is projective relative to D.*

PROOF. We have seen that as an $R(G \times G)$-module, B is projective relative to $\Delta(G)$. Thus by Higman's criterion (Proposition 3.6.4), H contains a defect group of B if and only if there is a map $\alpha \in \mathrm{End}_{R\Delta(H)}(RG)$ such that $\mathrm{Tr}_{\Delta(H),\Delta(G)}(\alpha)$ is the identity map on $B = e.RG$. If α is such a map then we have

$$\begin{aligned} e = (\mathrm{Tr}_{\Delta(H),\Delta(G)}(\alpha))(e) &= \sum_{(g,g)\in\Delta(G)/\Delta(H)} (g,g)\alpha((g^{-1},g^{-1})(e)) \\ &= \sum_{g\in G/H} g\alpha(e)g^{-1} = \mathrm{Tr}_{\Delta(H),\Delta(G)}\alpha(e) \end{aligned}$$

and so $e \in RG^{\Delta(G)}_{\Delta(H)}$. Conversely if $e = \mathrm{Tr}_{\Delta(H),\Delta(G)}(a) \in RG^{\Delta(G)}_{\Delta(H)}$ then we define a map $\alpha : RG \to RG$ by $\alpha(x) = a.x$. Since $a \in RG^{\Delta(H)}$ we have $\alpha \in \mathrm{End}_{R\Delta(H)}(RG)$, and for $x \in e.RG$ we have

$$\begin{aligned} (\mathrm{Tr}_{\Delta(H),\Delta(G)}(\alpha))(x) &= \sum_{(g,g)\in\Delta(G)/\Delta(H)} (g,g)\alpha((g^{-1},g^{-1})(x)) \\ &= \sum_{g\in G/H} g.a.g^{-1}xg.g^{-1} = (\mathrm{Tr}_{\Delta(H),\Delta(G)}(a)).x = ex = x. \end{aligned}$$

Finally, if M is an RG-module in the block B, then e acts as the identity on M, and so the above element $a = \alpha(e)$ acts as an endomorphism of M whose transfer is the identity. Again applying Higman's criterion, we see that M is projective relative to D. □

REMARK. In fact there is always an indecomposable module in B whose vertex is exactly D, as we shall see when we come to discuss Brauer's second main theorem in Section 6.3.

EXAMPLE. The block of RG containing the trivial RG-module R is called the **principal block**. Since a Sylow p-subgroup of G is a vertex of the trivial module, it is also a defect group of the principal block, by the above proposition. The principal block is usually denoted $B_0 = B_0(G)$.

COROLLARY 6.1.3. *If B is a block of kG and $\hat{B}$ is the corresponding block of $\mathcal{O}G$ then B and $\hat{B}$ have the same defect groups.*

PROOF. If D is a defect group of B then by the proposition the corresponding central idempotent e lies in $(kG)^{\Delta(G)}_{\Delta(D)}$. Thus the central idempotent $\hat{e}$ corresponding to $\hat{B}$ lies in $(\mathcal{O}G)^{\Delta(G)}_{\Delta(D)} + \mathfrak{p}Z(\mathcal{O}G)$. So by Rosenberg's Lemma 1.7.10 we have either $\hat{e} \in (\mathcal{O}G)^{\Delta(G)}_{\Delta(D)}$ or $\hat{e} \in \mathfrak{p}Z(\mathcal{O}G)$. But $\mathfrak{p}Z(\mathcal{O}G) \subseteq \mathfrak{p}G \subseteq J(\mathcal{O}G)$ so the latter is impossible. Conversely if $\hat{e} \in (\mathcal{O}G)^{\Delta(G)}_{\Delta(D)}$ then $e \in (kG)^{\Delta(G)}_{\Delta(D)}$. □

6.2. The Brauer map

From now on, we restrict our attention to the (diagonal) action of G on G by conjugation, and drop the symbol Δ used in the last section. The following lemma is our starting point for the discussion of the Brauer map and Brauer's three main theorems. The first of these gives a one–one correspondence between blocks of RG with defect group D and blocks of $RN_G(D)$ with defect group D. First note that we are assuming that R is either $\mathcal{O}$ or k, so that by Corollary 6.1.3 it suffices to consider the case $R = k$.

LEMMA 6.2.1. *Suppose D is a p-subgroup of G. With the notation of Definition 3.6.2, we have*
(i) $(kG)^D = kC_G(D) \oplus \sum_{D'<D} (kG)^D_{D'}$
(ii) $(kG)^{N_G(D)} = (kC_G(D))^{N_G(D)} \oplus \sum_{D \not\leq Q \leq N_G(D)} (kG)^{N_G(D)}_Q$
in each case as a direct sum of a subring and a two-sided ideal. Each summand in (ii) *is contained in the corresponding summand in* (i).

PROOF. The space $(kG)^D$ has as a basis the orbit sums of D on elements of G. The orbits of length one span $kC_G(D)$, and the sums of orbits of length greater than one are transfers from proper subgroups of D. The intersection of $kC_G(D)$ and $\sum_{D'<D}(kG)^D_{D'}$ is zero since k has characteristic p.

Similarly, for the second statement we split up the orbits of $N_G(D)$ into those on which D acts trivially and those on which D has no fixed points.

In each case, the right-hand summand is a two-sided ideal by Lemma 3.6.3 (i) and (ii). □

REMARK. The above lemma does not hold with k replaced by $\mathcal{O}$ since the two pieces on the right hand sides do not intersect in zero. However, it follows from the discussion in the last section that there is a one-one correspondence between blocks of kG and blocks of $\mathcal{O}G$ preserving defect groups, so that if we wish to count blocks then it suffices to work over k.

PROPOSITION 6.2.2. *Suppose D is a normal p-subgroup of G. Then every idempotent in $Z(kG)$ lies in $kC_G(D)$. In particular, if $C_G(D) \leq D$ them kG has only one block.*

PROOF. Since D is a p-subgroup, it acts trivially on every simple kG-module (its fixed points form a non-zero invariant submodule, by Lemma 3.14.1), and hence on $kG/J(kG)$. Since transfer from a proper subgroup

$D' < D$ to D is zero on a module on which D acts trivially, we have $\sum_{D'<D}(kG)^D_{D'} \subseteq J(kG)$. It now follows from Lemma 6.2.1 (i) that every idempotent in $(kG)^D$ lies in $kC_G(D)$. If $C_G(D) \leq D$ then $C_G(D)$ is a p-group, and so $kC_G(D)$ has only the identity as an idempotent, and hence kG has only one block. □

DEFINITION 6.2.3. We define the **Brauer map**

$$\mathrm{Br}_D : (kG)^D \to kC_G(D)$$

to be the projection onto the first factor in the decomposition given in Lemma 6.2.1 (i). Since the second factor in this decomposition is a two-sided ideal, this map is a ring homomorphism. Note also that by Lemma 6.2.1 (ii), the kernel of Br_D on $(kG)^{N_G(D)}$ is equal to $\sum_{D \not\leq Q \leq N_G(D)}(kG)_Q^{N_G(D)}$.

In a sense, we are only really interested in the map

$$\mathrm{Br}_D : Z(kG) \to Z(kC_G(D))$$

obtained by restricting the above map to $Z(kG) = (kG)^G \subseteq (kG)^D$, and its effect on idempotents. The point of the extended definition will become clearer in Lemma 6.2.5.

If H is a subgroup of G with $DC_G(D) \leq H \leq N_G(D)$ then by the above proposition, every idempotent in $Z(kH)$ lies in $kC_G(D)$. Let $e \in Z(kH)$ be a primitive idempotent corresponding to a block b of kH, and $1 = e_1 + \cdots + e_s$ be a decomposition of 1 as a sum of primitive orthogonal idempotents in $Z(kG)$ corresponding to the block decomposition $kG = B_1 + \cdots + B_s$. Then in $Z(kH)$ we have

$$e = e.\mathrm{Br}_D(1) = e.\mathrm{Br}_D(e_1) + \cdots + e.\mathrm{Br}_D(e_s).$$

Since e is primitive, we have $e = e.\mathrm{Br}_D(e_i)$ for some i, and $e.\mathrm{Br}_D(e_j) = 0$ for $j \neq i$. We define the Brauer correspondent b^G of b to be the block B_i of kG. In general, the Brauer correspondence is not a one–one correspondence, but in case $H = N_G(D)$, Brauer's first main theorem, which we prove next, states that it is a one-one correspondence between blocks with defect group D.

LEMMA 6.2.4. *A block idempotent $e \in Z(kG)$ has defect group D if and only if $e \in (kG)^G_D$ and $\mathrm{Br}_D(e) \neq 0$.*

PROOF. By Proposition 6.1.2, e has defect group D if and only if $e \in (kG)^G_D$ and $e \notin (kG)^G_{D'}$ for $D' < D$. By Rosenberg's Lemma 1.7.10, the latter condition is equivalent to $e \notin \sum_{D'<D}(kG)^G_{D'}$. The lemma now follows from the fact that the kernel of Br_D on $(kG)^G_D$ is $\sum_{D'<D}(kG)^G_D$. □

LEMMA 6.2.5. *We have a commutative diagram*

$$\begin{array}{ccc} (kG)^D & \xrightarrow{\mathrm{Br}_D} & kC_G(D) \\ \downarrow{\scriptstyle \mathrm{Tr}_{D,G}} & & \downarrow{\scriptstyle \mathrm{Tr}_{D,N_G(D)}} \\ (kG)^G_D & \xrightarrow{\mathrm{Br}_D} & kC_G(D)^{N_G(D)}_D \end{array}$$

in which the bottom map is surjective. Note that we have $(kG)^G_D \subseteq (kG)^G = Z(kG)$ *and* $kC_G(D)^{N_G(D)}_D \subseteq Z(kC_G(D))$.

PROOF. By Lemma 3.6.3 (iv), for $x \in (kG)^D$ we have

$$\mathrm{Br}_D(\mathrm{Tr}_{D,G}(x)) = \mathrm{Br}_D(\sum_{N_G(D)gD} \mathrm{Tr}_{N_G(D)\cap {}^gD, N_G(D)}(gx)) = \mathrm{Br}_D(\mathrm{Tr}_{D,N_G(D)}(x))$$

since the Brauer map is defined in such a way that it vanishes on the remaining terms. Comparing parts (i) and (ii) of Lemma 6.2.1, we see that Br_D commutes with $\mathrm{Tr}_{D,N_G(D)}$, and so the above diagram commutes. Since the top and right hand maps are surjective, it follows that the bottom map is also surjective. □

THEOREM 6.2.6 (Brauer's first main theorem). *The map* Br_D *establishes a one–one correspondence, called the* **Brauer correspondence**, *between block idempotents in* $Z(kG)$ *with defect group* D *and block idempotents in* $Z(kN_G(D))$ *with defect group* D. *If* b *is a block of* $kN_G(D)$ *with defect group* D *then* b^G *is the corresponding block of* kG *with defect group* D.

PROOF. By Lemma 6.2.4, a block idempotent $e \in Z(kG)$ has defect group D if and only if $e \in (kG)^G_D$ and $\mathrm{Br}_D(e) \neq 0$. By Lemma 6.2.5, the map $\mathrm{Br}_D : (kG)^G_D \to kC_G(D)^{N_G(D)}_D$ is onto, and by Proposition 6.2.2, every block idempotent in $Z(kN_G(D))$ with defect group D lies in $kC_G(D)^{N_G(D)}_D$. □

Brauer's first main theorem allows us to extend the above notation b^G as follows. If $DC_G(D) \leq H \leq G$ (but H not necessarily contained in $N_G(D)$) and b is a block of H with defect group D, then by the first main theorem, there is a unique block b' of $N_H(D)$ with $(b')^H = b$. We then write b^G for the block $(b')^G$. It is easy to see that if $H \leq K \leq G$ then $(b^K)^G = b^G$.

Another way to view the Brauer correspondence, due to Alperin, is as follows.

LEMMA 6.2.7. *Suppose* $DC_G(D) \leq H \leq G$ *and* b *is a block of* kH *with defect group* D. *Then* b^G *is the unique block* B *of* kG *such that* b *is a summand of the restriction* $B\downarrow_{H\times H}$ *as a* $k(H \times H)$*-module.*

PROOF. We first show that kG has a unique block B with b a summand of $B\downarrow_{H\times H}$ As in Proposition 6.1.1, we have

$$kG\downarrow_{H\times H} = k_{\Delta(G)}\uparrow^{G\times G}\downarrow_{H\times H} = \bigoplus_{(1,g)\in H\times H\backslash G\times G/\Delta(G)} k_{(H\times H)\cap^{(1,g)}\Delta(G)}\uparrow^{H\times H}$$

and

$$(H \times H) \cap {}^{(1,g)}\Delta(G) = {}^{(1,g)}\Delta(H \cap {}^{g^{-1}}H).$$

If ${}^{(1,g)}\Delta(H \cap {}^{g^{-1}}H) \geq {}^{(h_1,h_2)}\Delta(D)$ then for $x \in D$ we have $h_1xh_1^{-1} = g^{-1}h_2xh_2^{-1}g$ and so $h_2^{-1}gh_1 \in C_G(D) \leq H$ and so $g \in H$. It follows that for $g \notin H$, no summand $k_{(H\times H)\cap^{(1,g)}\Delta(G)}\uparrow^{H\times H}$ has vertex containing $\Delta(D)$. The identity double coset corresponds to the summand kH, and so $kG\downarrow_{H\times H}$ has only one summand isomorphic to b. It follows that there is a unique block of kG whose restriction to $H \times H$ has a summand isomorphic to b.

To prove that this summand is b^G, we argue as follows. It suffices to prove this in case $H \leq N_G(D)$, since applying this case twice yields the general case. But in this case, if B is a summand of kG such that b is not a summand of $B\downarrow_{H\times H}$, and e and e' are the central idempotents in kH and kG corresponding to b and B respectively, then the projection of ee' onto kH as a summand of kG is zero, so $e\mathrm{Br}_D(e') = \mathrm{Br}_D(ee') = 0$ and $B \neq b^G$. □

6.3. Brauer's second main theorem

The following is Nagao's module theoretic version of Brauer's second main theorem.

THEOREM 6.3.1 (Nagao). *Let $e \in Z(kG)$ be a central idempotent, let D be a p-subgroup of G, and let K be a subgroup with $C_G(D) \leq K \leq N_G(D)$. If M is a kG-module with $M = e.M$ then*

$$M\downarrow_K = \mathrm{Br}_D(e).M\downarrow_K \oplus M'$$

where M' is a direct sum of modules projective relative to p-subgroups Q with $D \not\leq Q \leq K$.

PROOF. We have $M' = (1 - \mathrm{Br}_D(e)).M_K$. Since e acts as the identity endomorphism of M, $e - \mathrm{Br}_D(e)$ acts as the identity endomorphism on M'. So by Lemma 6.2.1 (ii) we have

$$1_{M'} \in \sum_{D\not\leq Q\leq N_G(D)} (M', M')_Q^{N_G(D)} \subseteq \sum_{D\not\leq Q\leq K} (M', M')_Q^K.$$

Thus by Rosenberg's Lemma 1.7.10 and Higman's criterion 3.6.4, each indecomposable summand of M' is projective relative to some p-subgroup Q with $D \not\leq Q \leq K$. □

COROLLARY 6.3.2. *Suppose D is a p-subgroup of G, M is an indecomposable kG-module with vertex D, and e is a primitive idempotent in $Z(kG)$. If we denote by M' the $N_G(D)$ module corresponding to M under the Green correspondence (see Theorem 3.12.2), then $e.M = M$ if and only if $\mathrm{Br}_D(e).M' = M'$.* □

COROLLARY 6.3.3. *Suppose $B = e.kG$ is a block with defect group D. Then there is an indecomposable trivial source module $M = e.M$ in the block B, with vertex exactly D.*

PROOF. Let M_0 be a projective indecomposable $k(N_G(D)/D)$-module regarded as an indecomposable $kN_G(D)$-module with vertex D, and chosen so that $\mathrm{Br}_D(e).M_0 = M_0$. Then by the above corollary the Green correspondent M of M_0 is a trivial source indecomposable kG-module with vertex D, and $e.M = M$. □

COROLLARY 6.3.4 (Blocks of defect zero). *If $B = e.kG$ is a block with defect group D, then the following are equivalent:*

(i) *$J(B) = 0$ (and hence B is a complete matrix ring over a division ring).*

(ii) *D is the trivial subgroup of G.*

(iii) *B contains a projective simple module.*

PROOF. By the above corollary and Proposition 6.1.2, D is the trivial subgroup if and only if every module in B has vertex the trivial subgroup, which is the same as being projective. By Lemma 1.2.4 this is equivalent to the condition that $J(B) = 0$. Now by Proposition 3.1.2 projective modules for kG are injective, and so by Proposition 1.8.5 this happens if and only if B contains a projective simple module. □

COROLLARY 6.3.5. *If B is a block of kG with defect group D, then B has finite representation type if and only if D is cyclic.*

PROOF. If D is cyclic, then every subgroup D' of D is cyclic and so kD' has finite representation type So by Proposition 6.1.2 there are only finitely many possible sources for modules in B, and therefore B has finite representation type. Conversely, if D is not cyclic then D has a quotient $\mathbb{Z}/p \times \mathbb{Z}/p$, so there are infinitely many indecomposable kD-modules with vertex exactly D. Let b be the Brauer correspondent of B, as a block of $N_G(D)$. Now if we induce the trivial kD-module to $N_G(D)$ then the resulting module $k(N_G(D)/D)$ has summands in every block of $kN_G(D)$ (since every simple $kN_G(D)$-module has D acting trivially). So given any indecomposable kD-module M with vertex D, $M\uparrow^{N_G(D)}$ has some summand M' in b. Since $M\uparrow^{N_G(D)}\downarrow_D$ is a sum of conjugates of M, the module M is a source of M'. Since each conjugacy class in $N_G(D)$ of kD-modules has finite cardinality, we obtain infinitely many non-isomorphic indecomposable modules in b with vertex D this way. By Corollary 6.3.2, the Green correspondents of these give infinitely many non-isomorphic indecomposable modules in B with vertex D. □

6.4. Clifford theory of blocks

In this section we examine the relationship between blocks and normal subgroups. We use this to establish the extended version of Brauer's first main theorem, and Brauer's third main theorem.

Let N be a normal subgroup of G, e a primitive central idempotent in kN, and $b = e.kN$ the corresponding block of kN. For any $g \in G$, geg^{-1} is again a primitive central idempotent in kN, and is hence either equal to or orthogonal to e. We define the **inertia group** $T(b)$ to be the subgroup of G consisting of those elements g with $geg^{-1} = e$. Thus

$$f = \sum_{g \in G/T(b)} geg^{-1}$$

is a central idempotent in kG.

The identity element of kG can be written as the sum of such elements f, one for each conjugacy class in kG of blocks of kN. Thus given any primitive central idempotent e' in kG corresponding to a block $B = e'.kG$, there exists a block b of kN as above, unique up to conjugacy, such that B is a summand of $f.kG$. This is equivalent to the condition that $e'f = e'$, which in turn happens if and only if $e'f \neq 0$ since e' is primitive. In the above situation, we say that the block B **covers** the block b. Clearly every block of kN is covered by some block of kG.

THEOREM 6.4.1. *Suppose B is a block of kG, and N is a normal subgroup of G.*

(i) *The blocks of kN covered by B form a single G-conjugacy class of blocks of kN.*

(ii) *Suppose $b = e.kN$ is a block of kN covered by B. Then some defect group $D(B)$ is contained in the inertia group $T = T(b)$.*

(iii) *Let $f = \sum_{g \in G/T} geg^{-1}$ as above. Then we have $f.kG \cong \mathrm{Mat}_n(e.kT)$ where $n = |G : T|$.*

(iv) *For some choice of $D(B)$ (but not necessarily for all choices of $D(B)$) the group $D(B) \cap N$ is a defect group of b.*

(v) *If $C_G(D(b)) \leq N$ then $b^G = B$ and B is the unique block covering b. Finally, if k is algebraically closed then $|T : D(B)N|$ is not divisible by p.*

PROOF. (i) This is clear from the definitions.

(ii) As a $k(N \times N)$-module we have $kG = \bigoplus_{g \in G/N} g.kN$, and so letting f be the idempotent $\sum_{g \in G/T} geg^{-1}$ as above, we have

$$f.kG = \bigoplus_{g \in G/N} g(f.kN) = \bigoplus_{g \in G/N} \bigoplus_{h \in G/T} g(he.kN.h^{-1}) = \bigoplus_{\substack{(g_1,g_2) \in \\ (G \times G)/(\Delta(T).N \times N)}} g_1 b g_2^{-1}.$$

The latter is just b regarded as a $k(\Delta(T).N \times N)$-module induced up to $G \times G$. It follows that as a $k(G \times G)$-module, every block B covering b is projective relative to $\Delta(T).N \times N$. Thus for some defect group $D(B)$ we have $\Delta(D(B)) \leq \Delta(T).N \times N$. Since $N \leq T$, it follows that $D(B) \leq T$.

(iii) We have

$$f.kG = \bigoplus_{(g_1,g_2) \in (G \times G)/(T \times T)} g_1(e.kT)g_2^{-1}.$$

Similarly, $e.kG$ can be written as an $e.kT$-module in the form

$$e.kG = \bigoplus_{g \in G/T} (e.kT)g^{-1}$$

so that it is free of rank $n = |G : T|$ over $e.kT$. Thus we have

$$\mathrm{End}_{e.kT}(e.kG)^{\mathrm{op}} \cong \mathrm{Mat}_n(e.kT).$$

We have an algebra homomorphism $f.kG \to \mathrm{End}_{e.kT}(e.kG)^{\mathrm{op}}$ given by right multiplication. This is injective since if $(ey)(fx) = 0$ for all $y \in kG$ then $fx = f.fx = \sum_{g \in G/T} geg^{-1}.fx = 0$. It is surjective since the matrix entry corresponding to sending $(e.kT)g_1^{-1}$ to $(e.kT)g_2^{-1}$ with right multiplication by $x \in b^T$ is achieved by the element $g_1xg_2^{-1} \in f.kG$.

(iv) Since B is a trivial source module with vertex $\Delta(D(B))$, the restriction to $\Delta(D(B))$ has the trivial module as a summand. Hence the restriction to $\Delta(D(B)) \cap (N \times N)$ also has the trivial module as a summand, and so some summand of $B\downarrow_{N\times N}$ has vertex containing $\Delta(D(B)) \cap (N \times N)$. But the restriction of B to $N \times N$ is a sum of modules of the form $g_1bg_2^{-1}$. The vertices of these modules are all $G \times G$-conjugate to $\Delta(D(b))$, and so some G-conjugate of $D(B) \cap N$ is contained in $D(b)$. Since b is a summand of $B\downarrow_{N\times N}$, it is projective relative to some G-conjugate of $\Delta(D(B) \cap N)$, and so we have equality.

(v) If $C_G(D(b)) \leq N$, then by Lemma 6.2.7 b^G is the only block B of kG such that b is a summand of $B\downarrow_{N\times N}$, and is hence the only block covering b.

Finally, suppose k is algebraically closed, and let S be a Sylow p-subgroup of T containing $D(B)$. As in the proof of (ii), we regard b as a module for $\Delta(T)(N \times N)$ whose restriction to $N \times N$ is indecomposable. Hence $b\downarrow_{\Delta(S)(N\times N)}$ is indecomposable and projective relative to $\Delta(D(B))(N \times N)$, so that by Green's Indecomposability Theorem 3.13.3 it is induced from a module for $\Delta(D(B))(N \times N)$. Since $b\downarrow_{\Delta(D(B))(N\times N)}$ is indecomposable the Mackey decomposition theorem implies that $\Delta(D(B))(N \times N) = \Delta(S)(N \times N)$, so that $D(B)N = SN$. This implies that $|T : D(B)N|$ is not divisible by p. □

LEMMA 6.4.2. *Suppose Q is a p-subgroup of G and $G = QC_G(Q)$. Then the natural map $\pi : kG \to kG/Q$ induces a one–one correspondence between blocks of kG with defect group D (which of course contain Q, for example by Proposition 6.1.1) and blocks of kG/Q with defect group D/Q.*

PROOF. By Proposition 6.2.2, every idempotent in $Z(kG)$ lies in $kC_G(Q)$. Conversely $Z(kC_G(Q)) \leq Z(kG)$ since $G = QC_G(Q)$, and so the primitive idempotents in $Z(kG)$ are the same as the primitive idempotents in $Z(kC_G(Q))$.

Now the map

$$\pi : kG \to kG/Q \cong kC_G(Q)/Z(Q)$$

maps $Z(kC_G(Q))$ surjectively onto $Z(kC_G(Q)/Z(Q))$, and has as its kernel $J(kQ).C_G(Q)$, which is nilpotent. So by the idempotent refinement theorem it induces a one–one correspondence between idempotents in $Z(kC_G(Q))$ and in $Z(kG/Q)$. Since the image under π of $(kG)_D^G$ is $(kG/Q)_{D/Q}^{G/Q}$ the statement about defect groups follows from Proposition 6.1.2. □

THEOREM 6.4.3 (Brauer's first main theorem, extended version).

Suppose that k is algebraically closed. Then there is a one–one correspondence between the following:

(i) *Blocks of kG with defect group D.*

(ii) *Blocks of $kN_G(D)$ with defect group D.*

(iii) *$N_G(D)$-conjugacy classes of blocks b of $kDC_G(D)$ with defect group D, such that $|T(b) : DC_G(D)|$ is not divisible by p.*

(iv) *$N_G(D)$-conjugacy classes of blocks b of $kDC_G(D)/D$ of defect zero, such that $|T(b) : DC_G(D)|$ is not divisible by p.*

The correspondence between (i), (ii) *and* (iii) *is given by the Brauer map $b \mapsto b^G$, while the correspondence between* (iii) *and* (iv) *is given by the natural map $\pi : kDC_G(D) \to kDC_G(D)/D$.*

PROOF. The correspondence between (i) and (ii) was shown in Theorem 6.2.6. The correspondence between (iii) and (iv) was shown in the above lemma (with $Q = D$). So it remains to discuss the correspondence between (ii) and (iii). We may thus assume that D is normal in G. We apply Theorem 6.4.1 with $N = DC_G(D)$. This says that if B is a block of kG with defect group D then the blocks of kN covered by B also have D as defect group and p does not divide $|T(b) : DC_G(D)|$. Conversely if b is a block of $DC_G(D)$ with defect group D and p does not divide $|T(b) : DC_G(D)|$ then $B = b^G$ is the unique block of kG which covers b. Moreover the defect group $D(b^G)$ is a p-subgroup of $T(b)$ which intersects $DC_G(D)$ in exactly D, so that since p does not divide $|T(b) : DC_G(D)|$, D is a defect group of b^G. □

It is worth making more explicit the structure of the blocks given in part (iii) of the above theorem.

PROPOSITION 6.4.4. *Suppose k is an algebraically closed field, and suppose B is a block of kG with normal defect group D and $G = DC_G(D)$. Then $B \cong \mathrm{Mat}_n(kD)$.*

PROOF. Since D acts trivially on simple kG-modules, and the block of defect zero of kG/D corresponding to B as in the last theorem has only one simple module by Corollary 6.3.4, it follows that B has only one simple module S. So to prove that $B \cong \mathrm{Mat}_n(kD)$ it suffices to show that B is Morita equivalent to kD.

If M is a kD-module then $M\uparrow^G$ is a kG-$kC_G(D)$-bimodule, with right $C_G(D)$-action given by $(g \otimes m)x = gx \otimes m$. So we have a functor ${}_{kD}\mathbf{mod} \to {}_{B}\mathbf{mod}$ given by $M \mapsto M\uparrow^G \otimes_{kC_G(D)} S$. Similarly, we regard the dual S^* of S as a kD-kG-bimodule with trivial left action and we have a functor ${}_{B}\mathbf{mod} \to {}_{kD}\mathbf{mod}$ given by $N \mapsto S^* \otimes_{kG} N$. Using the identity $S^* \otimes_{kC_G(D)} S \cong \mathrm{Hom}_{kC_G(D)}(S, S) \cong k$ it is easy to see that these functors give an equivalence of categories. □

THEOREM 6.4.5 (Brauer's third main theorem). *Suppose H is a subgroup of G with $DC_G(D) \leq H$, and b is a block of kH with defect group D. Then b^G is the principal block $B_0(kG)$ if and only if b is the principal block $B_0(kH)$.*

PROOF. If we apply Nagao's Theorem 6.3.1 to the trivial module, we find that if $b = B_0(kH)$ then $B = B_0(kG)$. Conversely, suppose $b^G = B_0(kG)$. We may suppose $H \le N_G(D)$, since applying this case twice gives the general case. Let D' be a defect group of $b^{N_G(D)}$. Then $D \le D' \le N_G(D)$ so that $C_G(D') \le C_G(D) \le N_G(D)$ and hence $D'C_G(D') \le N_G(D)$. By the extended first main theorem there is a unique $N_G(D')$-conjugacy class of blocks b' of $kD'C_G(D')$ with $(b')^G = B_0(kG)$. Since $B_0(kD'C_G(D'))$ is such a block and is stable under $N_G(D')$-conjugation, this is the unique such block. But there is also a block b' of $kD'C_G(D')$ with $(b')^{N_G(D)} = b^{N_G(D)}$. For such a block we also have $(b')^G = B_0(kG)$ and hence $b' = B_0(kD'C_G(D'))$. Now by the first part of the proof we have $b^{N_G(D)} = B_0(kN_G(D))$. So we may assume that D is normal in G.

We now apply Theorem 6.4.1 with $N = DC_G(D)$. Let b' be a block of kN covered by b. By part (iv) of this theorem D is a defect group of b'. Since $C_G(D) \le N$, by part (v) we have $(b')^H = b$ and so $(b')^G = B_0(kG)$. But also $B_0(kN)^G = B_0(kG)$, so that since $B_0(kN)$ is stable under G-conjugation we have $b' = B_0(kN)$ and hence $b = (b')^H = B_0(kH)$. □

6.5. Blocks of cyclic defect

The situation of a block whose defect groups are cyclic is one which is very well understood. The case of a cyclic defect group of order p was originally described by Brauer. Using ideas of Green and Thompson, the general case was analysed by Dade. We shall only describe that part of the theory which has to do with the modular representations. We shall not describe the ordinary characters or decomposition numbers.

Suppose B is a block of kG whose defect group D is cyclic of order p^n. By Proposition 6.1.2 every indecomposable module in B has vertex contained in D. Since a cyclic group has only finitely many indecomposable modules, there are only finitely many sources and hence only finitely many indecomposables in B. Thus B has finite representation type.

Let Q be the unique subgroup of D of order p, so that $N_G(Q) \ge N_G(D)$. By Brauer's first main theorem, there is a unique block b of $N_G(Q)$ with $b^G = B$. By Theorem 6.4.1, there is a unique $N_G(Q)$-conjugacy class of blocks b_1 of $C_G(Q)$ (note that $Q \le C_G(Q)$) with $b_1^{N_G(Q)} = b$ (so that $b_1^G = B$). Let $T = T(b_1) \le N_G(Q)$ be the inertia group of b_1, and set $e = |T : C_G(Q)|$, the **inertial index** of B (this is not the usual definition, but it is equivalent and more suitable for our purposes). We shall not use the letter e to stand for an idempotent during this section, so there should be no notational confusion. Note that e divides $|N_G(Q) : C_G(Q)|$, which in turn divides $p - 1$ since $N_G(Q)/C_G(Q)$ is isomorphic to a group of automorphisms of Q.

We next analyse the Green correspondence between modules for G and $N_G(Q)$.

LEMMA 6.5.1. *Green correspondence between G and $N_G(Q)$ sets up a one–one correspondence between non-projective modules in B and non-projective modules in b, in such a way that*

$$M\downarrow_{N_G(Q)} = f(M) \oplus (\text{projective}) \oplus (\text{modules not in } b)$$
$$M'\uparrow^G = g(M') \oplus (\text{projective}).$$

If M_1 and M_2 are modules in B then

$$(M_1, M_2)^{G,1} \cong (f(M_1), f(M_2))^{N_G(Q),1}.$$

(see Definition 3.6.2 for the notation).

PROOF. By Proposition 6.1.2 every non-projective module in B has vertex D' with $1 < D' \leq D$. Since Q is a characteristic subgroup of D' this means that $N_G(D') \leq N_G(Q)$ and so we may apply Green correspondence. By Nagao's theorem a module with vertex D' lies in B if and only if its Green correspondent lies in b.

The theorem now follows from Theorem 3.12.2 once we have evaluated the sets of subgroups $\mathcal{X}$ and $\mathcal{Y}$. Since Q is the unique minimal subgroup of D, if $x \notin N_G(Q)$ then we have ${}^xD \cap D = 1$, and so $\mathcal{X} = \{1\}$. Similarly $\mathcal{Y}$ consists of subgroups of $N_G(Q)$ not containing Q, so that any non-projective indecomposable $kN_G(Q)$-module with vertex in $\mathcal{Y}$ does not lie in b by Proposition 6.1.2. □

We shall also need the following lemma.

LEMMA 6.5.2. *Suppose Λ is a finite dimensional algebra with the property that every projective indecomposable Λ-module and every injective indecomposable Λ-module is uniserial. Then every indecomposable Λ-module is uniserial, and in particular is a quotient of a projective indecomposable.*

PROOF. Suppose M is indecomposable and S is a simple submodule of M. Let M' be a submodule of M which is maximal subject to the condition $S \cap M' = 0$. Then M/M' has S as its socle since otherwise we could enlarge M'. So the injective hull of M/M' is the injective hull of S and hence uniserial, and so M/M' is uniserial. So the projective cover P of M/M' is also uniserial. The map $P \to M/M'$ lifts to a map $P \to M$ whose image M'' is a uniserial submodule of M containing S and hence intersects M' trivially. Thus $M = M' \oplus M''$, and since $M'' \neq 0$ we have $M' = 0$. □

We now analyse blocks of cyclic defect with inertial index one. This is an easy case to understand, and acts as a model for the arguments in the general case.

PROPOSITION 6.5.3. *Suppose B is a block of kG with cyclic defect group D of order p^n and with inertial index one. Then there is only one simple module S in B, and the projective cover of S is uniserial of length p^n.*

PROOF. We first prove that if B has only one simple module S then the projective cover of S is uniserial. For this purpose we use the fact that B has finite representation type.

Denote by P_S the projective cover of S. We have $\mathrm{Rad}(P_S)/\mathrm{Rad}^2(P_S) \neq 0$, since otherwise S is projective so that by Corollary 6.3.4 S lies in a block of defect zero. If $\mathrm{Rad}(P_S)/\mathrm{Rad}^2(P_S)$ is a sum of at least two copies of S then P_S has a quotient M with $\mathrm{Rad}(M) \cong S \oplus S$ and $M/\mathrm{Rad}(M) \cong S$. Thus the basic algebra $\mathrm{End}_{kG}(P_S)^{\mathrm{op}}$ of B is a quotient of the algebra $\mathrm{End}(M)^{\mathrm{op}}$. Setting $\Delta = \mathrm{End}_{kG}(S)^{\mathrm{op}}$ we have $\mathrm{End}_{kG}(M)^{\mathrm{op}} \cong \Delta[X,Y]/(X^2, XY, Y^2)$. This algebra has infinitely many non-isomorphic indecomposable modules. This may be seen as follows. The algebra $\Delta[X,Y]/(X^2,Y^2)$ is a self-injective algebra of dimension four over Δ, and the modules $\Omega^n k$ are all non-isomorphic indecomposables on which XY acts as zero. Since B has only finitely many indecomposables, we deduce that $\mathrm{Rad}(P_S)/\mathrm{Rad}^2(P_S) \cong S$. It follows that P_S is uniserial, since by induction on r, if $\mathrm{Rad}^i(P_S)/\mathrm{Rad}^{i+1}(P_S) \cong S$ then $\mathrm{Rad}^i(P_S)/\mathrm{Rad}^{i+2}(P_S)$ is a quotient of $P_S/\mathrm{Rad}^2(P_S)$ and so we have either $\mathrm{Rad}^{i+1}(P_S) = 0$ or $\mathrm{Rad}^{i+1}(P_S)/\mathrm{Rad}^{i+2}(P_S) \cong S$.

We now prove the proposition by induction on the order of G, and we begin with the case where the subgroup Q of order p in D is central in G (note that the case $D = 1$ was dealt with in Corollary 6.3.4). Applying Lemma 6.4.2, we see that the natural map $\pi : kG \to kG/Q$ induces a one–one correspondence between blocks of kG with defect group D and blocks of kG/Q with defect group D/Q. If $\bar{B}$ is the block of kG/Q corresponding to B then the inertial index of $\bar{B}$ is again one, and since G/Q is smaller than G the inductive hypothesis shows that $\bar{B}$ has only one simple module S. Since Q acts trivially on simple B-modules, B also has only one simple module, namely S again. Now if g is a generator of Q then as a kG-module we have

$$(g-1)^j kG/(g-1)^{j+1} kG \cong kG/Q$$

for $j = 0, 1, \ldots, p-1$, and $(g-1)^p = 0$. Thus $(g-1)^j B/(g-1)^{j+1} B \cong \bar{B}$ and so the length of the projective cover of S as a B-module is p times what it is as a $\bar{B}$-module. This completes the proof in case $Q \le Z(G)$.

Next, we treat the case where Q is normal but not central. In this case we apply Theorem 6.4.1 to the normal subgroup $C_G(Q)$, which is equal to T since $e = 1$. We see that if b_1 is a block of $kC_G(Q)$ covered by B then $B \cong \mathrm{Mat}_m(b_1)$, where $m = |G : T|$. Thus B is Morita equivalent to b_1 and the result follows in this case.

Finally, if Q is not normal in G then we apply Green correspondence between G and $N_G(Q)$, as in the Lemma 6.5.1. If B has more than one simple module, say S and S' are simple B-modules, then the Green correspondents $f(S)$ and $f(S')$ are uniserial modules for $N_G(Q)$, and so one is a quotient of the other, say $f(S')$ is a quotient of $f(S)$. Now it follows from Proposition 3.6.6 that a homomorphism from one module to another lies in the image of $\mathrm{Tr}_{1,G}$ if and only if it factors through the projective cover of the second module. Since $f(S)$ is not projective, the surjection $f(S) \to f(S')$ does not lie in $(f(S), f(S'))_1^{N_G(Q)}$. Hence we have $(f(S), f(S'))^{N_G(Q),1} \neq 0$ so that by Lemma 6.5.1 $(S, S')^{G,1} \neq 0$. Since S and S' are distinct simple modules this is absurd, and so B has only one simple module. Since the length of the projective cover of this simple module is one more than the number of non-projective indecomposables in B, it also follows from the Green correspondence that this is the same as the length for the corresponding block of $N_G(Q)$. □

The next easiest case to consider is the one in which Q is normal in G. To avoid complications we assume that k is algebraically closed.

PROPOSITION 6.5.4. *Suppose k is algebraically closed, and B is a block of kG with cyclic defect group D of order p^n and inertial index e. Suppose further that the subgroup Q of D of order p is normal in G. Then there are e simple modules in B, all of the same dimension over k. These simple modules may be labelled $S_1, \dots, S_e$ in such a way that the projective cover P_j of S_j is uniserial of length p^n with $\mathrm{Rad}^i(P_j)/\mathrm{Rad}^{i+1}(P_j) \cong S_{i+j}$, with the subscripts being taken modulo e.*

B is Morita equivalent to the group algebra of a split extension with normal subgroup D and complement cyclic of order e acting faithfully on D.

PROOF. Let b_1 be a block of $C_G(Q)$ covered by B, and let $T = T(b_1)$ be the inertia group. Then by Theorem 6.4.1 B is isomorphic to a complete matrix algebra over b_1^T. It thus suffices to prove the proposition with $G = T$. We may apply the last proposition to b_1 to see that it has a unique simple module S, and its projective cover is uniserial of length n.

We first claim that S extends to a simple kG-module in exactly e ways, and that these are all the simple modules in B (note that this is not true unless $G = T$). Since b_1 is stable under conjugation by G, so is the simple module S. Thus if $g \in G$ generates the cyclic group $G/C_G(Q)$ we have $g \otimes S \cong S$. Let θ be an element of $\mathrm{End}_k(S)$ such that $g \otimes s \mapsto \theta(s)$ is such a $kC_G(Q)$-module isomorphism. In other words, $\theta(g^{-1}hg(s)) = h(\theta(s))$ for $h \in C_G(Q)$. Then $g^j \otimes s \mapsto \theta^j(s)$ gives an isomorphism $g^j \otimes S \cong S$. Since $g^e \in C_G(Q)$, $g^e(s) \mapsto \theta^e(s)$ is a $kC_G(Q)$-module endomorphism of S. Since k is algebraically closed, it follows that for some $\mu \in k$ we have $g^e(s) = \mu\theta^e(s)$. Since e is coprime to p, there are exactly e distinct choices for an element $\lambda \in k$ with $\lambda^e = \mu$. For each such choice, we may extend S to a simple kG-module by letting g act as the endomorphism $\lambda\theta$. Distinct choices of λ give non-isomorphic extensions of S to a kG-module, since an isomorphism between two such extensions restricts to give an isomorphism on $C_G(Q)$, which may therefore be taken to be the identity map.

If we let $S_1, \dots, S_e$ be the extensions of S to kG-modules, then

$$\mathrm{Hom}_{kG}(S_j, S\uparrow^G) \cong \mathrm{Hom}_{kC_G(Q)}(S, S) \cong k$$

and so by counting dimensions we have $S\uparrow^G = S_1 \oplus \cdots \oplus S_e$. Now if M is any simple kG-module then

$$\mathrm{Hom}_{kG}(M, S\uparrow^G) \cong \mathrm{Hom}_{kC_G(Q)}(M\downarrow_{C_G(Q)}, S) \neq 0$$

and so M must be one of the S_j.

Now by the Eckmann–Shapiro lemma

$$\mathrm{Ext}^1_{kG}(S_j, S_1 \oplus \cdots \oplus S_e) \cong \mathrm{Ext}^1_{kC_G(Q)}(S, S) \cong k$$

so that there is a unique value of i for which $\mathrm{Ext}^1_{kG}(S_j, S_i) \cong k$, and for the remaining values of i we have $\mathrm{Ext}^1_{kG}(S_j, S_i) = 0$. Since B is a block, all simple modules are connected by some chain of extensions, and it follows that we may label the simple modules in such a way that $\mathrm{Ext}^1_{kG}(S_j, S_{j+1}) \cong k$, where j is taken modulo e. Thus if we denote by P_j the projective cover of S_j we have $\mathrm{Rad}(P_j)/\mathrm{Rad}^2(P_j) \cong S_{j+1}$. It follows that P_j is uniserial with either $\mathrm{Rad}^i(P_j)/\mathrm{Rad}^{i+1}(P_j) \cong S_{i+j}$ or $\mathrm{Rad}^i(P_j) = 0$, since by induction on i, if $\mathrm{Rad}^i(P_j)/\mathrm{Rad}^{i+1}(P_j) \cong S_{i+j}$ then $\mathrm{Rad}^i(P_j)/\mathrm{Rad}^{i+2}(P_j)$ is a quotient of $P_{i+j}/\mathrm{Rad}^2(P_{i+j})$ and so either $\mathrm{Rad}^{i+1}(P_j) = 0$ or $\mathrm{Rad}^{i+1}(P_j)/\mathrm{Rad}^{i+2}(P_j) \cong S_{i+j+1}$. The restriction of P_j to $C_G(Q)$ is some multiple of P_S; this multiple has to be one since $P_S\uparrow^G$ is projective and hence equal to the sum of the P_j. Thus the length of P_j is the same as the length of P_S, namely p^n.

Finally, to see that B is Morita equivalent to the group algebra of a split extension of D by $\mathbb{Z}/e$, we notice that the Ext-quiver is an oriented cycle of length e with relations saying that any composite of p^n successive arrows is zero, so this determines the basic algebra by the method of Section 4.1. Since the group algebra of the split extension is a block of the type being considered, it has the same Morita type. □

THEOREM 6.5.5. *Suppose k is algebraically closed, and B is a block of kG with cyclic defect group D of order p^n and inertial index e. Then there are e simple modules and $e.p^n$ indecomposable modules in B. The stable Auslander–Reiten quiver of B is a finite tube $(\mathbb{Z}/e)A_{p^n-1}$. The algebra B is a Brauer tree algebra (see Section 4.18) for a Brauer tree with e edges and exceptional multiplicity $(p^n-1)/e$.*

PROOF. Let Q be the subgroup of D of order p, and let b be the Brauer correspondent of B in $N_G(Q)$. Then by the previous proposition there are e simple modules in b, and their projective covers are uniserial of length p^n. Thus every indecomposable in b is uniserial and a quotient of a projective indecomposable. Each projective indecomposable has $p^n - 1$ non-projective quotients, and so b has $e(p^n - 1)$ non-projective indecomposables. It now follows from the Green correspondence (Lemma 6.5.1) that B also has $e(p^n - 1)$ non-projective indecomposables. So we must show that B has e simple modules, and hence also e projective indecomposables.

We first claim that if S and S' are simple B-modules then the Green correspondents $f(S)$ and $f(S')$ have non-isomorphic heads. For otherwise, since they are uniserial modules, one is a quotient of the other, and just as in the proof of Proposition 6.5.3 this implies that $0 \neq (f(S), f(S'))^{N_G(Q),1} \cong (S, S')^{G,1}$ which is absurd. Thus there are at most e simple modules in B.

Conversely if S_0 and S_0' are simple modules in b, we claim that the same simple B-module S cannot appear in the top radical layer of the Green correspondents $g(S_0)$ and $g(S_0')$. For if S appears in $g(S_0)/\mathrm{Rad}\, g(S_0)$ then $0 \neq (g(S_0), S)^{G,1} \cong (S_0, f(S))^{N_G(Q),1}$ so that S_0 is a submodule of $f(S)$, and similarly S_0' is also a submodule of $f(S)$ and so $S_0 \cong S_0'$. So there are at least e simple modules in B.

It follows from Lemma 6.5.1, Corollary 6.3.2 and Theorem 4.12.11 that the induction to G of an almost split sequence of modules in b is a sum of an almost split sequence of modules in B and a split sequence of projective modules. Thus the stable Auslander–Reiten quiver of B is isomorphic to that for b. It is easy to see from the structure of the projective modules in b given in the previous proposition that the irreducible maps between modules in b are the injections and surjections with simple kernels and cokernels, so that the stable Auslander–Reiten quiver of b, and hence also for B, is a tube of type $(\mathbb{Z}/e)A_{p^n-1}$. In particular there are at most $2e$ almost split sequences in B with the property that the middle term has only one non-projective summand. Thus we may apply Theorem 4.18.3 to deduce that B is a Brauer graph algebra. Since B has finite representation type, we may then apply Theorem 4.18.4 to deduce that B is a Brauer tree algebra. The number of edges in the tree is equal to the number of isomorphism classes of simple modules, namely e. Following Alperin [**3**], the multiplicity of the exceptional vertex may be determined by looking at the determinant of the Cartan matrix as follows. It follows from Lemma 6.5.1 that Green correspondence gives an isomorphism between the cokernels of the Cartan homomorphisms for b and for B (cf. Corollary 5.3.5 and Theorem 5.9.3). So the determinant of the Cartan matrix of B is equal to that of b, namely $\det(I + ((p^n - 1)/e)J) = p^n$, where I is an $e \times e$ identity matrix and J is a matrix of the same size with all entries equal to one. We claim that for a Brauer tree with e edges and exceptional multiplicity m the determinant of the Cartan matrix is $em + 1$, which gives the value of m as $(p^n - 1)/e$ as required. We prove this by induction on e. We first treat the case in which the tree is a star; in other words there is at most one vertex of valency greater than one. If the exceptional vertex is the one with valency greater than one, then as above the determinant is $\det(I + mJ) = em + 1$, while if the exceptional vertex has valency one then the determinant is

$$\det \begin{pmatrix} m+1 & 1 & \cdots & 1 \\ 1 & 2 & & 1 \\ \vdots & & & \vdots \\ 1 & 1 & \cdots & 2 \end{pmatrix} = em + 1.$$

Finally, if the tree is not a star then there exists an edge E such that neither of the vertices at its ends has valency one. Denote by L and R the trees to the left and right of E, intersecting in exactly E, and with union the whole tree. Denote by L_0 and R_0 the trees obtained from L and R by removing the edge E. We choose the labelling so that the exceptional vertex, if there is one, lies in L. Denote by l and r the numbers of edges in L and R, so that $l+r=e+1$. The Cartan matrix has the form

$$C = \left(\begin{array}{c|c|c} X & & 0 \\ \hline & z & \\ \hline 0 & & Y \end{array}\right)$$

with a single entry z in the overlap, equal to either 2 or $m+1$, depending on whether the exceptional vertex is the left-hand vertex of E or not. Denote by X_0 and Y_0 the submatrices of X and Y obtained by removing the row and column containing the overlap. Then X, Y, X_0 and Y_0 are the Cartan matrices of L, R, L_0 and R_0. By the inductive hypothesis we know the determinants of X, Y, X_0 and Y_0.

Expanding $\det(C)$ about the row containing the overlap, we see that

$$\det(C) = \det(X)\det(Y_0) + \det(X_0)\det(Y) - z.\det(X_0)\det(Y_0).$$

If $z=2$ this equals

$$(lm+1)r + ((l-1)m+1)(r+1) - 2((l-1)m+1)r = em+1,$$

while if $z=m+1$ it equals

$$(lm+1)r + ((l-1)m+1)(rm+1) - (m+1)((l+1)m+1)r = em+1.$$

This completes the calculation of the determinant of C. □

6.6. Klein four defect groups

In this section we show how the methods of the last section can be pushed to determine the structure of blocks whose defect groups are Klein four groups. In an extraordinary series of papers, Erdmann [**101, 102, 103, 104, 105**] has taken this method to its natural conclusion by completing the analysis of tame blocks of group algebras; namely those whose defect groups are dihedral, semidihedral or generalised quaternion. For all other possible defect groups the representation type is wild, and so one does not expect an analysis of almost split sequences to determine the algebra structure.

Suppose k is an algebraically closed field of characteristic two, and suppose B is a block of kG whose defect group D is a Klein four group $\mathbb{Z}/2\times\mathbb{Z}/2$. By Brauer's first main theorem, there is a unique block b of $N_G(D)$ with $b^G = B$. By the extended first main theorem there is a unique $N_G(D)$-conjugacy class of blocks b_1 of $kC_G(D)$ (note that $D \le C_G(D)$) with $b_1^{N_G(D)} = b$ (so that $b_1^G = B$). Let $T = T(b_1) \le N_G(D)$ be the inertial group of b_1, and set $e = |T : C_G(D)|$, the inertial index of B. Since $N_G(D)/C_G(D)$ is isomorphic to a subgroup of $\mathrm{Aut}(D)$, which has order six, and e is coprime to $p=2$, it follows that $e=1$ or 3. We begin with the case $e=1$.

THEOREM 6.6.1. *Suppose k is an algebraically closed field of characteristic two, and B is a block of kG with Klein four defect group D and inertial* *index $e = 1$. Then there is a unique simple module S in B, and if P_S is its projective cover then $\mathrm{Rad}(P_S)/\mathrm{Soc}(P_S) \cong S \oplus S$. For some $n \geq 1$ we have $B \cong \mathrm{Mat}_n(kD)$.*

PROOF. We begin with the case where D is central in G. In this case Proposition 6.4.4 shows that $B \cong \mathrm{Mat}_n(kD)$. Next, if D is normal but not central then by Theorem 6.4.1 there is a unique conjugacy class of blocks b of $C_G(D)$ covered by B. Since $e = 1$ the inertial group of b is $C_G(D)$ and so $B \cong \mathrm{Mat}_m(b)$, where $m = |G : C_G(D)|$.

To examine the case where D is not normal in G, we first examine the stable Auslander–Reiten quiver of kG-modules. Let b be the Brauer correspondent of B as a block of $kN_G(D)$. Referring to the examples as the end of Section 4.17, we see that the stable quiver of b-modules consists of a component of type $\mathbb{Z}\tilde{A}_{12}$ and an infinite set of 1-tubes, each fixed by Ω. Since every proper subgroup of D is cyclic, all but a finite number of these modules have vertex D, and the remaining modules lie in 1-tubes. It thus follows from Proposition 4.12.11 and Corollary 6.3.2 that the stable quiver of B-modules also consists of a component of type $\mathbb{Z}\tilde{A}_{12}$ and an infinite set of 1-tubes, each fixed by Ω.

Now suppose P_S is the projective cover of a simple B-module S. If P_S is attached at the end of a 1-tube then $\mathrm{Rad}(P_S) \cong P_S/\mathrm{Soc}(P_S)$. It is easy to see that this implies P_S is uniserial with all composition factors isomorphic to S and hence S is the only simple module in a block of finite representation type. This is absurd, and so P_S is connected to the component of type $\mathbb{Z}\tilde{A}_{12}$. This implies that $\mathrm{Rad}(P_S)/\mathrm{Soc}(P_S) \cong S \oplus S$. So S is the only simple module in B, and by Morita theory $B \cong \mathrm{Mat}_n(\mathrm{End}_B(P_S)^{\mathrm{op}})$. Let α be an endomorphism of P_S taking the top composition factor to one of the summands of the middle, and β be an endomorphism taking it to the other summand. Since the image M of α is a module with a resolution by projective modules all of dimension four, it follows from Auslander's Theorem 4.14.2 that M does not lie in the $\mathbb{Z}\tilde{A}_{12}$ component and so it satisfies $M \cong \Omega(M)$, so that $\alpha^2 = 0$. Similarly $\beta^2 = 0$ and $\alpha\beta$ is some non-zero multiple of $\beta\alpha$. But B, and hence also $\mathrm{End}_B(P_S)$, is a symmetric algebra. If λ is a linear map as in the definition of a symmetric algebra, then $\lambda(\alpha\beta) = \lambda(\beta\alpha)$. Since λ cannot vanish on the left ideal generated by $\alpha\beta$, we have $\alpha\beta = \beta\alpha$ and so $\mathrm{End}_B(P_S) \cong k\langle\alpha, \beta\rangle/(\alpha^2, \beta^2, \alpha\beta - \beta\alpha) \cong kD$. □

THEOREM 6.6.2. *Suppose k is an algebraically closed field of characteristic two, and B is a block of kG with Klein four defect group D normal in G, and inertial index $e = 3$. Then there are three simple modules, S_1, S_2, S_3 in B with projective covers P_1, P_2, P_3. We have $\mathrm{Rad}(P_j)/\mathrm{Soc}(P_j) \cong S_{j-1} \oplus S_{j+1}$ where the indices are taken modulo three. For some $n \geq 1$ we have $B \cong \mathrm{Mat}_n(kA_4)$, where A_4 is the alternating group of degree four.*

PROOF. Let b be a block of $kC_G(D)$ covered by B. Then $T(b) = G$, $|G : C_G(D)| = 3$, and by the previous theorem we have $b \cong \mathrm{Mat}_n(kD)$. Let S be the simple module in b. Then S is stable under conjugation by G, and so by the same argument as we used in the proof of Proposition 6.5.4, we see that S extends to give three non-isomorphic B-modules S_1, S_2, S_3, and this is a complete list of simple B-modules. Also $P_S\uparrow^G \cong P_1 \oplus P_2 \oplus P_3$ and $P_j\downarrow_{C_G(D)} \cong P_S$. Thus $\mathrm{Rad}(P_j)/\mathrm{Soc}(P_j)$ is isomorphic to a direct sum of two simples. Since the three simples are related by tensoring with the three one dimensional modules for $kG/C_G(D) = k(\mathbb{Z}/3)$, the whole picture is invariant under the substitution $S_j \mapsto S_{j+1}$. Now the fact that the determinant of the Cartan matrix is a power of two (see Corollary 5.7.2 or Theorem 5.9.3) leaves only one possibility, namely $\mathrm{Rad}(P_j)/\mathrm{Soc}(P_j) \cong P_{j-1} \oplus P_{j+1}$. We now compute the basic algebra by the method of quivers with relations (Section 4.1). The Ext-quiver is as follows:

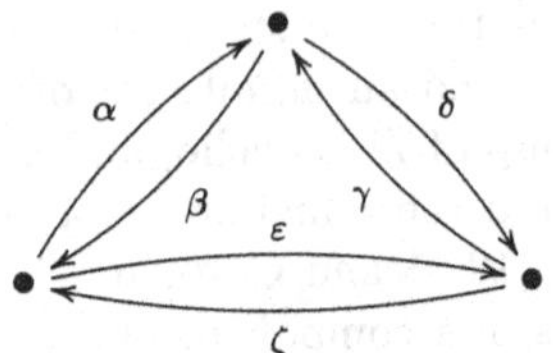

The relations are $0 = \delta\alpha = \beta\gamma = \alpha\zeta = \varepsilon\beta = \zeta\delta = \gamma\varepsilon$, and there are non-zero constants μ_1, μ_2, μ_3 with $\beta\alpha = \mu_1\zeta\varepsilon$, $\varepsilon\zeta = \mu_2\delta\gamma$ and $\gamma\delta = \mu_3\alpha\beta$. By replacing (say) ζ and δ by non-zero multiples, we may assume that $\mu_1 = \mu_2 = 1$. Now B is a symmetric algebra, and hence so is the basic algebra. If λ is a linear map as in the definition of symmetric algebra then

$$0 \neq \lambda(\alpha\beta) = \lambda(\beta\alpha) = \lambda(\zeta\varepsilon) = \lambda(\varepsilon\zeta) = \lambda(\delta\gamma) = \lambda(\gamma\delta) = \mu_3\lambda(\alpha\beta)$$

and so $\mu_3 = 1$. Thus the basic algebra is completely determined. Now kA_4 is an example of such a block B, and is basic. Since S_1, S_2, S_3 have the same dimension, n say, Morita theory implies that B is isomorphic to the endomorphism ring of a direct sum of copies of the regular representation of kA_4, i.e., $B \cong \mathrm{Mat}_n(kA_4)$. □

THEOREM 6.6.3. *Suppose k is an algebraically closed field of characteristic two, and B is a block of kG with Klein four defect group D of inertial index $e = 3$. Then there are three simple B-modules S_1, S_2, S_3 with projective covers P_1, P_2, P_3. One of the following possibilities holds:*

(i) *B is Morita equivalent to the group algebra kA_4. In this case*

$$\mathrm{Rad}(P_j)/\mathrm{Soc}(P_j) \cong S_{j-1} \oplus S_{j+1},$$

where the indices are taken modulo three.

(ii) *B is Morita equivalent to the principal block $B_0(kA_5)$ of the group algebra of the alternating group A_5. In this case, after re-indexing if necessary we have*

$$\begin{aligned}
\mathrm{Rad}(P_1)/\mathrm{Soc}(P_1) &= \mathrm{Uni}(S_2, S_1, S_3) \oplus \mathrm{Uni}(S_3, S_1, S_2)\\
\mathrm{Rad}(P_2)/\mathrm{Soc}(P_2) &= \mathrm{Uni}(S_1, S_3, S_1)\\
\mathrm{Rad}(P_3)/\mathrm{Soc}(P_3) &= \mathrm{Uni}(S_1, S_2, S_1).
\end{aligned}$$

Here, $\mathrm{Uni}(A, B, \ldots)$ *denotes a uniserial module with composition factors* $A, B, \ldots$ *starting from the top.*

PROOF. The proof of this theorem is a variant of the ideas in Theorem 4.18.3 in which the almost split sequences considered have non-isomorphic ends. The problem with applying the theorem as it stands is that there are infinitely many 1-tubes, as we shall see.

We use Green correspondence to examine the stable Auslander–Reiten quiver of B-modules. Let b be the block of $kN_G(D)$ corresponding to B by Brauer's first main theorem, and b_1 a block of $kC_G(D)$ covered by b. By the previous theorem, b is Morita equivalent to the group algebra kA_4 of the alternating group of degree four, and so by the example at the end of Section 4.17, the stable quiver of b-modules consists of one component $\mathbb{Z}\tilde{A}_5$, two 3-tubes, and an infinite set of 1-tubes, and Ω fixes each component setwise. The non-projective modules in b with vertex a proper subgroup of D lie at the ends of 1-tubes. Moreover, if M is an indecomposable non-projective module in B with vertex a proper subgroup D' of D, then by Nagao's Theorem, the Green correspondent $f(M)$ is a module lying in $b_1^{N_G(D')}$. This is a block with defect group D and inertial index one, and so by Theorem 6.6.1, $f(M) \cong \Omega f(M)$ and hence $M \cong \Omega(M)$. It now follows from Proposition 4.12.11 that the stable quiver of B-modules consists of one component $\mathbb{Z}\tilde{A}_5$, two 3-tubes, and an infinite set of 1-tubes, and Ω fixes each component setwise. The non-projective indecomposable modules with vertex a proper subgroup of D lie in 1-tubes.

We shall show that for each simple B-module S we can produce two almost split sequences satisfying the following conditions:

(a) The middle term has at most one non-projective summand.

(b) The left and right hand terms are non-isomorphic.

Such an almost split sequence cannot lie in a $\mathbb{Z}\tilde{A}_5$ component or a 1-tube, so it must lie at the end of a 3-tube. Since there are only six modules lying at the ends of 3-tubes, this will say there are at most three simple modules.

Suppose S is a simple module with projective cover P_S. If S lies in a tube it lies at the end, and so $\mathrm{Rad}(P_S)/\mathrm{Soc}(P_S)$ is indecomposable. Conversely, if $\mathrm{Rad}(P_S)/\mathrm{Soc}(P_S)$ is indecomposable, then the almost split sequences terminating in $\overline{U} = P_S/\mathrm{Soc}(P_S)$ and $\overline{V} = S$ (notation as in Section 4.18) satisfy (a) by Proposition 4.12.9, and S lies at the end of a tube. Since $S \not\cong \Omega(S)$, these almost split sequences satisfy (b), and S lies at the end of a 3-tube.

On the other hand, if $\mathrm{Rad}(P_S)/\mathrm{Soc}(P_S)$ is decomposable, then it has exactly two summands U and V, and S lies in the $\mathbb{Z}\tilde{A}_5$ component. By Lemma 4.18.2, the almost split sequences terminating in $\overline{U}$ and $\overline{V}$ satisfy (a). If $\Omega^2(\overline{U}) \cong \overline{U}$, then since every periodic module has period one or three, we have $\Omega(\overline{U}) \cong \overline{U}$, i.e., $\underline{V} \cong \overline{U}$. Since $\mathrm{Rad}(P_S)$, U, V and $P_S/\mathrm{Soc}(P_S)$ lie in the $\mathbb{Z}\tilde{A}_5$ component, there is an irreducible morphism $\Omega(V) \xrightarrow{\alpha} U$, and $\Omega(V) \not\cong \mathrm{Rad}(P_S)$. Since $\Omega(V)$ has more composition factors than U, α is surjective and has $\mathrm{Soc}\,\Omega(V) \cong V/\mathrm{Rad}(V) \cong \overline{U}/\mathrm{Rad}(\overline{U}) \cong S$ in its kernel.

But $\Omega(V)/\operatorname{Soc}\Omega(V) \cong V \oplus S$ and so either $U \cong V$ or $U \cong S$. Since neither of these holds we have $\Omega^2(\overline{U}) \not\cong \overline{U}$ and so the almost split sequences terminating in $\overline{U}$ and $\overline{V}$ satisfy (b).

We now know that there are at most three simple B-modules. Next, we note that not all simple B-modules can be periodic, since otherwise by the Horseshoe Lemma 2.5.1, all finitely generated B-modules would have resolutions by projective modules of bounded dimension. But then the $\mathbb{Z}\tilde{A}_5$ component would contradict Auslander's Theorem 4.14.2.

Let S_1 be a non-periodic simple module whose projective cover has as few composition factors as possible. Then S_1 lies in the $\mathbb{Z}\tilde{A}_5$ component, and hence so does $\Omega(S_1)$. So $\operatorname{Rad}(P_1)/\operatorname{Soc}(P_1) \cong U_1 \oplus V_1$ with U_1 and V_1 non-zero indecomposable. An examination of the action of Ω on the $\mathbb{Z}\tilde{A}_5$ component shows that there have to be almost split sequences $0 \to \Omega(U_1) \to V_1 \oplus S_1 \to \Omega^{-1}(U_1) \to 0$ and $0 \to \Omega(V_1) \to U_1 \oplus S_1 \to \Omega^{-1}(V_1) \to 0$ with the possible addition of a projective summand in the middle. We divide into two cases according to whether such a projective summand appears in at least one of these sequences.

First suppose that one of these sequences has a projective summand, say $0 \to \Omega(U_1) \to V_1 \oplus S_1 \oplus P_2 \to \Omega^{-1}(U_1) \to 0$. In this case we have $U_2 \cong V_1$, $V_2 \cong S_1$ and $S_2 \cong U_1$, and so there is an almost split sequence $0 \to \Omega(V_1) \to S_1 \oplus S_2 \to \Omega^{-1}(V_1) \to 0$ with the possible addition of a projective summand in the middle. But this almost split sequence does not make sense without a projective summand in the middle, so in fact it is $0 \to \Omega(V_1) \to S_1 \oplus S_2 \oplus P_3 \to \Omega^{-1}(V_1) \to 0$. Thus there are exactly three simple modules and $\operatorname{Rad}(P_j)/\operatorname{Soc}(P_j) \cong P_{j-1} \oplus P_{j+1}$ (indices modulo three). We now argue as in the proof of the previous theorem to show that the basic algebra of B is isomorphic to kA_4.

Now suppose that we have almost split sequences $0 \to \Omega(U_1) \to V_1 \oplus S_1 \to \Omega^{-1}(U_1) \to 0$ and $0 \to \Omega(V_1) \to U_1 \oplus S_1 \to \Omega^{-1}(V_1) \to 0$ without projective summands in the middle. If the projective cover of U_1 has the projective cover of a non-periodic simple as a summand then $\Omega(U_1)$ has more composition factors than $V_1 \oplus S_1$, so this cannot happen. Therefore there is at least one periodic simple, say S_2, which lies at the end of a 3-tube. The modules at the end of this 3-tube are therefore $\operatorname{Rad}(P_2) = \underline{U}_2$, $P_2/\operatorname{Soc}(P_2) = \overline{U}_2$ and $S_2 = \underline{V}_2 = \overline{V}_2$.

Now consider the four modules $\overline{U}_1$, $\overline{V}_1$, $\underline{U}_1$ and $\underline{V}_1$. Since these all lie at the ends of 3-tubes, either two of them are isomorphic, or one of them is isomorphic to one of S_2, $\overline{U}_2$ or $\underline{U}_2$. In the first case, since $\overline{V}_1$ is not isomorphic to either $\overline{U}_1$ or $\underline{U}_1$, without loss of generality we have $\overline{V}_1 \cong \underline{V}_1$. This implies that V_1 is uniserial with all composition factors isomorphic to S_1. Since there is an injective irreducible map $\Omega(U_1) \to V_1$, $\Omega(U_1)$ has the same property, contradicting the fact that the projective cover of U_1 does not have P_1 as a summand. Thus the second case holds, and by dualising all arguments if necessary we may assume without loss of generality that $\overline{U}_1 \cong \underline{U}_2$.

Hence $\underline{V}_1 \cong \Omega(\overline{U}_1) \cong \overline{U}_2$ and so $U_2/\mathrm{Rad}(U_2) \cong \mathrm{Soc}(U_2) \cong S_1$, and hence $\mathrm{Ext}^1_B(S_1, S_2) \cong k$. Now $\Omega(\underline{V}_1) \cong \Omega(\overline{U}_2) \cong S_2$ and hence $\Omega(V_1)$ is a non-split extension $0 \to S_2 \to \Omega(V_1) \to S_1 \to 0$. Now, $\mathrm{Soc}^2(U_1) \cong \mathrm{Soc}^2(\overline{U}_1) \cong \mathrm{Soc}^2(\underline{U}_2)$ is also a non-split extension $0 \to S_2 \to \mathrm{Soc}^2(U_1) \to S_1 \to 0$, so the irreducible map $\Omega(V_1) \to U_1$ is injective with simple cokernel. We claim that this simple cokernel is not isomorphic to S_1 or S_2. If it were isomorphic to S_1 then $U_1 = \mathrm{Uni}(S_1, S_1, S_2)$ and $V_1/\mathrm{Soc}(V_1) \cong S_2$ so that there would be no surjective map from P_1 to U_1, which contradicts the fact that U_1 has a projective cover. If it were isomorphic to S_2 then $\mathrm{Rad}(P_1)/\mathrm{Rad}^2(P_1) \cong S_2 \oplus S_2$, which would contradict the fact that $\mathrm{Ext}^1_B(S_1, S_2) \cong k$. Thus there is a third simple module S_3, and we have $U_1 = \mathrm{Uni}(S_3, S_1, S_2)$, $U_2 = \mathrm{Uni}(S_1, S_3, S_1)$ and $V_1 = \mathrm{Uni}(S_2, S_1, S_3)$. Since the projective cover of U_1 does not have the projective cover of a non-periodic simple as a summand, P_3 is attached at the end of the other 3-tube. It is now easy to see that $\underline{U}_3 \cong \overline{U}_1$, so that $U_3 = \mathrm{Uni}(S_1, S_2, S_1)$. We have proved that B is a Brauer graph algebra, and so finally the basic algebra may be determined by the method of quivers and relations as explained in Section 4.18. As usual, the last scalar is determined using the fact that B is a symmetric algebra.

Since $A_5 \cong SL_2(4)$, the principal block of kA_5 has a two dimensional simple module whose restriction to a Sylow 2-subgroup is indecomposable of even dimension and hence periodic, so this principal block is an example of a block of type (ii). □

REMARK. Part of this theorem, namely the fact that there are three simple modules in B, may be given an alternative proof using generalised decomposition numbers, see Brauer [**36**] (Vol. III, 20–52: Some Applications of the Theory of Blocks of Characters of Finite Groups, IV). The above proof in fact shows something stronger, namely that any finite dimensional symmetric algebra Λ whose stable Auslander–Reiten quiver is isomorphic to that of kA_4 is in fact Morita equivalent to one of the two algebras listed, and in particular has exactly three simple modules.

Auslander has conjectured that if Λ and Γ are any finite dimensional algebras such that ${}_\Lambda\underline{\mathbf{mod}}$ is equivalent to ${}_\Gamma\underline{\mathbf{mod}}$ (such algebras are said to be **stably equivalent**) then Λ and Γ have the same number of simple modules (other than those lying in summands isomorphic to complete matrix algebras). This may be related to a conjecture of Alperin, which says that if B is a block of kG with abelian defect group D and Brauer correspondent b as a block of $N_G(D)$, then the number of simple modules in B is equal to the number of simple modules in b. A theorem of Knörr implies that in this situation D is a vertex of every simple module in B, and so this is a special case of a more general conjecture, usually called Alperin's conjecture. This says that if B is a block of kG with defect group D (not necessarily abelian), then the number of simple modules in B is equal to the sum over all conjugacy classes of p-subgroups P of G of the number of projective simple modules for $N_G(P)/P$ which when viewed as modules for $N_G(P)$ lie in a block b for which

$b^G = B$. In other words, the number of simple modules in B, which may be thought of as a global invariant, is equal to the number of simple Green correspondents of modules in B, which may be calculated locally. This is considered by many to be one of the most important conjectures in modular representation theory at this time.

Bibliography

[1] M. Aigner. *Combinatorial theory.* Grundlehren der Mat. Wiss. 234, Springer-Verlag, Berlin/New York, 1979.

[2] J. L. Alperin. *Diagrams for modules.* J. Pure Appl. Algebra 16 (1980), 111–119.

[3] J. L. Alperin. *Local representation theory.* Cambridge Studies in Advanced Mathematics 11, Cambridge Univ. Press, 1986.

[4] J. L. Alperin and M. Broué. *Local methods in block theory.* Ann. of Math. 110 (1979), 143–157.

[5] J. L. Alperin and D. W. Burry. *Block theory with modules.* J. Algebra 65 (1980), 225–233.

[6] M. F. Atiyah and I. G. Macdonald. *Introduction to commutative algebra.* Addison-Wesley, Reading, Mass. (1969).

[7] M. Auslander. *Applications of morphisms determined by objects.* Proc. Conf. on Representation Theory, Philadelphia 1976. Marcel Dekker (1978), 245–327.

[8] M. Auslander. *Existence theorems for almost split sequences.* Oklahoma Ring Theory Conference, March 1976, Marcel Dekker (1977), 1–44.

[9] M. Auslander and J. F. Carlson. *Almost-split sequences and group rings.* J. Algebra 103 (1986), 122–140.

[10] M. Auslander and I. Reiten. *Representation theory of Artin algebras, III: almost split sequences.* Comm. in Algebra 3 (3) (1975), 239–294.

[11] M. Auslander and I. Reiten. *Representation theory of Artin algebras, IV: invariants given by almost split sequences.* Comm. in Algebra 5 (5) (1977), 443–518.

[12] V. A. Bašev. *Representations of the group $Z_2 \times Z_2$ in a field of characteristic* 2. (Russian), Dokl. Akad. Nauk. SSSR 141 (1961), 1015–1018.

[13] W. Baur. *Decidability and undecidability of theories of abelian groups with predicates for subgroups.* Compositio Math. 31 (1975), 23–30.

[14] R. Bautista. *On algebras of strongly unbounded representation type.* Comment. Math. Helvetici 60 (1985), 392–399.

[15] D. J. Benson. *Lambda and psi operations on Green rings.* J. Algebra 87 (1984), 360–367.

[16] D. J. Benson. *Some recent trends in modular representation theory.* Proc. Rutgers Group Theory Year, 1983–1984, ed. M. Aschbacher et al. Cambridge Univ. Press, 1984.

[17] D. J. Benson. *Modular representation theory: New trends and methods.* Lecture Notes in Mathematics 1081, Springer-Verlag. Berlin/New York 1984.

[18] D. J. Benson. *Modules for finite groups: representation rings, quivers and varieties.* Representation Theory II, Groups and Orders. Proceedings, Ottawa 1984. Lecture Notes in Mathematics 1178, Springer-Verlag, Berlin/New York 1986.

[19] D. J. Benson. *Representation rings of finite groups.* Representations of Algebras, ed. P. J. Webb, Durham 1985. L.M.S. Lecture Note Series 116, Cambridge Univ. Press, 1986.

[20] D. J. Benson and J. F. Carlson. *Nilpotent elements in the Green ring.* J. Algebra 104 (1986), 329–350.

[21] D. J. Benson and J. F. Carlson. *Diagrammatic methods for modular representations and cohomology.* Comm. in Algebra 15 (1987), 53–121.

[22] D. J. Benson and J. H. Conway. *Diagrams for modular lattices.* J. Pure Appl. Algebra 37 (1985), 111–116.

[23] D. J. Benson and R. A. Parker. *The Green ring of a finite group.* J. Algebra 87 (1984), 290–331.

[24] G. M. Bergman. *Modules over coproducts of rings.* Trans. Amer. Math. Soc. 200 (1974), 1–32.

[25] S. Berman, R. Moody and M. Wonenburger. *Certain matrices with null roots and finite Cartan matrices.* Indiana Math. J. 21 (1972), 1091–1099.

[26] I. N. Bernstein, I. M. Gel'fand and V. A. Ponomarev. *Coxeter functors and Gabriel's theorem.* (Russian) Uspekhi Mat. Nauk 28 (1973), Russian Math. Surveys 29 (1973), 17–32. Also London Math. Soc. Lecture Note Series 69, Representation Theory, Cambridge Univ. Press, 1982.

[27] C. Bessenrodt. *On blocks of finite lattice type.* Arch. Math. (Basel) 33 (1979/80), 334–337.

[28] C. Bessenrodt. *On blocks of finite lattice type, II.* Integral representations and applications (Oberwolfach, 1980), Lecture Notes in Mathematics 882 (1981), 390–396.

[29] C. Bessenrodt. *A criterion for finite module type.* Proc. Amer. Math. Soc. 85 (1982), 520–522.

[30] C. Bessenrodt. *Indecomposable lattices in blocks with cyclic defect groups.* Comm. in Algebra 10 (2) (1982), 135–170.

[31] C. Bessenrodt. *Vertices of simple modules over p-solvable groups.* J. London Math. Soc. (2), 29 (1984), 257–261.

[32] C. Bessenrodt. *The Auslander–Reiten quiver of a modular group algebra revisited.* Math. Zeit. 206 (1991), 25–34.

[33] V. M. Bondarenko. *Representations of dihedral groups over a field of characteristic 2.* (Russian), Mat. Sbornik 96 (1975), 63–74; translation: Math. USSR Sbornik 25 (1975), 58–68.

[34] V. M. Bondarenko and Yu. A. Drozd. *Representation type of finite groups.* Zap. Naučn. Sem. Leningrad (LOMI) 57 (1977), 24–41.

[35] A. K. Bousfield and D. M. Kan. *Homotopy limits, completions and localizations.* Lecture Notes in Mathematics 304, Springer-Verlag, Berlin/New York, 1972.

[36] R. Brauer. *Collected papers, Vols. I–III.* M.I.T. Press, 1980, ed. P. Fong and W. Wong.

[37] S. Brenner. *Modular representations of p-groups.* J. Algebra 15 (1970), 89–102.

[38] S. Brenner. *On four subspaces of a vector space.* J. Algebra 29 (1974), 587–599.

[39] S. Brenner and M. C. R. Butler. *The equivalence of certain functors occurring in the representation theory of Artin algebras and species.* J. London Math. Soc. 14 (1976), 183–187.

[40] M. Broué and L. Puig. *Characters and local structure in G-algebras.* J. Algebra 63 (1980), 306–317.

[41] M. Broué and L. Puig. *A Frobenius theorem for blocks.* Invent. Math. 56 (1980), 117–128.

[42] M. Broué and L. Puig. *On the fusion in local block theory.* Preprint, 1981.

[43] M. Broué and G. R. Robinson. *Bilinear forms on G-algebras.* J. Algebra 104 (1986), 377–396.

[44] W. Burnside. *Theory of groups of finite order.* 2nd ed., Cambridge Univ. Press, 1911.

[45] D. W. Burry. *A strengthened theory of vertices and sources.* J. Algebra 59 (1979), 330–344.

[46] D. W. Burry. *Scott modules and lower defect groups.* Comm. in Algebra 10 (17) (1982), 1855–1872.

[47] D. W. Burry and J. F. Carlson. *Restrictions of modules to local subgroups.* Proc. Amer. Math. Soc. 84 (1982), 181–184.

[48] M. C. R. Butler. *On the classification of local representations of finite abelian p-groups.* Carleton Math. Lecture Notes 9, Carleton Univ., Ottawa, Ontario, 1974.

[49] M. C. R. Butler. *The 2-adic representations of Klein's four group.* Proc. 2nd Int. Conf. on Group Theory, Canberra, 1973, Lecture Notes in Mathematics 372, Springer-Verlag, Berlin/New York 1974.

[50] M. C. R. Butler and M. Shahzamanian. *The construction of almost split sequences III. Modules over two classes of tame local algebras.* Math. Ann. 247 (1980), 111–122.

[51] J. F. Carlson. *The modular representation ring of a cyclic 2-group.* J. London Math. Soc. (2), 11 (1975), 91–92.

[52] J. F. Carlson. *Endo-trivial modules over (p,p) groups.* Illinois J. Math. 24 (1980), 287–295.

[53] H. Cartan and S. Eilenberg. *Homological Algebra.* Princeton University Press, 1956.

[54] S. B. Conlon. *Certain representation algebras.* J. Austr. Math. Soc. 5 (1965), 83–99.

[55] S. B. Conlon. *The modular representation algebra of groups with Sylow 2-subgroups* $Z_2 \times Z_2$. J. Austr. Math. Soc. 6 (1966), 76–88.

[56] S. B. Conlon. *Structure in representation algebras.* J. Algebra 5 (1967), 274–279.

[57] S. B. Conlon. *Relative components of representations.* J. Algebra 8 (1968), 478–501.

[58] S. B. Conlon. *Decompositions induced from the Burnside algebra.* J. Algebra 10 (1968), 102–122.

[59] S. B. Conlon. *Modular representations of $C_2 \times C_2$.* J. Austr. Math. Soc. 10 (1969), 363–366.

[60] W. W. Crawley-Boevey. *Functorial filtrations and the problem of an idempotent and a square zero matrix.* J. London Math. Soc. (2), 38 (1988), 385–402.

[61] W. W. Crawley-Boevey. *Functorial filtrations II: clans and the Gel'fand problem.* J. London Math. Soc. (2), 40 (1989), 9–30.

[62] W. W. Crawley-Boevey. *Functorial filtrations III: semidihedral algebras.* J. London Math. Soc. (2), 40 (1989), 31–39.

[63] W. W. Crawley-Boevey. *On tame algebras and BOCS's.* Proc. London Math. Soc 56 (1988), 451–483.

[64] C. W. Curtis and I. Reiner. *Representation theory of finite groups and associative algebras.* Wiley-Interscience 1962.

[65] C. W. Curtis and I. Reiner. *Methods in representation theory, Vol. I.* J. Wiley and Sons, 1981.

[66] C. W. Curtis and I. Reiner. *Methods in representation theory, Vol. II.* J. Wiley and Sons, 1987.

[67] E. C. Dade. *Endo-permutation modules over p-groups, I.* Ann. of Math. 107 (1978), 459–494.

[68] E. C. Dade. *Endo-permutation modules over p-groups, II.* Ann. of Math. 108 (1978), 317–346.

[69] E. C. Dade. *Algebraically rigid modules.* Proc. ICRA II (Ottawa 1979), Lecture Notes in Mathematics 832, 195–215, Springer-Verlag, Berlin/New York 1980.

[70] P. Deligne. *Séminaire de Géométrie Algébrique du Bois-Marie SGA $4\frac{1}{2}$.* Lecture Notes in Mathematics 569, Springer-Verlag, Berlin/New York, 1977.

[71] T. tom Dieck. *Idempotent elements in the Burnside ring.* J. Pure Appl. Algebra 10 (1977), 239–247.

[72] T. tom Dieck. *Transformation groups and representation theory.* Lecture Notes in Mathematics 766, Springer-Verlag, Berlin/New York 1979.

[73] T. tom Dieck. *Transformation groups.* De Gruyter Studies in Mathematics 8, Walter de Gruyter, Berlin/New York 1987.

[74] E. Dieterich. *Representation types of group rings over complete discrete valuation rings.* Integral representations and their applications, Oberwolfach 1980. Lecture Notes in Mathematics 882, 369–389, Springer-Verlag, Berlin/New York 1981.

[75] E. Dieterich. *Construction of Auslander–Reiten quivers for a class of group rings.* Math. Zeit. 183 (1983), 43–60.

[76] E. Dieterich. *Representation types of group rings over complete discrete valuation rings II.* Orders and their applications, Oberwolfach 1984. Lecture Notes in Mathematics 1142, 112–125, Springer-Verlag, Berlin/New York 1985.

[77] J. Dieudonné. *Sur la réduction canonique des couples de matrices.* Bull. Soc. Math. France 74 (1946), 130–146.

[78] V. Dlab and C. M. Ringel. *Indecomposable representations of graphs and algebras.* Memoirs of the Amer. Math. Soc. (6) 173, 1976.

[79] V. Dlab and C. M. Ringel. *The representations of tame hereditary algebras.* In representation theory of algebras, Proc. of the Philadelphia Conference (ed. by R. Gordon), Marcel Dekker, New York/Basel (1978).

[80] V. Dlab and C. M. Ringel. *On modular representations of A_4.* J. Algebra 123 (1989), 506–522.

[81] P. W. Donovan. *Dihedral defect groups.* J. Algebra 56 (1979), 184–206.

[82] P. W. Donovan and M.-R. Freislich. *Representable functions on the category of modular representations of a finite group with cyclic Sylow subgroups.* J. Algebra 32 (1974), 356–364.

[83] P. W. Donovan and M.-R. Freislich. *Representable functions on the category of modular representations of a finite group with Sylow subgroup $C_2 \times C_2$.* J. Algebra 32 (1974), 365–369.

[84] L. Dornhoff. *Group representation theory, part B.* Marcel Dekker, New York, 1972.

[85] A. Dress. *A characterization of solvable groups.* Math. Zeit. 110 (1969), 213–217.

[86] A. Dress. *Notes on the theory of representations of finite groups I: The Burnside ring of a finite group and some AGN-applications.* Lecture notes, Bielefeld, Dec. 1971.

[87] A. Dress. *Operations in representation rings.* Representation Theory of Finite Groups and related topics, Wisconsin, 1970. Proc. Symp. Pure Math. 21 (1971), 39–45.

[88] A. Dress. *Contributions to the theory of induced representations.* Algebraic K-theory, Proc. Conf. Seattle 1972, 182–240. Lecture Notes in Mathematics 342, Springer-Verlag, Berlin/New York 1973.

[89] A. Dress. *On relative Grothendieck rings.* Representations of Algebras, Ottawa 1974. Lecture Notes in Mathematics 488, Springer-Verlag, Berlin/New York 1975.

[90] A. Dress. *Modules with trivial source, modular monomial representations and a modular version of Brauer's induction theorem.* Abh. Math. Sem. Univ. Hamburg 44 (1975), 101–109.

[91] Yu. A. Drozd. *On tame and wild matrix problems.* Matrix problems, Kiev (1977), 104–114 (in Russian).

[92] Yu. A. Drozd. *Tame and wild matrix problems.* Representations and quadratic forms 39–74, 154, Akad. Nauk. Ukrain. SSR, Inst. Mat., Kiev, 1979 (in Russian).

[93] B. Eckmann. *Der Cohomologie-Ring einer beliebigen Gruppe.* Comment. Math. Helv. 18 (1946), 232–282.

[94] S. Eilenberg and T. Nakayama. *On the dimensions of modules and algebras, V. Dimensions of residue rings.* Nagoya Math. J. 11 (1957), 9–12.

[95] K. Erdmann. *Blocks and simple modules with cyclic vertices.* Bull. London Math. Soc. 9 (1977), 216–218.

[96] K. Erdmann. *Principal blocks of groups with dihedral Sylow 2-subgroups.* Comm. in Algebra 5 (1977), 665–694.

[97] K. Erdmann. *Blocks whose defect groups are Klein four groups.* J. Algebra 59 (1979), 452–465.

[98] K. Erdmann. *On 2-blocks with semidihedral defect groups.* Trans. Amer. Math. Soc. 256 (1979), 267–287.

[99] K. Erdmann. *Blocks whose defect groups are Klein 4-groups: a correction.* J. Algebra 76 (1982), 505–518.

[100] K. Erdmann. *On modules with cyclic vertices in the Auslander–Reiten quiver.* J. Algebra 104 (1986), 289–300.

[101] K. Erdmann. *Algebras and dihedral defect groups.* Proc. London Math. Soc. (3) 54 (1987), 88–114.

[102] K. Erdmann. *Algebras and semidihedral defect groups I.* Proc. London Math. Soc. (3) 57 (1988), 109–150.

[103] K. Erdmann. *Algebras and semidihedral defect groups II.* Proc. London Math. Soc. 60 (1990), 123–165.

[104] K. Erdmann. *Algebras and quaternion defect groups I.* Math. Ann. 281 (1988), 545–560.

[105] K. Erdmann. *Algebras and quaternion defect groups II.* Math. Ann. 281 (1988), 561–582.

[106] K. Erdmann and G. O. Michler. *Blocks with dihedral defect groups in solvable groups.* Math. Zeit. 154 (1977), 143–151.

[107] W. Feit. *The representation theory of finite groups.* North-Holland, 1982.

[108] P. Freyd. *Abelian categories.* Harper and Row, New York 1964.

[109] P. Gabriel. *Unzerlegbare Darstellungen, I.* Manuscripta Math. 6 (1972), 71–103.

[110] P. Gabriel. *Indecomposable representations, II.* Symp. Math. Inst. Nazionale Alta Mat. (Rome), 11 (1973), 81–104.

[111] P. Gabriel. *Représentations indécomposables.* Séminaire Bourbaki 444 (1973–4), Lecture Notes in Mathematics 431, Springer-Verlag, Berlin/New York, 1975.

[112] P. Gabriel. *Auslander–Reiten sequences and representation-finite algebras.* Representation theory I, Ottawa 1979. Lecture Notes in Mathematics 831, Springer-Verlag, Berlin/New Nork 1980.

[113] P. Gabriel and Ch. Riedtmann. *Group representations without groups.* Comm. Math. Helvetici 54 (1979), 240–287.

[114] I. M. Gel'fand and V. A. Ponomarev. *Problems of linear algebra and classification of quadruples of subspaces in a finite-dimensional vector space.* Hilbert space operators and operator algebras, 163–237; Colloquia Mathematica Societatis János Bolyai, 5. Tihany, Hungary, 1970; North Holland, Amsterdam/London 1972.

[115] D. Gluck. *Idempotent formula for the Burnside algebra with applications to the p-subgroup simplicial complex.* Ill. J. Math. 25 (1981), 63–67.

[116] J. A. Green. *On the indecomposable representations of a finite group.* Math. Zeit. 70 (1959), 430–445.

[117] J. A. Green. *The modular representation algebra of a finite group.* Ill. J. Math. 6 (4) (1962), 607–619.

[118] J. A. Green. *A transfer theorem for modular representations.* J. Algebra 1 (1964), 73–84.

[119] J. A. Green. *Some remarks on defect groups.* Math. Zeit. 107 (1968), 133–150.

[120] J. A. Green. *A transfer theorem for modular representations.* Trans. Amer. Math. Soc. 17 (1974), 197–213.

[121] T. Hannula, T. Ralley and I. Reiner. *Modular representation algebras.* Bull. Amer. Math. Soc. 73 (1967), 100–101.

[122] D. Happel, U. Preiser and C. M. Ringel. *Vinberg's characterization of Dynkin diagrams using subadditive functions with applications to DTr-periodic modules.* Representation theory II, Ottawa 1979. Lecture Notes in Mathematics 832, Springer-Verlag, Berlin/New York 1980.

[123] M. Harada and Y. Sai. *On categories of indecomposable modules, I.* Osaka J. Math. 7 (1970), 323–344.

[124] A. Heller and I. Reiner. *Indecomposable representations.* Illinois J. Math. 5 (1961), 314–323.

[125] D. G. Higman. *Indecomposable representations at characteristic p.* Duke Math. J. 21 (1954), 377–381.

[126] P. J. Hilton and U. Stammbach. *A course in homological algebra.* Graduate Texts in Mathematics 4, Springer-Verlag, Berlin/New York, 1971.

[127] J. E. Humphreys. *Introduction to Lie algebras and representation theory.* Graduate Texts in Mathematics 9, Springer-Verlag, Berlin/New York 1972.

[128] N. Jacobson. *Lectures in Abstract Algebra I, II, III.* Graduate Texts in Mathematics 30, 31, 32, Springer-Verlag, Berlin/New York 1975–6.

[129] A. V. Jakovlev. *A classification of the 2-adic representations of a cyclic group of order eight.* (Russian), Investigations on the theory of representations. Zap. Naučn. Sem. Leningrad Otdel. Mat. Inst. Steklov (LOMI) 28 (1972), 93–129.

[130] G. J. Janusz. *Faithful representations of p groups at characteristic p.* Representation Theory of Finite Groups and related topics, Wisconsin, 1970. Proc. Symp. Pure Math. 21 (1971), 89–90.

[131] S. A. Jennings. *The structure of the group ring of a p-group over a modular field.* Trans. Amer. Math. Soc. 50 (1941), 175–185.

[132] H. Kawai, S. Kawata and T. Okuyama. *On a conjecture of D. Benson.* J. Algebra 121 (1989), 244–247.

[133] S. Kawata. *Auslander–Reiten quivers and Green correspondence.* Proceedings of the 21st symposium on ring theory (Hirosaki, 1988), 64–69, Okayama Univ., Okayama, 1989.

[134] S. Kawata. *The Green correspondence and Auslander–Reiten sequences.* J. Algebra 123 (1989), 1–5.

[135] S. Kawata. *Module correspondence in Auslander–Reiten quivers for finite groups.* Osaka J. Math. 26 (1990), 671–678.

[136] M. A. Kervaire. *Opérateurs d'Adams en théorie des représentations linéaires des groupes finis.* Enseign. Math. (2) 22 (1976), 1–28.

[137] R. Knörr. *Blocks, vertices and normal subgroups.* Math. Zeit. 148 (1976), 53–60.

[138] R. Knörr. *Semisimplicity, induction, and restriction for modular representations of finite groups.* J. Algebra 48 (1977), 347–367.

[139] R. Knörr. *On the vertices of irreducible modules.* Ann. of Math. (2) 110 (1979), 487–499.

[140] R. Knörr. *Virtually irreducible modules.* Proc. London Math. Soc 59 (1989), 99–132.

[141] D. Knutson. *λ-rings and the representation theory of the symmetric group.* Lecture Notes in Mathematics 308. Springer-Verlag, Berlin/New York 1973.

[142] A. I. Kokorin and V. I. Mart'yanov. *Universal extended theories.* Algebra, Irkutsk (1973), 107–114.

[143] C. Kratzer et J. Thévenaz. *Fonctions de Möbius d'un groupe fini et anneau de Burnside.* Comment. Math. Helvetici 59 (1984), 425–438.

[144] B. Külshammer. *A remark on conjectures in modular representation theory.* Arch. Math. 49 (1987), 396–399.

[145] T. Y. Lam. *A theorem on Green's modular representation ring.* J. Algebra 9 (1968), 388–392.

[146] T. Y. Lam and I. Reiner. *Relative Grothendieck groups.* J. Algebra 11 (1969), 213–242.

[147] P. Landrock. *The Cartan matrix of a group algebra modulo any power of its radical.* Proc. Amer. Math. Soc. 88 (1983), 205–206.

[148] P. Landrock. *Finite group algebras and their modules.* London Math. Soc. Lecture Note Series 84, Cambridge Univ. Press 1984.

[149] S. Mac Lane. *Homology.* Springer-Verlag, Berlin/New York 1974.

[150] J. M. Maranda. *On p-adic integral representations of finite groups.* Canad. J. Math. 5 (1953), 344–355.

[151] J. W. Milnor. *On axiomatic homology theory.* Pacific J. Math. 12 (1962), 337–341.

[152] H. Nagao. *A proof of Brauer's theorem on generalized decomposition numbers.* Nagao Math. J. 22 (1963), 73–77.

[153] H. Nagao and Y. Tsushima. *Representations of finite groups.* Academic Press, 1988.

[154] L. A. Nazarova. *Representations of tetrads.* Izv. Akad. Nauk SSSR Ser. Mat. 31 (1967), 1361–1378 (= Math. USSR Izv. 1, 1305–1321 (1969).)

[155] L. A. Nazarova and A. V. Rojter. *Categorical matrix problems and the Brauer–Thrall conjecture.* Preprint, Kiev (1973). German version in Mitt. Math. Sem. Giessen 115 (1975), 1–153.

[156] T. Okuyama. *On the Auslander–Reiten quiver of a finite group.* J. Algebra 110 (1987), 420–424.

[157] T. Okuyama. *Subgroups and almost split sequences of a finite group.* J. Algebra 110 (1987), 420–424.

[158] M. F. O'Reilly. *On the modular representation algebra of metacyclic groups.* J. London Math. Soc. 39 (1964), 267–276.

[159] M. F. O'Reilly. *On the semisimplicity of the modular representation algebra of a finite group.* Ill. J. Math. 9 (1965), 261–276.

[160] M. Prest. *Model theory and modules.* L.M.S. Lecture Note Series 130, Cambridge Univ. Press, 1988.

[161] L. Puig. *Pointed groups and construction of characters.* Math. Z. 176 (1981), 265–292.

[162] L. Puig. *Local fusion in block source algebras.* J. Algebra 104 (1986), 358–369.

[163] L. Puig. *Nilpotent blocks and their source algebras.* Invent. Math. 93 (1988), 77–116.

[164] L. Puig. *Pointed groups and construction of modules.* J. Algebra 116 (1988), 7–129.

[165] D. G. Quillen. *On the associated graded ring of a group ring.* J. Algebra 10 (1968), 411–418.

[166] I. Reiner. *Nilpotent elements in rings of integral representations.* Proc. Amer. Math. Soc. 17 (1966), 270–274.

[167] I. Reiner. *Integral representation algebras.* Trans. Amer. Math. Soc. 124 (1966), 111–121.

[168] I. Reiten. *Almost split sequences.* Workshop on permutation groups and indecomposable modules, Giessen, September 1975.

[169] I. Reiten. *Almost split sequences for group algebras of finite representation type.* Trans. Amer. Math. Soc. 335 (1977), 125–136.

[170] J.-C. Renaud. *The decomposition of products in the modular representation ring of a cyclic group of prime power order.* J. Algebra 58 (1979), 1–11.

[171] J.-C. Renaud. *Recurrence relations in a modular representation algebra.* Bull. Austr. Math. Soc. 26 (2) (1982), 215–219.

[172] J. Rickard. *Morita theory for derived categories.* J. London Math. Soc. 39 (1989), 436–456.

[173] Ch. Riedtmann. *Algebren, Darstellungsköcher, Überlagerungen und zurück.* Comm. Math. Helvetici 55 (1980), 199–224.

[174] C. M. Ringel. *The representation type of local algebras.* Representations of algebras, Ottawa 1974. Lecture Notes in Mathematics 488, Springer-Verlag, Berlin/New York 1975.

[175] C. M. Ringel. *The indecomposable representations of the dihedral 2-groups.* Math. Ann. 214 (1975), 19–34.

[176] C. M. Ringel. *Report on the Brauer–Thrall conjectures.* Representations of algebras, Ottawa 1979. Lecture Notes in Mathematics 831, Springer-Verlag, Berlin/New York 1980.

[177] C. M. Ringel. *Tame algebras and integral quadratic forms.* Lecture Notes in Mathematics 1099, Springer-Verlag, Berlin/New York 1984.

[178] K. W. Roggenkamp. *Almost split sequences for group rings.* Mitt. Math. Sem. Giessen 121 (1976), 1–25.

[179] K. W. Roggenkamp. *The construction of almost split sequences for integral group rings and orders.* Comm. in Algebra 5 (13) (1977), 1363–1373. Correction: Comm. in Algebra 6 (17) (1978), 1851.

[180] K. W. Roggenkamp and J. W. Schmidt. *Almost split sequences for integral group rings and orders.* Comm. in Algebra 4 (1976), 893–917.

[181] A. V. Rojter. *The unboundedness of the dimension of the indecomposable representations of algebras that have a finite number of indecomposable representations.* Izv. Acad. Nauk SSSR 32 (1968), 1275–1282.

[182] R. Schultz. *Tensoring Auslander–Reiten sequences with modules.* J. Pure Appl. Algebra 52 (1988), 147–152.

[183] R. Schwänzl. *Koeffizienten im Burnsidering.* Arch. Math. 29 (1977), 621–622.

[184] R. Schwänzl. *On the spectrum of the Burnside ring.* J. Pure Appl. Algebra 15 (1979), 181–185.

[185] L. L. Scott. *Modular permutation representations.* Trans. Amer. Math. Soc. 175 (1973), 101–121.

[186] L. L. Scott. *The modular theory of permutation representations.* Representation Theory of Finite Groups and related topics, Wisconsin, 1970. Proc. Symp. Pure Math. 21 (1971), 137–144.

[187] S. Shelah. *Infinite abelian groups, Whitehead problem and some constructions.* Israel J. Math. 18 (1974), 243–256.

[188] P. Sin. *The Green ring and modular representations of finite groups of Lie type.* J. Algebra 123 (1989), 185–192.

[189] L. Solomon. *The Burnside algebra of a finite group.* J. Combin. Theory 2 (1967), 603–615.

[190] E. H. Spanier. *Algebraic topology.* McGraw-Hill, New York 1966.

[191] B. Srinivasan. *Representations of finite Chevalley groups.* Lecture Notes in Mathematics 764, Springer-Verlag, Berlin/New York, 1979.

[192] U. Stammbach. *On the principal indecomposables of a modular group algebra.* J. Pure Appl. Algebra 30 (1983), 69–84.

[193] R. G. Swan. *Induced representations and projective modules.* Ann. of Math. 71 (3) (1960), 552–578.

[194] M. E. Sweedler. *Hopf algebras.* Benjamin, New York, 1969.

[195] J. Thévenaz. *G-algebras, Jacobson radical and almost split sequences.* Invent. Math. 93 (1988), 131–159.

[196] J. Thévenaz. *Some remarks on G-functors and the Brauer morphism.* J. Reine Angew. Math. 384 (1988), 24–56.

[197] K. Uno. *On the sequences induced from Auslander–Reiten sequences.* Osaka J. Math. 24 (1987), 409–415.

[198] K. Uno. *Auslander–Reiten sequences for certain group modules.* J. Algebra 121 (1989), 239–243.

[199] K. Uno. *Relative projectivity and extendibility of Auslander–Reiten sequences.* Osaka J. Math. 25 (1988), 499–518.

[200] E. B. Vinberg. *Discrete linear groups generated by reflections.* Izv. Akad. Nauk. SSSR 35 (1971), 1072–1112 (= Math. USSR Izvestija 5 (1971), 1083–1119).

[201] W. D. Wallis. *Decompositions of representation algebras.* J. Austral. Math. Soc. 10 (1969), 395–402.

[202] W. D. Wallis. *Direct summands in representation algebras.* Representation Theory of Finite Groups and related topics, Wisconsin, 1970. Proc. Symp. Pure Math. 21 (1971), 165–167.

[203] P. J. Webb. *Restricting ZG-lattices to elementary abelian subgroups.* Integral representations and their applications, Oberwolfach 1980. Lecture Notes in Mathematics 882, 423–429, Springer-Verlag, Berlin/New York 1981.

[204] P. J. Webb. *The Auslander–Reiten quiver of a finite group.* Math. Zeit. 179 (1982), 97–121.

[205] P. J. Webb. *On the orthogonality coefficients for character tables of the Green ring of a finite group.* J. Algebra 89 (1984), 247–263.

[206] A. Wiedemann. *The Auslander–Reiten-graph of blocks with cyclic defect two.* Integral representations and their applications, Oberwolfach 1980. Lecture Notes in Mathematics 882, 397–410, Springer-Verlag, Berlin/New York 1981.

[207] T. Yoshida. *Idempotents in Burnside rings and Dress induction theorem.* J. Algebra 80 (1983), 90–105.

[208] J. R. Zemanek. *Nilpotent elements in representation rings.* J. Algebra 19 (1971), 453–469.

[209] J. R. Zemanek. *Nilpotent elements in representation rings.* Representation Theory of Finite Groups and related topics, Wisconsin, 1970. Proc. Symp. Pure Math. 21 (1971), 173–174.

[210] J. R. Zemanek. *Nilpotent elements in representation rings over fields of characteristic 2.* J. Algebra 25 (1973), 534–553.

FURTHER REFERENCES (ADDED SINCE THE FIRST EDITION):

[211] J. L. Alperin. *Loewy structure of permutation modules for p-groups.* Quart. J. Math. Oxford 39 (1988), 129–133.

[212] J. L. Alperin. *Trees and Brauer trees.* Discrete Math. 83 (1990), 127–128.

[213] M. Auslander, I. Reiten and S. O. Smalø. *Representation theory of Artin algebras.* Cambridge Studies in Advanced Mathematics 36, Cambridge Univ. Press, 1995.

[214] L. Barker. *Induction, restriction and G-algebras.* Comm. in Algebra 22 (1994), 6349–6383.

[215] L. Barker. *Blocks of endomorphism algebras.* J. Algebra 168 (1994), 728–740.

[216] L. Barker. *Modules with simple multiplicity modules.* J. Algebra 172 (1995), 152–158.

[217] L. Barker. *G-algebras, Clifford theory, and the Green correspondence.* J. Algebra 172 (1995), 335–353.

[218] C. Bessenrodt. *Modular representation theory for blocks with cyclic defect groups via the Auslander–Reiten quiver.* J. Algebra 140 (1991), 247–262.

[219] C. Bessenrodt. *Endotrivial modules and the Auslander–Reiten quiver.* Representation theory of finite groups and finite-dimensional algebras (Bielefeld, 1991), 317–326, Progr. Math., 95, Birkhäuser, Basel, 1991.

[220] M. Broué. *On Scott modules and p-permutation modules: an approach through the Brauer homomorphism.* Proc. Amer. Math. Soc. 93 (1985), 401–408.

[221] M. Broué. *Blocs, isométries parfaites, catégories dérivées.* C. R. Acad. Sci. Paris Sér. I Math. 307 (1988), 13–18.

[222] M. Broué. *Isométries parfaites, types de blocs, catégories dérivées.* Astérisque 181–182 (1990), 61–92.

[223] M. Cabanes and C. Picaronny. *Types of blocks with dihedral or quaternion defect groups.* J. Fac. Sci. Univ. Tokyo 39 (1992), 141–161.

[224] A. H. Clifford. *Representations induced in an invariant subgroup.* Ann. of Math. 38 (1937), 533–550.

[225] E. C. Dade. *Blocks with cyclic defect groups.* Ann. of Math. 84 (1966), 20–48.

[226] K. Erdmann. *Blocks of tame representation type and related algebras.* Lecture Notes in Mathematics 1428. Springer-Verlag, Berlin/New York 1990.

[227] K. Erdmann. *On the vertices of modules in the Auslander–Reiten quiver of p-groups.* Math. Zeit. 203 (1990), 321–334.

[228] K. Erdmann. *Representations of groups and Auslander–Reiten theory.* Mitt. Math. Ges. DDR No. 3 (1990), 17–30.

[229] K. Erdmann. *On Auslander–Reiten components for wild blocks.* Representation theory of finite groups and finite-dimensional algebras (Bielefeld, 1991), 371–387, Progr. Math., 95, Birkhäuser, Basel, 1991.

[230] K. Erdmann. *On Auslander–Reiten components for group algebras.* J. Pure Appl. Algebra 104 (1995), 149–160.

[231] K. Erdmann and A. Skowroński. *On Auslander–Reiten components of blocks and self-injective biserial algebras.* Trans. Amer. Math. Soc. 330 (1992), 165–189.

[232] Y. Fan. *Permutation modules, p-permutation modules and Conlon species.* Sci. China Ser. A 34 (1991), 1290–1301.

[233] A. Franchetta. *Almost split sequences and tensor product over some group algebras.* Boll. Un. Math. Ital. B (1983), 255–268.

[234] P. Gabriel and A. V. Roiter. *Representations of finite-dimensional algebras.* Algebra VIII, Encyclopædia of Mathematical Sciences 73, Kostrikin and Shafarevich (Eds.), Springer-Verlag, Berlin/New York 1992.

[235] O. Garotta. *On Auslander–Reiten systems.* Astérisque No. 181-182 (1990), 191–194.

[236] O. Garotta. *Suites presque scindées d'algèbres intérieures.* Seminaire sur les groupes finis IV, 137–237. Publ. Math. Univ. Paris VII, 34, 1994.

[237] J. A. Green. *Multiplicities, Scott modules and lower defect groups.* J. London Math. Soc. (2) 28 (1983), 282–292.

[238] H. Heldner. *Nilpotent elements of order 3 in the Green ring of the dihedral-2-groups.* Arch. Math. (Basel) 63 (1994), 193–196.

[239] T. Inoue and Y. Hieda. *A note on Auslander–Reiten quivers for integral group rings.* Osaka J. Math. 32 (1995), 483–494.

[240] S. Kawata. *The modules induced from a normal subgroup and the Auslander–Reiten quiver.* Osaka J. Math. 27 (1990), 265–269.

[241] S. Kawata. *On Auslander–Reiten components for certain group modules and an additive function.* Proceedings of the 26th Symposium on Ring Theory (Tokyo, 1993), 61–68, Okayama Univ., 1993.

[242] S. Kawata. *On Auslander–Reiten components for certain group modules.* Osaka J. Math. 30 (1993), 137–157.

[243] S. Kawata. *On Auslander–Reiten components for group algebras of finite groups.* Representation theory of finite groups and algebras (Kyoto, 1993). Sūrikaisekikenkyūsho Kōkyūroku No. 877 (1994), 34–40.

[244] M. Kleiner. *On the almost split sequences for relatively projective modules over a finite group.* Proc. Amer. Math. Soc. 116 (1992), 943–947.

[245] R. Knörr. *On the indecomposability of induced modules.* J. Algebra 104 (1986), 261–265.

[246] R. Knörr. *A remark on Brauer's $k(B)$-conjecture.* J. Algebra 131 (1990), 444–454.

[247] R. Knörr. *Auslander–Reiten sequences and a certain ideal in mod-FG.* Rostock. Math. Kolloq. No. 49 (1995), 89–97.

[248] S. Koshitani and B. Külshammer. *A splitting theorem for blocks.* Osaka J. Math. 33 (1996), 343–346.

[249] B. Külshammer. *Morita equivalent blocks in Clifford theory of finite groups.* Astérisque 181–182 (1990), 209–215.

[250] B. Külshammer. *Modular representations of finite groups: conjectures and examples.* Darstellungstheorietage Jena 1996, 93–125, Sitzungsber. Math.-Naturwiss. Kl., 7, Akad. Gemein. Wiss. Erfurt, Erfurt, 1996.

[251] B. Külshammer and L. Puig. *Extensions of nilpotent blocks.* Invent. Math. 102 (1990), 17–71.

[252] B. Külshammer and G. R. Robinson. *Characters of relatively projective modules, II.* J. Algebra 135 (1990), 19–56.

[253] M. Linckelmann. *Derived equivalence for cyclic blocks over a P-adic ring.* Math. Zeit. 207 (1991), 293–304.

[254] M. Linckelmann. *A derived equivalence for blocks with dihedral defect groups.* J. Algebra 164 (1994), 244–255.

[255] M. Linckelmann. *On stable and derived equivalences of blocks and algebras.* Representation theory of groups, algebras and orders (Constanţa, 1995). An. Ştiinţ. Univ. Ovidius Constanţa Ser. Mat. 4 (1996), 74–97.

[256] P. A. Linnell. *The Auslander–Reiten quiver of a finite group.* Arch. Math. (Basel) 45 (1985), 289–295.

[257] S. Martin. *Periodic modules in the Auslander–Reiten quiver for some group algebras.* Comm. Algebra 18 (1990), 4087–4102.

[258] T. Okuyama and K. Uno. *On vertices of Auslander–Reiten sequences.* Bull. London Math. Soc. 22 (1990), 153–158.

[259] T. Okuyama and K. Uno. *On the vertices of modules in the Auslander–Reiten quiver. II.* Math. Zeit. 217 (1994), 121–141.

[260] L. Puig. *Vortex et sources des foncteurs simples.* C. R. Acad. Sc. Paris Sér. I Math. 306 (1988), 223–226.

[261] L. Puig. *On Thévenaz' parametrization of interior G-algebras.* Math. Zeit. 215 (1994), 321–335.

[262] J. Rickard. *Derived categories and stable equivalence.* J. Pure Appl. Algebra 61 (1989), 303–317.

[263] J. Rickard. *Derived equivalences as derived functors.* J. London Math. Soc. (2) 43 (1991), 37–48.

[264] J. Rickard. *Lifting theorems for tilting complexes.* J. Algebra 142 (1991), 383–393.

[265] J. Rickard. *Splendid equivalences: derived categories and permutation modules.* Proc. London Math. Soc. (3) 72 (1996), 3331–358.

[266] G. R. Robinson. *On characters of relatively projective modules.* J. London Math. Soc. (2) 36 (1987), 44–58.

[267] P. Schmid. *Clifford theory and tensor induction.* Arch. Math. (Basel) 54 (1990), 549–557.

[268] P. Schmid. *Tensor induction of projective representations.* J. Algebra 147 (1992), 442–449.

[269] A. Shalev. *Graded permutation modules and a theorem of Quillen.* Proc. Amer. Math. Soc. 120 (1994), 333–337.

[270] J. Thévenaz. *Extensions of group representations from a normal subgroup.* Comm. in Algebra 11 (1983), 391–425.

[271] J. Thévenaz. *Lifting idempotents and Clifford theory.* Comment. Math. Helv. 58 (1983), 86–95.

[272] J. Thévenaz. *Relative projective covers and almost split sequences.* Comm. in Algebra 13 (1985), 1535–1554.

[273] J. Thévenaz. *G-algebras and modular representation theory.* Oxford University Press, 1995.

[274] J. G. Thompson. *Vertices and sources.* J. Algebra 6 (1967), 1–6.

[275] K. Uno. *On the vertices of modules in the Auslander–Reiten quiver.* Math. Zeit. 208 (1991), 411–436.

[276] A. Watanabe. *On nilpotent blocks of finite groups.* J. Algebra 163 (1994), 128–134.

[277] C. A. Weibel. *An introduction to homological algebra.* Cambridge studies in advanced mathematics 38. Cambridge Univ. Press, 1994.

[278] A. Wiedemann. *The Auslander–Reiten graph of integral blocks with cyclic defect two and their integral representations.* Math. Zeit. 179 (1982), 407–429.

[279] T. Yoshida. *On a theorem of Benson and Parker.* J. Algebra 115 (1988), 442–444.

[280] T. Yoshida. *On the unit groups of Burnside rings.* J. Math. Soc. Japan 42 (1990), 31–64.

[281] A. Zimmermann. *Structure of blocks with cyclic defect groups and Green correspondence.* Representation theory of groups, algebras, and orders (Constanţa, 1995). An. Ştiinţ. Univ. Ovidius Constanţa Ser. Mat. 4 (1996), no. 1, 24–73.

[282] A. Zimmermann. *Two sided tilting complexes for Green orders and Brauer tree algebras.* J. Algebra 187 (1997), 446–473.

Index

Printed in the United States
By Bookmasters